Cosmic Radio Waves

Cosmic Radio Waves

I. S. Shklovsky

Sternberg Astronomical Institute, Moscow

TRANSLATED BY

RICHARD B. RODMAN AND **CARLOS M. VARSAVSKY**

Harvard College Observatory

HARVARD UNIVERSITY PRESS · Cambridge, Massachusetts · 1960

London: Oxford University Press

Printed in the United States of America

To the memory of

GRIGORY ABRAMOVICH SHAJN

Preface to the English Edition

This book was first published in Moscow, in 1956. It constitutes an exposition of a very young branch of science, one that has burgeoned with extraordinary speed. Therefore, in preparing the revisions of the text pursuant to its translation into English, considerable care has been taken to ensure that the book shall reflect the progress made over the past two years in the study of cosmic radio waves. The basic content and the conclusions of the book have nevertheless remained substantially unaltered.

Thanks to intensive work by investigators in various countries, the status of cosmic radio waves has now become more or less clear. Perhaps this may serve to justify the appearance of the book as it is now brought to the attention of American readers.

America is the homeland of radio astronomy: Jansky's and, later, Reber's work inaugurated the field of cosmic radio-wave research. At the present time the American radio astronomers occupy a leading position in the continuing development of this field. The author therefore takes especial pleasure in the translation of his book in America. He hopes that this monograph will prove useful to a wide community of physicists and astronomers, who may desire to study the fascinating new science of radio astronomy.

I am very grateful to the translators, R. B. Rodman and C. M. Varsavsky, for the high quality of their work. I also wish to express my thanks to Dr. D. S. Heeschen of the National Radio Astronomy Observatory and Dr. G. B. Field of the Princeton University Observatory, who have read the manuscript in translation, and have communicated some new results announced at the 1958 Paris symposium on radio astronomy.

I. S. SHKLOVSKY

Sternberg Astronomical Institute,
  Moscow
September 1958

Preface to the Russian Edition

In December 1956 radio astronomy marks its silver jubilee. For the quarter century following Jansky's pioneering work, radio astronomy enriched both physics and astronomy with a series of discoveries whose significance can scarcely be overestimated. While the young science hardly developed at all during the first ten or fifteen years of its "life," we could, in the postwar period, witness a striking acceleration of research. This can naturally be explained by the great advances in radiophysics and radar.

We are aware that radiophysical methods of investigation have recently been introduced into a wide variety of scientific fields: astronomy is no exception. Furthermore this "incursion" of radiophysics into astronomy represents perhaps the most effective demonstration of the power and possibilities of modern radiophysics. Today radio astronomy employs the most sensitive receivers and the largest antenna systems. The tasks with which radio astronomy comes into contact stimulate, in their turn, researches in radiophysics. It is the demands of radio astronomy which above all dictate the exploitation, by modern radiophysics, of large antennas, interferometers with exceptionally high resolving power, and receivers of the highest sensitivity and stability. In this way, radio astronomy is already exerting an influence—a feedback, as it were—on the radiophysics which engendered it.

Nevertheless it must not be forgotten that radio astronomy is above all a branch of *astronomy*. In the course of its long centuries of development, the most ancient of the natural sciences has made use of various methods of investigation. Thus, for example, in the second half of the nineteenth century spectroscopic methods were employed in astronomy to an ever-increasing extent. To most astronomers of that period these methods seemed no less new and unusual than do radiophysical methods at the present time. Vigorously developed at the close of the nineteenth and particularly at the beginning of the twentieth century, spectroscopy transformed the old science of the sky. Can we today conceive of astronomy apart from spectroscopic methods of investigation?

Over the last decade, photoelectric methods have entered astronomy to a considerable degree; in this area outstanding results have been attained and, what is more important, very broad perspectives have been opened up. There are firm grounds for supposing that in the near future the use of electronic methods at the telescope will acquire exceptional significance for astronomy; they will lead to a sharp increase in the sensitivity of receivers of radiation, which may of itself entail quite limitless consequences.

An important trend in the development of modern astronomy is the extension of the limits of the spectral region within which one may examine the radiation of cosmic objects. For example, research has been carried out in recent years on the infrared radiation of the planets, raising anew the question of the nature of their atmospheres. Investigation of the ultraviolet and x-radiation of the sun by means of high-altitude rocketry has been of the utmost importance.

In this way the rise of radio astronomy—as one of the manifestations of a basic trend in contemporary astronomy—proves to be quite a natural development. And however diverse, however singular may be the methods of astronomical research, however little they may resemble one another, it must always be kept in mind that they are but branches of one and the same venerable old tree.

Radio astronomy is valuable above all because it materially enriches our understanding of the universe. However, in order to interpret the results of radio-astronomical research, it is absolutely essential to make use of the data of optical astronomy. Only very close cooperation and interdependence between optical and radio astronomy can lead to genuine progress. One must always remember that radio astronomy has not sprung up from a desert: that, over the long years of development of optical astronomy, there has accumulated an immense store of information on the physical nature of the various celestial objects.

Radio astronomy has at the present time attained such a level of development that it has already become impossible to expound its principles adequately in a single monograph. We have consequently excluded from consideration all questions dealing with radio waves from the sun; these problems merit a special monograph. Similarly, we shall not consider the matter of lunar radio waves; nor shall we deal with radar astronomy, which comprises an extensive and rather detached branch of radio astronomy. The subject of the present monograph will be the problem of cosmic radio waves—that is to say, the radiation from objects found far beyond the solar system. It is in this field that radio astronomy has achieved its most conspicuous results, changing all our habitual ideas in a fundamental way.

It has by now become clear that the development of radio astronomy in this last direction is exerting a powerful influence on such important problems of contemporary natural science as the nature and origin of primary cosmic rays, cosmology, and the series of basic astronomical problems connected with the investigation of interstellar matter and the dynamics of our stellar system. A principal aim of this monograph is the analysis of the relations between radio astronomy on the one hand, and astrophysics, cosmology, the problem of the origin of cosmic rays on the other. We shall examine this broad range of questions from a unified point of view. At the same time we have attempted to give as full an account as possible of the results and the observational methods.

The present monograph represents an endeavor to review the development of a quarter century of research in the domain of cosmic radio waves. During the next year or two, powerful new radio telescopes will be constructed in various countries, so that undoubtedly a number of interesting and important new facts will soon be revealed. There is no question but that, in the light of this material, individual conclusions of our monograph will demand revision, perhaps even radically. This is unavoidable for so rapidly evolving a science as radio astronomy. It is our conviction, however, that the basic findings of this book will hardly suffer substantive change.

I. S. SHKLOVSKY

Moscow
July 1956

Contents

Foreword

Too few scientists of the Western World read the Russian language with a fluency sufficient to absorb with ease the contents of books written by Soviet scientists. Since many important scientific developments have originated in the past and are continuing to come from the U.S.S.R., there is a real demand for translations into the English language of some of the more significant books and papers. The recent book by Shklovsky is in this class, and scientists, young and old, will be grateful to Mr. Rodman and Mr. Varsavsky for having prepared a translation of this important volume. It is a fortunate circumstance that Dr. Shklovsky has been able to bring his text up to date before publication, because in a fast-moving science like radio astronomy a book is in danger of becoming out of date in many parts in one or two years. This is a new book, rather than a translation.

Dr. Shklovsky has personally made some mighty contributions to the development of radio astronomy. His name will always be associated in our minds with one of the most fruitful suggestions of the past decade—that of the importance of the "synchrotron process" of relativistically accelerated electrons moving in interstellar magnetic fields. We think in terms of his work whenever we touch upon problems relating to radio radiation from the galactic halo—or anything involving the sources of emission in the radio continuum. From him have come as well some of the most useful researches regarding monochromatic radio radiations, and the suggestions made by himself and other Soviet astronomers regarding the nature and functions of the galactic halo are currently receiving much attention. Dr. Shklovsky is a brilliant theoretical astrophysicist of the younger generation, whose mind is clear and critical—and who is forever searching for fresh avenues of approach.

The book deals with cosmic radio waves exclusively and hence there is no place in it for radio studies of the sun, moon, and planets and for radar work in the solar system. This limitation is a good one, for it makes it possible for the author to cover in a comprehensive manner the whole of his field.

The principal contributions to radio astronomy that have come from the U.S.S.R. have been along theoretical lines. In his first chapter, Dr. Shklovsky is wise not to attempt a new variety of the presentation of observational techniques and it is with pleasure that an astronomer writing from Australia notes how prominently the writings of the radiophysicists of the C.S.I.R.O. in Sydney have influenced the author's approach to practical problems. In the second chapter, the author gives a comprehensive semihistorical survey of the basic observational material; for many years to come this chapter will provide a fine introduction for the novice in the field and a convenient summary of references for the old hands. The later chapters of the book are the ones that will be read with greatest interest by the active research workers in radio astronomy, for here we enter upon studies which are in the author's own field; I enjoyed personally the privilege of getting a bit of an insight into the working of the Shklovsky mind. He rightly stresses the point that radio and optical astronomy are not really separate fields. All astronomers should learn to use both techniques, in many cases applying them side by side for the solution of a given problem. Of special interest to many astrophysicists will be Shklovsky's far-sighted insistence on an intimate connection between sources of cosmic radio waves and the origin of cosmic rays, with, in the author's opinion, the supernovae playing a major role in producing the needed plasma clouds with associated interstellar magnetic fields.

One of the most useful aspects of the book is that it focuses attention in a precise manner upon the major problems in urgent need of further study. This is especially so in the concluding chapter, which deals with cosmology. But the same spirit is found throughout the volume, whether the author is dealing with monochromatic radio radiations, or with emission nebulae and the role they play in the spiral pattern of our galaxy, with the gaseous near-spherical halo of our galaxy, or with the interstellar gas in the galactic nucleus. This is a book that deserves to be read and pondered—and the world of science is indebted to the author, the translators, and the publishers for making it available to all.

BART J. BOK

Mount Stromlo Observatory
Canberra, Australia
March 14, 1960

Cosmic Radio Waves

I

Receivers and Antennas in Radio Astronomy

1. THE MEASUREMENT OF COSMIC NOISE

The primary task of observational radio astronomy is the measurement of the sources and intensities of the radiation from cosmic objects, in those portions of the spectrum which lie within the radio range. In addition, the polarization characteristics of cosmic radio waves are often measured, most frequently for solar radio waves.

It seems to us expedient to give here definitions of the basic quantities which are measured in radio-astronomical investigations. Let a plane linearly polarized electromagnetic wave be emitted by a source of radio waves and pass through some point of space; at this point there arises a varying electric and magnetic field. We recall that the flux of radiation is given by the Poynting vector $\mathbf{S}$. If the electromagnetic wave is monochromatic, that is, if the intensities $\mathbf{E}$ of the electric field and $\mathbf{H}$ of the magnetic field vary according to the expressions

$$\mathbf{E} = \mathbf{E}_0 \cos (2\pi\nu t + \alpha), \quad \mathbf{H} = \mathbf{H}_0 \cos (2\pi\nu t + \alpha),$$

then the energy flux is determined by the formula

$$S = cE_0^2/8\pi.$$

In the general case,

$$S = c\langle E^2 \rangle /4\pi;$$

this last formula gives the energy flux integrated over the entire spectrum.

When we analyze a beam of radiation in optical astronomy with spectroscopic apparatus, we can obtain the spectral distribution of the beam, that is, either a function $S(\nu)$ or a function $S(\lambda)$. The flux of radiation in the frequency interval from ν to $\nu + d\nu$ will then be $S(\nu)d\nu$, with $\int S(\nu)d\nu = S$. Analogous expressions are obtained on a wavelength scale. We note, however, that the procedure customarily employed by astrophysicists for resolving "white" radiation into a spectrum is far from an elementary one. Only a series of simplifying assumptions, specifically adapted to optical astronomy, allows us to

1

retain the elementary definition of spectral resolution with an accuracy sufficient for practical purposes. An analysis of the concept of *spectral flux density* will assist us in understanding its specific character and its possibilities for radio-astronomical methods.

Radiation arriving from some source represents, in general, an electromagnetic field, which may be described by a very rapidly and irregularly varying function $\varphi(t)$ of time. In order to simplify the problem we shall, without loss of generality, consider only one component of polarization. The function $\varphi(t)$ does not of itself give us information on the spectral composition of the radiation, for it describes the *time* dependence of the electromagnetic field vectors. If we had

$$\varphi(t) \propto \cos(2\pi \nu t + \alpha),$$

that is, if $\varphi(t)$ were a periodic function of time, then the spectral character of the radiation (the frequency ν) would be assigned by the same expression. But it is a familiar fact that monochromatic electromagnetic oscillations do not exist in nature; for example, spectral lines have a certain finite width for various reasons. In radio astronomy $\varphi(t)$ is, as a rule, an irregularly varying function of time, or "noise" (except for the hydrogen radio line at $\lambda = 21$ cm). In this event $\varphi(t)$ yields *all* frequencies when it is analyzed as a Fourier integral; hence we are then dealing with a continuous spectrum.

Since $\langle \varphi(t) \rangle = 0$, that is, since the permanent noise component vanishes in the mean, the total flux may be defined as

$$S = \langle [\varphi(t)]^2 \rangle ; \tag{1-1}$$

here the coefficient of proportionality depends on the choice of units, and we have taken it to be unity.

We consider the interval τ of time from $t = 0$ to $t = \tau$, and analyze $\varphi(t)$ as a Fourier integral. The Fourier coefficients will then be given by

$$A_\tau(\nu) = \int_0^\tau \varphi(t) e^{2\pi i \nu t}\, dt. \tag{1-2}$$

We carry out the same decomposition for the time interval from $t = \tau$ to $t = 2\tau$, obtaining another set of coefficients $A_\tau'(\nu)$; for the time interval from $t = 0$ to $t = 2\tau$, let

$$A_{2\tau}(\nu) = A_\tau(\nu) + A_\tau'(\nu). \tag{1-3}$$

Analogously we may define the quantities $A_\tau''(\nu), \ldots, A_\tau^{(n)}(\nu)$, and the sums $A_{3\tau}(\nu), \ldots, A_{(n+1)\tau}(\nu)$. As the time interval $0 < t < (n + 1)\tau$ increases without bound, the sums $A_{(n+1)\tau}(\nu)$ do not tend to a definite limit $A(\nu)$, since the Fourier coefficients defined above are complex numbers of independent phase. Consequently it is not physically meaningful to speak of the Fourier analysis of a noise spectrum extending over an infinite time.

We consider now the mean square Fourier coefficients, which are formed by taking the mean over all the quantities $|A_\tau(\nu)|^2$. By a theorem from statistics, the relative variance of the true value of $|A_\tau(\nu)|^2$, that is, the quantity

$$\frac{|A_\tau(\nu)|^2 - \langle|A_\tau(\nu)|^2\rangle}{\langle|A_\tau(\nu)|^2\rangle},$$

will be proportional to $\tau^{-1/2}$. The quantities $|A_\tau(\nu)|^2$ will be almost strictly proportional to τ for sufficiently large τ, and their relative variance will be small. It follows that there exists a limit

$$\lim_{\tau\to\infty} \frac{1}{\tau}|A_\tau(\nu)|^2 = S(\nu). \tag{1-4}$$

One may show that $\int S(\nu)d\nu = S$; hence the quantity $S(\nu)$, as defined by Eq. (1-4), represents the spectral flux density.

Radio "telescopes" are employed in radio astronomy for receiving and measuring cosmic radio waves. The essential aspects of a radio telescope are outlined in Fig. 1. Energy of cosmic radio waves, collected by the antenna, is delivered to the input of the receiver (usually a superheterodyne) by a wave guide or a coaxial cable. The electromagnetic oscillations are then amplified considerably (by, say, 10^8) and rectified. A measuring device is placed at the output of the radio telescope. For calibration, a source of radiation of known power, such as a noise generator, may be connected at the receiver input. Figure 2 illustrates schematically the transformations suffered by electromagnetic oscillations when accepted by the receiver.

The electromagnetic field of a wave falling upon the antenna corresponds to the function $\varphi(t)$ which we have defined above, a function changing very rapidly and irregularly with time. It may be regarded as a superposition of harmonics of continuously varying amplitude and phase. All types of antennas respond differently to incident electromagnetic fields of different spectral composition; in other words, anten-

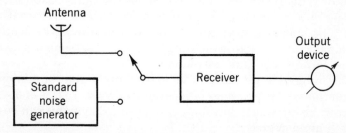

Fig. 1. The essential parts of a radio telescope (Pawsey and Bracewell 1955).

nas produce a "filtering" of frequencies. But the principal filtering of frequencies is produced by the receiver, which selectively amplifies the frequencies in some rather narrow frequency interval $\Delta\nu$ about ν. This *physical* operation is utilized in a fundamental way in resonance phenomena; it corresponds to our *mathematical* operation of deriving Fourier coefficients. Rectification in the receiver performs the operation of obtaining the squares $|A_\tau(\nu)|^2$ of the moduli of the Fourier coefficients. The time τ during which sample signals, or "pulses," of cosmic radio waves are analyzed (that is, the time of integration or smoothing) is determined by the time constant of the recording device at the output (for example, the recording pen).

According to Eq. (1-4), defining the spectral flux density, an infinitely long "smoothing time" τ is required to obtain $S(\nu)$ from the observations. Since in practice the time τ is finite (usually from 1 to 100 sec), the quantities $S(\nu)$, obtained for various pulses of uniform duration τ, will necessarily differ from one another; thus the recording device will produce a record exhibiting dispersion. We will, so to speak, observe "fluctuations" in the spectral flux density. These fluctuations are a consequence of our definition of the spectral density. An increase in the length of time during which each pulse is observed (and hence an increase in the time constant of the recording device) may decrease the magnitude of the fluctuations; we show below that this will lead to an increase in the sensitivity of the receiver.

There is, however, an additional factor which determines the magnitude of the fluctuations shown by the recording device: this is the frequency bandwidth $\Delta\nu$ within which the receiver amplifies noise. It can be shown that

$$\Delta R/R = \alpha(\Delta\nu \cdot \tau)^{-1/2}, \tag{1-5}$$

where ΔR is the standard deviation of the readings R of the recording device, and α is a dimensionless factor of the order of unity, depending on the construction of the receiver. Inspection of Fig. 2 will clarify the physical meaning of Eq. (1-5). The current in the receiver will, as the figure indicates, have constant phase and amplitude after selective amplification in the frequency interval $\Delta\nu$ only for a time interval of the order of $1/\Delta\nu$. If the recording device has the time constant τ, then it averages $n = \tau \cdot \Delta\nu$ elementary pulses of current over that time interval. The standard deviation of the readings must evidently be proportional to $n^{-1/2}$ from which we have Eq. (1-5), one of the most important formulas of radio astronomy.

A basic problem in practical radio astronomy is that of minimizing the fluctuations in the readings of recording devices. To this end, either $\Delta\nu$ or τ must be increased; but for a variety of reasons this can be done only to a limited degree.

In optical astronomy, as we have pointed out above, an elementary

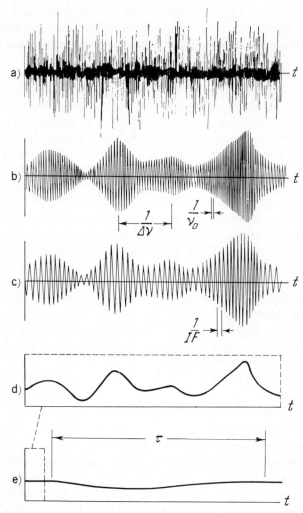

Fig. 2. Transformation of cosmic radio waves by the receiver. The figures successively show the time dependence for electromagnetic oscillations (a) as received, (b) after amplification at high frequency, (c) after amplification at an intermediate frequency, (d) after rectification, (e) after smoothing by the recording device (Pawsey and Bracewell 1955).

determination of the spectral flux density usually suffices. The fluctuations in the readings of the recording device, which are inherent in the very nature of the concept of "spectral flux density," are usually immaterial. Thus, for example, if a monochromator (such as a polarizing interference filter) cuts out a 1-A segment from the spectrum, the corresponding frequency interval will be $\Delta\nu \approx 10^{11}$ c/s, and with $\tau = 0.1$ sec, we have $\Delta R/R \approx 10^{-5}$, a very small quantity. With photographic methods of recording the radiation, the frequency interval $\Delta\nu$ will usually be still greater. On the other hand, in photoelectric photometry and spectrophotometry considerable fluctuations may arise from the spectral resolution procedure itself.

We shall estimate the magnitude $\Delta R/R$ of the fluctuations under the conditions of radio-astronomical observation. The passband of the receiver will usually be of the order of 1 to 10 Mc/s wide; with a time constant $\tau \approx 100$ sec, we have $\Delta R/R \approx 10^{-4}$. It is to be noted that the flux of cosmic radio waves is very small, as a rule; it frequently constitutes only a fraction of a percent of the intrinsic noise level in the receiver. In this event the readings of the recording device will be determined by the internal noise level, and the fluctuations will be centered upon this level. Evidently, if a "useful" signal from a cosmic source induces a deflection δR in the recording device, then the condition for detectability of the signal will be

$$\delta R > \Delta R; \qquad (1\text{-}6)$$

unless this condition is fulfilled, it will not be possible to distinguish the signal from the background of the fluctuations due to the internal noise of the apparatus.

We shall examine this important matter in more detail. We replace the antenna by a resistance at temperature T, connected to the receiver. As is shown in radiophysics, the power generated by the thermal fluctuations and incident at the receiver input will then be given by

$$P_0 = kT \cdot \Delta\nu, \qquad (1\text{-}7)$$

where $\Delta\nu$ is the bandwidth. In every receiver the role of this resistance is played by the so-called *input resistance* of the receiver. If an antenna is connected to the receiver input, then under ideal conditions (to which we shall refer at the beginning of Sec. 2) it will transmit to the receiver only the radiant energy falling upon it. In this event the power incident at the receiver input will be equal to P.

One should bear in mind that various kinds of losses are unavoidable in actual antennas, as well as in the lines connecting antenna to receiver. For this reason a "spurious" power, generated in the antenna channel, will be delivered to the receiver input, in addition to the "useful" power P. By taking appropriate steps, the "spurious" power may be rendered sufficiently small and may be allowed for when reducing the observational results. Furthermore, the thermal motions of the electrons within the resistors give rise to the "shot effect" in the receiver itself, generating a power P_i in the tubes. This power has the character of a random noise; hence its spectrum is continuous, and is similar to the spectrum of the useful signal P. The noise is termed the *internal noise* of the receiver; it arises in each stage of amplification, and is further amplified in all the succeeding stages. Therefore the noise arising in the first stage of amplification will be amplified in the succeeding stages in just the same way as the "external" noise (where the latter here includes the useful signal). The total receiver noise is thus given by $P_0 + P_i$, where P_0 is the power of the noise in the input resistance of the receiver, and P_i is the noise generated in the remainder of the receiver.

After the "external noise" and the "internal noise" are amplified G times within the receiver (where G is the *gain*, the coefficient of amplification of the receiver), the ratio of the useful signal to the noise at the receiver output will be

$$\frac{\delta R}{R} = \frac{GP}{GP_0 + GP_i}.$$

Then the signal-to-noise ratio at the receiver input divided by the signal-to-noise ratio at the receiver output will be equal to

$$N = \frac{P/P_0}{P/(P_0 + P_i)} = \frac{P_0 + P_i}{P_0}. \tag{1-8}$$

The quantity N, depending only on the inherent properties of the receiver, is called the *noise factor* of the receiver. It follows from Eqs. (1-8) and (1-7) that the power P_i of the internal receiver noise is

$$P_i = (N - 1)P_0 = (N - 1)kT \cdot \Delta \nu; \tag{1-9}$$

the temperature T is usually taken to be $290°K$. For an ideal "noise-free" receiver, $P_i = 0$ and $N = 1$.

Assume that the "useful" signal represents the power P of radio waves emitted by some source and absorbed by the antenna; here $P = P(\nu)\Delta\nu$, where $\Delta\nu$ is the width of the passband of the receiver. (A rigorous definition of P will be given in Sec. 2.) Then in accordance with Eqs. (1-6), (1-5), (1-7), and (1-9), the condition that a weak cosmic radio-wave signal be observable may, to within a factor of the order of unity, be written

$$P(\nu)_{\min} \geq NkT(\Delta\nu \cdot \tau)^{-1/2}. \tag{1-10}$$

As a result, the smaller the noise factor of the receiver, the smaller will be the quantity $P(\nu)_{\min}$; in this way the noise factor defines the sensitivity of the receiver. For a good receiver, $N \approx 3$ to 10 in the meter and decimeter range, with the smaller noise factor being attained in the meter range. In the centimeter range, $N \approx 5$ to 10 for good receivers.

We have already pointed out that in attempting to increase the sensitivity of the receiver it is not possible to increase $\Delta\nu$ and τ without limit. For example, interferometric systems are employed in a variety of ways for investigating cosmic radio waves. It is naturally essential to admit only sufficiently monochromatic radio waves to such a system, for otherwise the interference pattern would deteriorate. For the same reason, when observing optical interference in white light one finds a coloration and deterioration of the interference rings. In practice this requires that $\Delta\nu$ shall not exceed several thousandths of the central frequency ν. For much work, however, one does not employ interferometric methods; in that event one may successfully construct receivers with a fairly wide passband, up to several tenths of ν.

There is, however, a limit of technical character to an unbounded increase in the bandwidths of receivers. For instance, in wide-band receivers the gain at each stage is comparatively small. This inevitably entails an increase in the internal noise of the receiver (the quantity N), for noise becomes important not only in the first stage, but in the following stages as well. Finally, it should be kept in mind that, when investigating cosmic radio waves, it is very desirable for the spectral flux density to remain fairly constant within the interval $\Delta\nu$; otherwise we shall not have adequate information on the spectral constitution of the cosmic radio waves received.

Upper bounds on the time constant of the recording device are similarly dictated by difficulties of a technical nature, primarily associated with the stability of the apparatus (for example, the constancy of the receiver's gain, the various kinds of "zero drift" in the measuring instrument, and the like). Instruments with a time constant of 100 to 150 sec are generally used today; but there are definite prospects in view that the "integration time" of recording devices may be substantially increased (for example, by an objective averaging of a number of readings of the device). Evidently, a considerable increase in the sensitivity of the receiving apparatus would be attained in just this fashion.

Special methods of reduction are necessary to measure the small power involved: the power level often approaches the threshold of the fluctuations in the apparatus. From the earliest development of radio astronomy wide use has been made of a *compensating method* of measurement, in which the noise level of the apparatus is compensated by a constant current from a local source. This permits the use of very sensitive recording devices. The accuracy of the method depends on the precision to which the receiver noise level is compensated; the latter is continually broken up by fluctuations in the noise level and, to a greater degree, by the various zero drifts, such as those arising from variations in the gain.

A *modulation method* developed by Dicke (1946) has been applied extensively in radio-astronomical measurement. The receiver input is alternately connected to the antenna that collects the cosmic radio

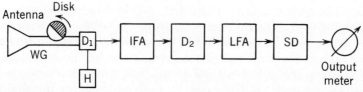

Fig. 3. Block diagram of the modulation radiometer: *WG,* wave guide; D_1, first detector; *H,* heterodyne; *IFA,* intermediate-frequency amplifier; D_2, second detector; *LFA,* low-frequency amplifier; *SD,* synchronous detector.

waves under investigation and to some equivalent resistor (such as another antenna) upon which the cosmic radiation is not incident. The frequency at which the receiver input is switched ranges from several tens to several hundreds of cycles per second. After passing through the receiving system, the useful signal is selectively amplified by a low-frequency amplifier tuned to the switching frequency. Figure 3 is a block diagram of Dicke's modulation radiometer. By using the modulation method one is able to measure cosmic noise of very low intensity. Indeed, modulation methods permit an approach to the theoretical limit of sensitivity as set by the fluctuations about the internal noise level of the receiver (where the noise level is determined by the time constant of the recording device and by the bandwidth).

A variant of the modulation method, known as the *frequency-modulation method,* is often employed in the study of monochromatic radio waves originating in the Galaxy. There also exist other methods of measuring weak cosmic noise against significantly more powerful, and not fully stable, background noise arising in the apparatus; but we shall not dwell upon them here.

The methods we have described permit the measurement of a signal of minimal power, close to the theoretical limit. If the bandwidth $\Delta \nu$ = 1 Mc/s and if the time constant $\tau = 100$ sec, then the ratio of the minimal power of a useful signal to the power of the noise generated in the apparatus will, for a receiver with a sufficiently small noise factor, become $(\Delta \nu \cdot \tau)^{-1/2} \approx 10^{-4}$ or 0.01 percent.

2. ANTENNAS AND INTERFEROMETERS IN RADIO ASTRONOMY

Antennas and receivers are the fundamental components of the various radio-telescope systems. In the preceding section we have given a brief survey of receivers, their functions, and their basic characteristics. In the present section we shall deal with the basic properties of antennas.

The purpose of the antenna of a radio telescope is to collect cosmic radio waves incident upon it from particular directions, having a particular state of polarization, and confined to a particular frequency range. The power collected by the antenna is partially or wholly delivered to the receiver input. In order that the energy selected by the antenna may be fully utilized, it is necessary that the energy flux in the line connecting antenna to receiver suffer no reflection as it passes along the line between them, and that there be no other losses. To this end, the resistance and reactance in the line and in the receiver must be suitably selected; that is, the antenna must be accurately *matched* to the receiver.

One of the properties of the various types of antennas is that most of them accept only one component of linear polarization of the incident radiation; for example, the half-wave dipole accepts only the component of polarization parallel to the axis of the dipole. Thus if the cosmic radio waves incident on such an antenna are unpolarized, the antenna will accept only one-half of the energy flux.

We have already pointed out that, in general, an antenna will "cut out" a more-or-less broad frequency band from the incident radiation flux. Compound antennas, of which the most elementary are the half-wave dipoles, are notable for the greatest selectivity; but even in this case the frequency band cut out by the antenna is significantly wider than the passband of the receiver. Much less selective ("wide-band") are the rhombic and helical antennas. The wide-band characteristics of paraboloidal reflectors are limited at high frequencies by the standards to which the reflector surface is manufactured. Another factor which governs the wide-band characteristics of paraboloids is the presence at their focus of the so-called *feed,* which is usually quite selective.

Let us assume that we have an extended source of radio waves with a given intensity distribution. The intensity is defined as usual in astrophysics: it is the energy passing through unit area in unit time, in unit frequency interval, and within unit solid angle. The spectral flux density is related to the intensity by the equation

$$F_\nu = \int I_\nu d\Omega. \tag{2-1}$$

In radio astronomy, spectral flux density is usually expressed in watts per square meter, per cycle per second [w m^{-2} (c/s)$^{-1}$]; and the intensity in watts per square meter, per cycle per second, per steradian [w m^{-2} (c/s)$^{-1}$ sterad^{-1}]. To convert these units into cgs units—and it is necessary to do this for theoretical calculations—they must be multiplied by 10^3.

Let θ and φ be the angular coordinates of a given element of a source of radio waves. Then the power absorbed by the antenna will be given by

$$P_{\text{abs}} = \tfrac{1}{2} \int A(\theta, \varphi) \cdot I_\nu(\theta, \varphi) \sin \theta d\theta d\varphi d\nu = \tfrac{1}{2} \int A(\theta, \varphi) \cdot I_\nu(\theta,\varphi)\, d\Omega d\nu. \tag{2-2}$$

The quantity $A(\theta, \varphi)$, which has dimensions of area, is called the *effective area* of the antenna. If the source has a sufficiently small angular dimension, then the spectral flux density may be observed; in this event,

$$P_{\text{abs}} = \tfrac{1}{2} F_\nu \cdot A(\theta, \varphi) \Delta\nu. \qquad (2\text{-}3)$$

The factor $\tfrac{1}{2}$ in Eqs. (2-2) and (2-3) arises from the fact that the antenna accepts only one component of polarization, while the radiation is assumed to be unpolarized.

The function $A(\theta, \varphi)$ is frequently represented graphically in polar coordinates, and is called the *directional diagram* of the antenna. A typical directional diagram appears in Fig. 4. Here, the portion where

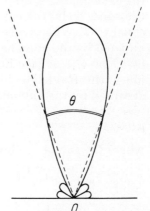

Fig. 4. Directional diagram of an antenna.

$A(\theta, \varphi)$ assumes its greatest value is called the *main lobe*. The minor lobes in the diagram are called the *side lobes;* furthermore, *back lobes* frequently occur as well. We shall designate by A_0 the maximum value of $A(\theta, \varphi)$, which is attained at $\theta = 0$. The mean value of the effective area of the antenna, taken over all directions, is

$$\langle A \rangle = \frac{1}{4\pi} \int A(\theta, \varphi) \, d\Omega. \qquad (2\text{-}4)$$

The dimensionless quantity

$$G(\theta, \varphi) = A(\theta, \varphi) / \langle A \rangle \qquad (2\text{-}5)$$

is called the *gain* of the antenna in the given direction. The maximum value of the gain is clearly

$$G_0 = A_0 / \langle A \rangle. \qquad (2\text{-}6)$$

It is shown in radiophysics that if a receiving antenna is used as a transmitting antenna the power radiated by it in each direction will have the same angular dependence $A(\theta, \varphi)$. It follows that the gain of an antenna is the ratio of the power radiated by it in a given direction to the power radiated in the same direction by an imaginary antenna whose directional diagram is spherical (that is, the radiation of the imaginary antenna is to be isotropic). We assume here that the total power radiated over all directions is the same for the real and the imaginary isotropic antennas.

Proceeding from simple thermodynamical considerations, we now derive an important relation between G and A. Let the resistance R be connected to the terminals of the antenna, and assume that all losses in the antenna are negligibly small. Furthermore, let an ideal black body at temperature T be placed at a given distance from the antenna, subtending the solid angle Ω at the antenna. As a result of the exchange of energy which takes place between the resistance and the black body, the equilibrium temperature of the resistance will be equal to T. When an equilibrium state has been attained, the energy radiated by the antenna and absorbed by the black body will be given by $kT\Delta\nu \cdot G\Omega/4\pi$ [see Eqs. (1-7) and (2-5)]. On the other hand, according to the Rayleigh–Jeans formula the power from the black body absorbed by the antenna will be equal to

$$\tfrac{1}{2} \cdot 2kT\Delta\nu \cdot \lambda^{-2}\Omega A,$$

where the factor $\tfrac{1}{2}$ allows for the acceptance by the antenna of only one component of polarization. Equating the absorbed energy to the energy radiated by the antenna, we obtain the desired relation:

$$G = 4\pi A\lambda^{-2}. \qquad (2\text{-}7)$$

Combining Eqs. (2-7), (2-5), and (2-4) we then have

$$\int A(\theta, \varphi)\, d\Omega = \lambda^2. \qquad (2\text{-}8)$$

The relations we have derived are valid for all types of antennas. Actual antennas are characterized by a quantity d, the *directivity*, related to the gain by the equation

$$G = \eta d, \qquad (2\text{-}9)$$

where η is a factor allowing for losses in the antenna. To characterize the directivity of an antenna one sometimes employs the "effective solid angle of the antenna," $\Omega_a = 4\pi/G$. The *beamwidth* is in wide usage as such a characteristic; this is the width of the main lobe of the directional diagram, the angle θ between the points of the diagram at which $A = \tfrac{1}{2}A_0$. In general, a plane section of the main lobe, taken perpendicular to the axis of the diagram, is elliptical. The width of the directional diagram along the two axes of this ellipse will be θ_A and θ_B, say. The values of the maximum gain G_0 and of θ_A and θ_B are related by the approximate expressions

$$G = 10/(\theta_A \theta_B), \qquad \theta_A, \theta_B \text{ in radians,}$$

$$G = 35\,000/(\theta_A \theta_B), \quad \theta_A, \theta_B \text{ in degrees.}$$

The sensitivity of an antenna is also determined by the *resolving power* of the radio telescope. The main lobe in the directional diagram corresponds to the primary maximum of a diffraction pattern, and the side lobes to the secondary maxima. If the angular distance between two sources is much less than the width of the main lobe, the sources are no longer resolvable with the given radio telescope.

One of the characteristic features of radio astronomy—the very low resolving power of radio telescopes as compared to optical telescopes —is of course a consequence of the great wavelength of the incident radiation. For example, if D is the diameter of a paraboloidal reflector, then $\theta = 1.03 \, \lambda/D$; even for a 250-ft paraboloid, the largest yet constructed, $\theta \approx 50'$ at $\lambda = 1$ m. Quite often the value of θ is of the order of several degrees, sometimes even $10°$. The application of interferometric methods considerably increases the resolving power of radio telescopes, but it nevertheless remains several orders of magnitude smaller than that of optical telescopes.

Observational results in radio astronomy are frequently expressed in thermal units, as equivalent temperatures; an analogous treatment is employed in optical astronomy. For example, the important concept of the "effective temperature" T_{eff} of a star does not at all signify that the star radiates as an ideal black body at temperature T_{eff}. The effective temperature is a convenient parameter describing the total radiation of the star, integrated over all frequencies. A similar interpretation obtains in connection with such terms as "color temperature" and "excitation temperature." The same concept may be expressed both in terms of equivalent temperatures and in terms of immediate observational results and measurements. The intensity of cosmic radio waves received is most frequently expressed as a *brightness temperature* T_b, defined by the relation

$$T_b = I_\nu \lambda^2 / 2k. \tag{2-10}$$

Evidently T_b is the temperature of an ideal black body whose radiant intensity in the frequency range $(\nu, \nu + d\nu)$ is the same as that of the observed source. The brightness temperature of a given source usually varies widely with frequency.

An important concept in radio astronomy is the *antenna temperature T_a*, defined by the expression

$$T_a = \frac{1}{k \cdot \Delta \nu} \cdot \tfrac{1}{2} \int A(\theta, \varphi) \cdot I_\nu(\theta, \varphi) d\Omega d\nu. \tag{2-11}$$

The term "antenna temperature" is clearly equivalent to "the power absorbed by the antenna." The physical meaning of the antenna temperature is as follows: if a radio telescope is surrounded on all sides by a surface representing a black body at temperature T_a, then the antenna will absorb the power $kT_a\Delta\nu$ (provided that no losses are suffered in the antenna and the surrounding medium). One may give another interpretation of the antenna temperature: let us assume that a matched impedance, heated to temperature T_a, is connected to the terminals of the antenna; then, because of the thermal motions of the electrons, it will deliver to the terminals the power $kT_a\Delta\nu$, which is equal to the power absorbed by the antenna when directed toward a source of radio waves.

It is important to note that the antenna temperature is not related to the temperature of the material of the antenna itself, nor to that of the surrounding medium. It depends only upon the temperature of the bodies whose radiation is being received. But we have adopted the brightness temperature as the parameter of this radiation; hence a relation of the form $T_a = f(T_b)$ must obtain. We may readily derive this relation. Equation (2-2), giving the power absorbed by the antenna, may be expressed in units of equivalent temperatures if we replace $I(\theta, \varphi)$ by $2kT_b(\theta, \varphi)\cdot\lambda^{-2}$ and make use of Eqs. (2-11) and (2-8). We shall then have:

$$T_a = \lambda^{-2} \int A(\theta, \varphi)\cdot T_b(\theta, \varphi)d\Omega$$
$$= \int T_b(\theta,\varphi)\cdot A(\theta, \varphi)d\Omega / \int A(\theta, \varphi)d\Omega. \quad (2\text{-}12)$$

If the intensity of the radio waves (or T_b) is constant within the main lobe, then, in Eq. (2-12), T_b may be taken outside the integral sign and

$$T_a = T_b. \quad (2\text{-}13)$$

If the angular dimension of the source of radio waves is small compared to the beamwidth, then we have approximately

$$T_a = T_b\Omega/\Omega_a, \quad (2\text{-}14)$$

where Ω is the solid angle subtended by a source at brightness temperature T_b, and $\Omega_a = 4\pi/G$ is the effective solid angle of the antenna. All the formulas we have cited are valid only for a loss-free radio-telescope system with matched components. It is not difficult to generalize them to the case in which losses cannot be neglected, but we shall not deal with the matter here.

Usually the power of a signal from a source of cosmic radio waves is expressed as an antenna temperature. The sensitivity of a radio telescope may consequently be defined as the minimum change in antenna temperature which can be recorded by a recording device placed at the output of the telescope. Equation (1-10) may therefore be rewritten as

$$\Delta T_{\mathrm{a}} \geq \alpha N T (\Delta \nu \cdot \tau)^{-1/2}, \tag{2-15}$$

where one usually sets $T = 290°\mathrm{K}$, and where α, as indicated previously, is a dimensionless parameter of the order of unity. For example if $N = 5$, $\Delta \nu = 1\,\mathrm{Mc/s}$, and $\tau = 100\,\mathrm{sec}$, then the minimum change in antenna temperature which can be detected by the radio telescope will be $\Delta T_{\mathrm{a}} \approx 0.1\,\mathrm{K\,deg}$.

The procedure for measuring the flux density of radio waves emanating from a discrete source consists in directing the radio telescope at the source, and in measuring the resulting deflections on the recording device from the level which would obtain in the absence of the source. If the receiving apparatus is calibrated, then the deflections may be converted into antenna temperature; knowing the gain of the antenna, one may now obtain the spectral flux density in absolute units, such as watts per square meter, per cycle per second.

To investigate the flux from weak sources, it is very desirable to employ an antenna of maximum effective area A. The sensitivity of the radio telescope is thereby increased insofar as the quantity $\frac{1}{2} F_{\nu} A$ is directly measured (that is, until the angular dimension of the main lobe becomes appreciably greater than that of the source); furthermore, a higher resolving power is achieved. The latter circumstance, in particular, diminishes the fluctuations in the intensity of the sky background.

Another important problem in radio-astronomical measurements is the derivation of the intensity distribution over an *extended* source of radio waves. It must be noted that all known sources of radio waves appear to be extended, so that this problem is a fundamental one in observational radio astronomy. In order to investigate intensity distribution, radio telescopes of sufficiently high resolving power are demanded—that is, with a sufficiently narrow beam.

The antenna in Fig. 5 exemplifies a design in which the resolving power is great enough for the intensity distribution over a number of extended sources to be observed successfully. This is the large radio telescope at the Pulkovo Observatory near Leningrad. Its reflector, a paraboloidal section of adjustable profile, consists of 90 plane panels measuring $1.5 \times 3\,\mathrm{m}$; the antenna opens out to a span of 120 m. At $\lambda = 3\,\mathrm{cm}$, the maximum angular size of the directional diagram is $1° \times 1.2$, but the vertical dimension is compressed for sources crossing the meridian at greater altitudes. Radio telescopes, however, are quite often found to have inadequate resolving power.

An immediate result of the observations is the antenna temperature T_{a}, which according to Eq. (2-12) is a mean of the brightness temperature T_{b} weighted according to the directional diagram. If the angular dimensions of the source are smaller than the beamwidth, the separate details of the distribution in brightness will not be revealed. This circumstance is illustrated in Fig. 6, which shows schematically the tracings made by the recording device as sources of radio waves, of angular dimensions as indicated in the lower part of the figure, traverse the directional diagram of the radio telescope. Curve A represents the case in which the angular dimensions of the source are negligibly small (a point source). Curve B corresponds to a source whose angular dimensions are approximately half the width of the main lobe. Figure 6 shows that these curves are practically indistinguishable. Curve C corresponds to a source whose angular diameter is greater than the width of the main lobe; even in this case, curve C is only 25 percent wider than curve A.

The problem of extracting the "true" distribution of brightness temperature $T_b(\theta, \varphi)$ from the "observed" distribution of antenna temperature $T_a(\theta, \varphi)$ is entirely analogous to the common spectroscopic problem of calculating the influence of the instrumental profile on the true profile of a spectral line. In the one-dimensional case, one may write the following well known integral equation, which may be regarded as expressing the inadequacy of the resolution:

$$T_a = \int\limits_{-\infty}^{+\infty} A(\xi - x)T_b(\xi)d\xi. \qquad (2\text{-}16)$$

Many methods have been devised for solving this equation. In studying the intensity distribution over the general field of galactic radio waves, some investigators have corrected for the inadequate resolving power of the antenna by solving Eq. (2-16). But in practice, the details in the intensity distribution are thereby lost if the dimensions of the source are equal to or smaller than the width of the main lobe. A considerably more effective way to investigate the intensity distribution within sources is to apply interference methods.

As we have already mentioned, interferometric methods of observation are employed to increase the resolving power of radio telescopes. These methods are in principle analogous to the celebrated method of Michelson for determining the angular dimensions of stars, which cannot be measured directly even with the largest modern optical telescopes. (Actually, the diameters of a few stars of maximum angular size can just be attained at the limit of the 200-in. Palomar reflector.) It is to be noted that in radio astronomy interference methods are applied more extensively than in optical astronomy. Many special instruments, known as *radio interferometers,* have already been constructed, with quite diverse properties.

Existing radio interferometers may be separated into two classes, the twin-wave and the multiple-wave. We first consider the basic characteristics of *twin-wave interferometers.* Two identical and similarly oriented antennas are joined by a line, the feeder, to a single

Fig. 5. A Pulkovo radio telescope, with adjustable reflecting panels and a movable focus.

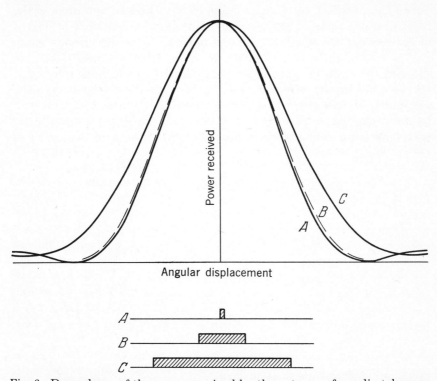

Fig. 6. Dependence of the power received by the antenna of a radio telescope on the angular dimensions of the source, denoted by the hatched rectangles below (Pawsey and Bracewell 1955).

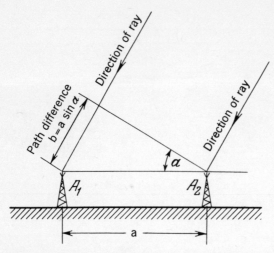

Fig. 7. Diagram of a two-antenna radio interferometer.

receiver (Fig. 7). If a plane wave incident upon them has a direction in space normal to the line A_1A_2 (the *baseline* of the interferometer), then the phase of the oscillations will be the same at A_1 as at A_2; hence the phase of the signals delivered by the two antennas to the receiver input will also be the same (assuming that the electric lengths of the feeders are equal). But if the normal to the direction in space makes the angle $\alpha \neq 90°$ with the base, then a phase difference $(2\pi/\lambda)a \sin \alpha$ will be introduced between the two signals arriving at the receiver input; here a is the *spacing*, the distance between the antennas. If the angle α is small, then the phase difference will be $ka\alpha$, where $k = 2\pi/\lambda$. Let P_0 denote the power from a single antenna, arriving at the receiver input. Then, for an arbitrary angle α,

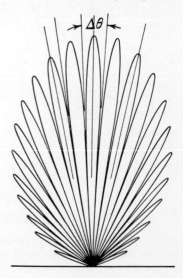

Fig. 8. A multilobed directional diagram.

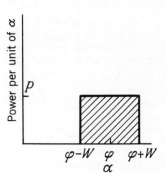

Fig. 9. Explanation of the method of deriving the angular dimensions of a source of rectangular intensity distribution (Pawsey and Bracewell 1955).

the power delivered to the receiver input by both antennas will have a mean value over a period which is given by

$$P = \tfrac{1}{2}P_0(2 \cos \tfrac{1}{2}ka\alpha)^2 = P_0(1 + \cos ka\alpha). \tag{2-17}$$

If the source of radio waves moves relative to the interferometer, the angle α will change and the received power P will oscillate between the values $2P_0$ and 0. This means that the directional diagram of the interferometer will have a multilobed pattern as shown in Fig. 8. The angular size of each lobe will be $\Delta\theta = \lambda/a$, which may be made very small by using a sufficiently large spacing. This circumstance permits of a substantial increase in the resolving power of radio telescopes.

With the aid of such a radio interferometer, one may determine the angular dimensions of sources of radio waves. To simplify the discussion we shall assume that the source has a rectangular shape, as shown in Fig. 9, and that the intensity within this rectangle is constant. If $pd\alpha$ denotes the signal delivered by a single antenna to the receiver input and determined by the radiation from the source within the angle $d\alpha$, then the signal resulting from both antennas of the interferometer will be given by

$$P = \int_{\varphi-W}^{\varphi+W} p(1 + \cos ka\alpha)\, d\alpha = 2pW + (2p/ka) \sin kaW \cos ka\varphi$$
$$= P_0[1 + (1/kaW) \sin kaW \cos ka\varphi], \tag{2-18}$$

where $2W$ is the width of the source, and $P_0 = 2pW$ is the signal from a single antenna determined by the radiation from the entire source. The angle φ varies as the source moves across the directional diagram of the interferometer; hence the signal at the receiver input will also fluctuate periodically about the constant value P_0 (Fig. 10). The ratio of the minimum to the maximum value of the signal is called the *percentage modulation,* and is equal to

$$R = \frac{1 - (\sin kaW)/kaW}{1 + (\sin kaW)/kaW}. \tag{2-19}$$

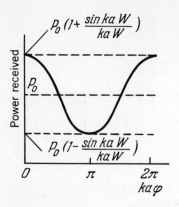

Fig. 10. Curve representing the power at the output of a radio interferometer as the source moves across the directional diagram (Pawsey and Bracewell 1955).

Clearly $0 < R < 1$. Measurement of the percentage modulation permits the determination of the angular dimension W of the source of radio waves.

We now consider the general case of an arbitrary intensity distribution over a source of finite angular dimensions (Fig. 11). Split the source into narrow strips, in each of which α may be regarded as constant. The position of each of these strips will be defined relative to an arbitrary line L across the source, for which $\alpha = \varphi$. Let β denote the angular distance of a strip from the line L. The signal arriving at the receiver input, and depending on radiation from a strip of width $d\beta$ located at distance β from the line L, will be given by $f(\beta)d\beta$, say. The resultant power delivered to the receiver input by the two antennas is now give by

$$
\begin{aligned}
P &= \int (1 + \cos ka\alpha) f(\alpha - \varphi) d\alpha \\
&= \int f(\alpha - \varphi) d\alpha + \int f(\alpha - \varphi) \cos ka\alpha \, d\alpha \qquad (2\text{-}20) \\
&= P_0 + F,
\end{aligned}
$$

since $\beta = \alpha - \varphi$. The term P_0 in Eq. (2-20) determines the power delivered to the receiver input by a single antenna and arising from the

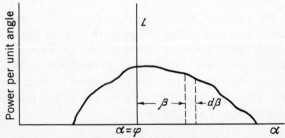

Fig. 11. Explanation of the method of deriving the angular dimensions of a source of arbitrary intensity distribution (Pawsey and Bracewell 1955).

radiation of the entire source. By Fourier integral theory the second
term in Eq. (2-20) may be written in the form

$$F = A(ka) \cdot \cos\left[ka\varphi - \psi(ka)\right], \qquad (2\text{-}21)$$

where

$$A(ka) \cdot \exp\left[i\psi(ka)\right] = \int\limits_{-\infty}^{+\infty} \exp\left[-ika\beta\right]f(\beta)d\beta.$$

Equation (2-20) is evidently insufficient to determine the unknown
intensity distribution $f(\beta)$, for one may always find a large number of
different functions having identical Fourier components at one frequency. The problem may, however, be solved if observations are carried out with various values of the quantity α, that is, if an interferometer with variable spacing is employed. As a result of such
observations, the unknown distribution $f(\beta)$ may be represented as a
Fourier integral:

$$f(\beta) = \frac{1}{\pi} \int\limits_{0}^{\infty} A(ka) \cdot \cos\left[ka\beta - \psi(ka)\right]d(ka). \qquad (2\text{-}22)$$

In practice, twin-wave interferometers are realized in the form of
either a sea interferometer or a two-antenna interferometer with a
horizontal base. In the *sea interferometer* (Fig. 12) the antenna is

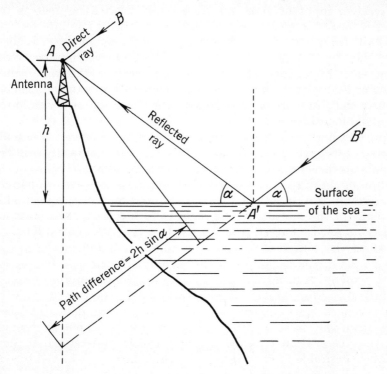

Fig. 12. Diagram of a sea-type radio interferometer.

mounted on a cliff at height h above the sea; the lobe is directed horizontally toward the point on the horizon where the source will rise or set. Interference is produced between the direct ray and the ray reflected from the surface of the sea. The optical analogue of this interferometric system is the Lloyd's mirror. The sea interferometer is equivalent to a two-antenna interferometer with a vertical base of length $a = 2h$, where the second antenna is taken as the mirror image of the actual antenna in the sea. Taking into account the loss of a half-wave upon reflection from the surface of the sea, we find that the signal at the receiver input experiences maxima at phase differences of $(4\pi h/\lambda) \sin \alpha = \pi, 3\pi, \ldots$ The signal at the receiver input will be equal to $2P_0[1 + (\sin kaW)/kaW]$, where $2W$ is the effective width of the source in a direction normal to the horizon, and P_0 is the signal from the same antenna when it is not operating as an interferometer.

The sea interferometer possesses a number of useful properties which render it especially convenient for the study of discrete sources. Very sharp lobes appear immediately, the largest lobe emerging first; if two close sources are to be resolved, this is very convenient since they will not rise or set simultaneously. The mean power delivered to the receiver input of a sea interferometer is twice as great as the power from the same antenna when it is not used as an interferometer; while for a two-antenna interferometer the mean power is given by P_0 [see Eq. (2-18)]. On the other hand, there are a number of deficiencies in the sea interferometer as compared with the two-antenna type. For example, the low elevation of sources above the horizon encumbers the observational results with the effects of refraction in the terrestrial ionosphere.

The baseline of a *two-antenna radio interferometer* is usually directed east–west. The main lobe of each of the identically oriented antennas is directed to some point on the meridian. Because of the rotation of the Earth, the sources of cosmic radio waves will sweep across the interferometer's directional diagram, which is located in the plane of the meridian. The record produced by a two-antenna interferometer is characterized by the fact that the central lobe, obtained as the source crosses the meridian, is found in the middle rather than at the beginning of the record, as is the case for the sea interferometer. A disadvantage of the two-antenna interferometer is the difficulty in separating the central lobe on the record from the adjacent ones, particularly when there is an abundance of lobes. Lobes from various sources become superimposed on one another, so that the observational results are not readily understood. In such work it is desirable to have fewer lobes; this may be achieved by employing sufficiently many antennas as interferometer elements. Then,

for a given spacing, a smaller number of lobes will be "inscribed" in
the directional diagram of each antenna (compare Fig. 8). But it
should be kept in mind that the number of lobes will be limited by
the finite bandwidth of the receiver. Only the central lobe will be the
same for all frequencies within the passband; lobes of higher order
will be displaced relative to one another for various frequencies. The
further they are from the central lobe, the greater will be this displacement. Finally, beginning with some particular lobe, the displacement of the lobes for the extreme frequencies within the passband
will begin to exceed the distance between adjacent lobes. This phenomenon is very like the coloring of the diffraction rings of a pattern
observed in white light.

For a two-antenna radio interferometer which is set up as described
above (a *meridian* radio interferometer), the right ascension of the
source is determined by the time of its meridian passage, that is, by
the time at which the central lobe of the interference pattern is observed. The declination of the source is determined by the speed with
which it crosses the meridian, according to the formula

$$\cos \delta = n\lambda/(a \sin H_n); \tag{2-23}$$

here n is the number of lobes observed between the time of transit of
the source and the given time t, and H_n is the hour angle of the source
at time t.

A significant improvement on the two-antenna radio interferometer was introduced by Ryle (1952). Figure 13 indicates the principle

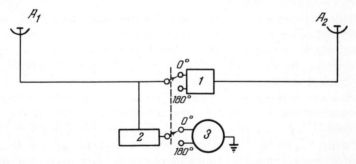

Fig. 13. Block diagram of the two-antenna Ryle interferometer: *1*, phase-
reversing switch; *2*, receiver; *3*, measuring device (Pawsey and Bracewell 1955).

of operation of such an interferometer. A switch is inserted in the line
connecting one of the antennas of the interferometer to the receiver
input; in one position of the contact, it introduces a supplementary
phase difference of 180° between the signals delivered by the antenna
to the receiver input, while at the other position of the contact it

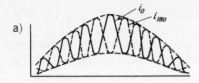

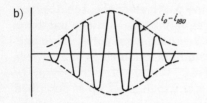

Fig. 14. Principle of the swept-lobe interferometer (Pawsey and Bracewell 1955).

introduces no phase difference. The broken line in Fig. 14a represents a record with the supplementary phase difference of 180°; the solid line, without the phase difference. In observing with the interferometer, the phase-reversing switch oscillates rapidly between its two positions. One should think of the directional diagram as being rapidly scanned in such a way that the lobes of the diagram jump back and forth over their own width. Interferometers constructed in this manner are therefore often called *swept-lobe* interferometers. The meter in the receiver output records the *difference* in current between the 0° and the 180° phase Fig. 14b; thus the method is seen to be a variant of the modulation method. A schematic tracing from the recorder is shown in Fig. 15.

This interferometer system has a decided advantage over the simple system since it retains all the advantages of the modulation method. An important feature is that it also eliminates, to a considerable degree, the effect of fluctuations in the general cosmic radiowave background, for only the *modulated* signal from the source is recorded at the receiver output. At the same time, when analyzing a single tracing obtained with a constant interferometer spacing *a,* it is not possible to secure from such a system any information on the angular dimensions of the source, since the percentage modulation of the interference pattern cannot be obtained. However, one may de-

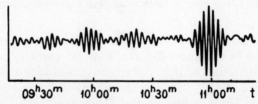

Fig. 15. A phase-switched interferometer record (Pawsey and Bracewell 1955).

rive the angular dimensions of the source by analyzing records obtained with a *variable* spacing.

There is a definite limit to the increasingly large interferometer spacing which would be required for a substantial rise in resolving power. The limit is determined by the attenuation of the signal from the antennas as it passes along the line to the receiver. As the length of this line (determined by the spacing) increases, the attenuation becomes more important. If there is considerable attenuation the signal will be very weak at the receiver input, and it will not be possible to distinguish it from the internal noise of the apparatus. To contend with this difficulty, supplementary amplifiers are installed in the line between the antennas and the receiver. By means of such systems it has now become possible to obtain a spacing as great as 10 km. It is also highly important to diminish the losses in the lines, a problem in the technology of the manufacture of feeders.

A *multiwave radio interferometer* of an interesting design has been constructed by Christiansen in Australia. This interferometer consists of a system of 32 reflectors, each of 1.7-m diameter, fed in phase, and placed at a constant distance apart along an east–west direction; the total length of the line is 217 m. Figure 16a shows a diagram of the apparatus and of the connection of the series of reflectors. In Fig. 16b is given a portion of the directional diagram of this interferometer at λ = 21 cm. This diagram is entirely analogous to the distribution of maxima formed by a diffraction grating. It consists of narrow (3′) primary maxima, separated by quite wide (1°.7) intervals; adjoining the

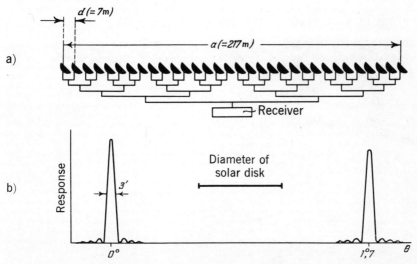

Fig. 16. The 32-reflector Christiansen radio interferometer: (*a*) feed arrangement; (*b*) a portion of the directional diagram (Pawsey and Bracewell 1955).

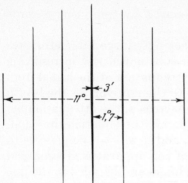

Fig. 17. Section by the celestial sphere of the diagram of the 32-reflector interferometer (Pawsey and Bracewell 1955).

primary maxima are considerably weaker secondary maxima. Figure 17 represents the section of the directional diagram of this interferometer by the celestial sphere. The size of the entire pattern (about $11°$ at $\lambda = 21$ cm) is determined by the directional diagram of an individual reflector. Figure 18 shows the series of 32 reflectors supplemented by a second series of 16 reflectors, placed perpendicular to the first series (Fig. 18). This permits a restriction of the directional diagram of the interferometer in the second coordinate, considerably extending its potentialities. It was planned to complete these series later on so as to secure a system in the form of a cross (see below). The interferometer has thus far been employed in the investigation of solar radio waves, and highly interesting results have been obtained with it. But it is doubtless possible to carry out research on galactic

Fig. 18. General view of the multiwave Christiansen interferometer.

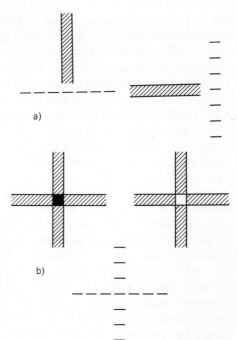

Fig. 19. Operation of the Mills cross: (*a*) an individual row of dipoles and its directional diagrams in the longitudinal and transverse directions; (*b*) arrangement of the dipoles in the form of a cross, and the directional diagrams of this cross (Pawsey and Bracewell 1955).

and extragalactic radio waves with this interferometer; it will probably be attempted in the near future.

In 1953 Mills proposed an essentially new antenna system; the operation of his instrument will be clear from Fig. 19. Consider an appropriately connected row of dipoles. The directional diagram of this assembly will be very narrow in the direction along the array, since it will be governed by the number of dipoles; while in the direction normal to the array it will be broad, since in this case it will be governed by the directional diagram of an individual dipole (see Fig. 19*a*, which shows the section of the diagram by the celestial sphere). Now consider another row of dipoles, perpendicular to the first; its directional diagram will be a narrow band perpendicular to the first. If the two antenna arrays are now connected in phase, then the amplitudes will combine within the central square; in the direction of the square, which comprises a very small fraction of the area of the sky whence radiation is being accepted, the sensitivity will increase by a factor of four (see Fig. 19*b*). If on the other hand the two dipole arrays are connected in opposite phase, then the sensitivity within the central square will vanish.

We now introduce a switching of the two antenna arrays from "in phase" to "antiphase" with a frequency of several tens or hundreds of cycles per second. The central square will then begin to "blink" with the same frequency; that is, the sensitivity in that direction will be alternately quite high and zero, while at the same time the sensitivity will be unaltered within the remainder of the large cross-shaped diagram. If within the region of the central square of the directional diagram there is located a source of cosmic radio waves, then the power proceeding from it to the receiver input will fluctuate periodically—that is, it will be modulated. Along with the power from this source, there will be delivered to the receiver input power from all sources located within the limits of the large cross-shaped diagram; the latter may significantly augment the power from the source located within the central square. To suppress this interfering radiation the receiver must be equipped with a standard modulation circuit. Then the constant radiation from all sources within the diagram, except for the central portion, will not pass through, and only radiation from the source within the central square will be observed. Consequently the resolving power of the Mills cross antenna system will be determined in *both* coordinates by the narrow portion of the directional diagram of each dipole array. In order to obtain such a narrow diagram by ordinary methods, it would be necessary to fill up an immense square with dipoles, at a suitable density; the side of the square would be equal to the length of the Mills cross dipole arrays.

Until recently, the solid angle of the main lobe of an antenna was always determined by its effective area. The Mills cross permits a radio telescope to be realized which incorporates, as it were, two opposite characteristics: an exceptionally high resolving power in both coordinates, and a relatively small area.

In Australia two crosses have been operating, at 22 and at 85 Mc/s. The resolving power at the latter frequency is 50', a record for so low a frequency. The length of each array (Fig. 20) is about 400 m. Some notable results have already been obtained with these antennas.

Because of the low cost of construction and the high resolving power of the cross, considerable notice has been taken of it by researchers. There is in prospect for the near future the construction of crosses extending up to 10 km or more. With such antennas, a resolving power of about 5' may be attained at a wavelength of 10 m, a matter of extreme interest. The main lobe of the cross may be directed to an arbitrary part of the sky, not very close to the horizon. It is desirable to increase the effective area of the crosses, so that objects of low energy flux may be observed to high accuracy. To this end, rather than use half-wave dipoles as elements of the cross, as Mills did, one might use more complex elements such as Yagi antennas, helical antennas, and even reflectors. For example, upon the base of the multiwave Australian interferometer which we discussed above there has been constructed a cross-shaped system, whose elements are reflectors of 1.7-m diameter. Each array in the cross extends for 400 m; the resolving power at 21 cm is about 2′.

In England an antenna system has recently been constructed whose directional diagram similarly provides a "pencil beam" (Blythe 1957*a*). This "moving-T antenna" (Fig. 21) consists of a line of dipoles forming a multiwave interferometer, together with a small movable antenna. In the same figure are represented a Mills cross, and a multi-wave interferometer in the form of a two-dimensional lattice. The directional diagrams of all these antennas are identical.

Observations with the new antenna system are carried out at different positions of the small antenna; after reduction, the results are equivalent to those which would be obtained with a single large pencil-beam antenna. As constructed at Cambridge, the pencil-beam system has a resolving power of 2°.2 at 38 Mc/s. Much valuable information has already been secured with it (see Sec. 5).

We conclude this chapter with a comparison of the characteristics of radio and optical telescopes.

As we have already stressed, the resolving power of optical telescopes is many orders of magnitude higher than for radio telescopes. Nevertheless in recent years radio astronomy has made a startling advance in this respect, by the wide application of interferometric methods. In particular instances the resolving power has already attained several tens of seconds of arc, which is higher than the resolving power of the unaided human eye!

The sensitivity of modern large radio telescopes is exceptionally great. At the limit of its resources, modern radio astronomy can detect and measure a spectral flux density in the meter wavelength

Fig. 20.   General view of a Mills cross.

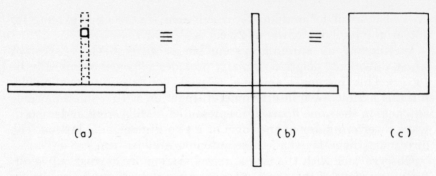

Fig. 21. Arrangements of antennas (a) in the moving-T system, (b) in a Mills cross, (c) for a conventional aperture giving equivalent resolution (Blythe 1957a).

range of $F_\nu^{\min} \approx 10^{-26}$ w m^{-2} (c/s)$^{-1}$. If we take the bandwidth of the receiver as ≈ 1 Mc/s, we find that the minimum flux of radio waves to which a large modern high-sensitivity radio telescope can still respond is $F_\nu^{\min}\Delta\nu \approx 10^{-17}$ erg cm^{-2} sec^{-1}. In the photographic range, a radiation flux of this order would correspond to an object of the 30th stellar magnitude! The most powerful existing telescope, the 200-in. reflector at the Palomar Mountain Observatory, can detect at its limit objects of $\approx 23^m.5$. Thus in absolute sensitivity radio telescopes already far surpass the optical; among all known receivers of radiation, they have no equal.

One might raise the objection that criteria of absolute sensitivity are immaterial, since the absolute radiating power of objects is many orders greater in the optical range than in the radio range. But this is not correct. In recent years objects have been detected whose radiation in the radio range, when expressed in absolute units, markedly exceeds their radiation in the optical range. Yet the optical brightness of these objects—peculiar galaxies—is immense. In such circumstances the extremely high absolute sensitivity of modern radio telescopes opens up rich perspectives for radio astronomy.

In radio astronomy we are always restricted to a quite limited frequency range, while in optical astronomy it is necessary to employ spectrographs or special filters to investigate a most important characteristic of radiation, its spectral composition. In the latter process a considerable loss of light inevitably arises. But in radio astronomy the problem of studying the spectral composition of radiation is solved in a very natural manner: every radio telescope is also a good monochromator.

A distinctive peculiarity of radiophysical methods is the ability to measure frequencies with extreme accuracy, unattainable in optical astronomy. This opens up the possibility of using the Doppler effect to study the diverse motions of cosmic objects, even those very distant from us. Thus far only the first steps have been taken in this direction, but again the perspectives here are most promising.

Thus, despite their comparatively low resolving power, radio telescopes surpass the optical in a number of basic characteristics. It must, however, be kept in mind that cosmic objects which radiate strongly in the radio range usually radiate little or not at all in the optical range, and conversely. As a result, radio-astronomical methods of investigation form an essential supplement to optical methods, without by any means replacing them. Cooperation between optical and radio-astronomical methods has already led to a significant enrichment of our understanding of the universe, an understanding which, in the future, will undoubtedly be enriched still further.

II

Cosmic Radio Waves:
Basic Observational Results

3. OBSERVATIONS AT DIFFERENT FREQUENCIES

In December 1931, Karl Jansky (1932) of the Bell Telephone Laboratories, while investigating atmospheric interference in radio reception at a wavelength of 14.7 m, discovered a source of interference whose intensity fluctuated with a period of 23^h56^m. Such a periodicity naturally led Jansky to the conclusion that the source was extraterrestrial, and in fact beyond the solar system. In the course of his investigation, using an antenna of relatively low directivity, he observed that the source of cosmic radio waves appeared to be in the Sagittarius region. As is well known, the center of the Galaxy also lies in this direction. More careful investigations carried out by Jansky (1933, 1935) during the following years showed that cosmic radio waves were concentrated along the Milky Way, and were most intense in the direction of the galactic center.

In the thirties, receiver sensitivity and antenna directivity were considerably lower than at the present time. Jansky was able to make his remarkable discovery, one which laid the foundations of radio astronomy, only because the intensity of cosmic radio waves attains a very high value at the frequencies at which he carried out his observations. The brightness temperature of cosmic radio waves in the region of the galactic center and at wavelengths of about 15 m is around $200\,000°$ K. The antenna temperature is of the same order since the angular dimensions of the emitting region are very large [see Eq. (2-14)]. It follows that even with a receiver noise factor $N \approx 50$ (and this would be a very poor receiver by present-day standards) the signal level would be several times as great as the noise level in the equipment.

The existence of intense sources of radio waves beyond the solar system was quite unexpected. One might assume that the radio emission from cosmic objects (for example, stars) would be more or less proportional to their optical emission. However, if that were the case, the observed radio flux from the sun would exceed the radio flux from

32

galactic and extragalactic objects by many orders of magnitude, since in the optical frequency range the solar radiation is about 10^8 times as great as that of all other objects in the universe combined (except the moon). Jansky was unable to detect solar radio waves, which were discovered only in 1944. This means that the flux of cosmic radio waves observed at the Earth is considerably greater than that arriving from the sun; it follows, then, that the ratio of the radiation fluxes of cosmic objects at radio frequencies is entirely different from the ratio observed at optical frequencies. For example, according to present-day data, at wavelengths ≈ 15 m the cosmic radio emission received is 1000 times as great as the solar.

Jansky's outstanding work was not developed further for several years. It was only in 1940 that the American radiophysicist Reber (1940) made the next important contribution to cosmic radio-wave research. He constructed a paraboloidal reflector of about 32-ft diameter and a receiver of relatively high sensitivity when operating at frequencies of about 160 Mc/s ($\lambda = 1.85$ m). The width of the directional diagram at this frequency was $10° \times 12°$. The reflector had a fixed mounting so that different parts of the sky drifted over its directional diagram on account of the Earth's rotation; the recording device was run continuously. Figure 22 shows some of the records obtained by Reber (1944). The maxima in these records correspond to the passage of a strip of the Milky Way across the main lobe of the

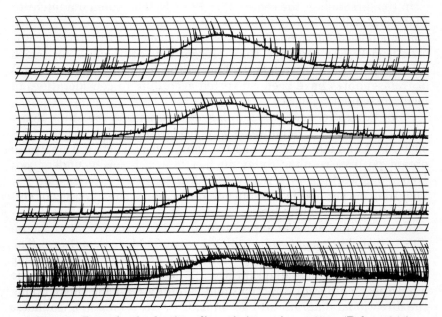

Fig. 22. Records of galactic radio emission at $\lambda = 1.85$ m (Reber 1944).

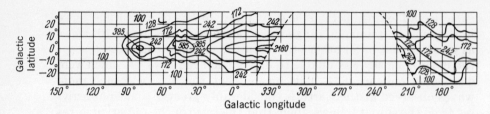

Fig. 23. Isophotes of galactic radio emission at $\lambda = 1.85$ m.

directional diagram. By changing the inclination of the antenna with respect to the horizon it was possible to observe "sections" of the celestial sphere at different declinations. After reducing the observations, Reber prepared a chart showing the brightness distribution of the cosmic radio waves (Fig. 23). Curves connecting different parts of the sky of the same intensity received the not very appropriate name of "isophotes." The broken lines in Fig. 23 indicate the limits of the survey.

The greatest intensity of radio waves was observed in the direction of the galactic center, where it reached a value of 1.48×10^{-20} w m^{-2} (c/s)$^{-1}$ sterad^{-1} or $4.5 \times 3.28 \times 10^{-18}$ cgs. From this value, the brightness temperature can be obtained from the relation

$$I_\nu = 2kT_b\lambda^{-2}.$$

In the direction of the center $T_b \approx 1200°$K. The minimum intensity that could still be detected was 7×10^{-22} w m^{-2} (c/s)$^{-1}$ sterad^{-1}, which corresponds to $\Delta T \approx 60$ K deg. As we can see, Reber's receiver was significantly less sensitive than present-day equipment, for which $\Delta T \approx 1$ K deg and even less. In Fig. 23 the brightness temperatures of the isophotes have been converted according to considerations to be discussed below, in Sec. 4.

Since he was very successful in his observations at 160 Mc/s, Reber attempted to detect radio waves at considerably higher frequencies—3300 Mc/s and 900 Mc/s—using the same reflector. He was unable to obtain positive results. Today such a failure is easily understood: Reber was working with receivers of low sensitivity, particularly at high frequencies, and in a frequency range where the intensity of cosmic radio waves is quite low.

Hey, Phillips, and Parsons (1946) in England investigated the intensity distribution of cosmic radio waves at 64 Mc/s ($\lambda = 4.7$ m). The antenna consisted of four Yagi arrays mounted on a freely rotating frame; the directional diagram was $12° \times 30°$. After measuring the intensities in different directions and carefully reducing them, taking the directional diagram into account, they drew emission isophotes covering the region $-30° < \delta < +60°$. These isophotes are shown in

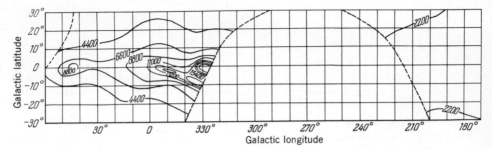

Fig. 24. Isophotes of galactic radio emission at $\lambda = 4.7$ m (Hey, Phillips, and Parsons 1946).

Fig. 24, where the brightness temperatures are given according to Piddington's reduction (see below). As seen from Fig. 24, the intensity distribution of cosmic radio waves at 64 Mc/s is fundamentally the same as at higher frequencies (Fig. 23).

At that time several surveys of cosmic radio waves were being carried out at different frequencies. For example, Sander (1946), working at 60 Mc/s, estimated the brightness temperature in the direction of the galactic equator to be 10 000° K, and in the direction of the galactic pole 1800° K, although this last figure was doubtful (5 percent of the power received by the antenna entered the side lobes). Moxon (1946) carried out a series of observations at 40, 90, and 220 Mc/s with the purpose of determining the frequency dependence of the intensity of cosmic radio waves (see Sec. 4). The maximum antenna temperature at 220 Mc/s was 350° K, while at 40 Mc/s it was 25 000° K.

Reber (1948), with his own reflector and a sensitive receiver (noise factor $N \approx 5$), investigated the intensity distribution of cosmic radio waves at 480 Mc/s ($\lambda = 62.5$ cm). Figure 25 shows the isophotes for 480 Mc/s drawn on the basis of Reber's observations. The isophotes show greater quantitative detail than isophotes at lower frequencies because of the relatively high directivity of the antenna (the width of the directional diagram was $2° \times 3°$).

Bolton and Westfold (1950) in Australia investigated in great detail

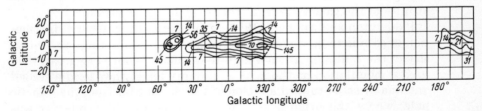

Fig. 25. Isophotes of galactic radio emission at $\lambda = 62.5$ cm (Reber 1948).

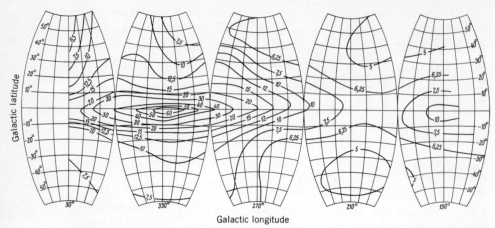

Fig. 26. Isophotes of galactic radio emission at $\lambda = 3$ m, in units of $100°$K (Bolton and Westfold 1950).

the intensity distribution of cosmic radio waves at 100 Mc/s ($\lambda = 3$ m). The antenna consisted of nine Yagi arrays combined in groups of three each. All the antennas were mounted on a structure that could move in altitude and azimuth. The half-width of the directional diagram of the system was 17°, and about two-thirds of the total energy incident on the antenna entered the main lobe. The side lobes were studied in detail and their influence was taken into account in the reduction of the observations. During the reduction process an attempt was made to include the insufficient resolving power of the antenna. Figure 26 shows the isophotes thus obtained by Bolton and Westfold for the frequency $\nu = 100$ Mc/s.

During the same year Allen and Gum (1950), also in Australia, investigated the intensity distribution of cosmic radio waves at 200 Mc/s ($\lambda = 1.5$ m). Their antenna consisted of four Yagi arrays. Since a standard zero point had not been established, they measured only the difference in brightness temperature between the region of the sky being investigated and a "cold" region surrounding the galactic pole. The results of their observations are shown in Fig. 27. The isophotes give the absolute value of the brightness temperature; according to Piddington's estimate (1951), this was taken as 71°K in the vicinity of the galactic pole, at the frequency in question.

Piddington and Minnett (1951) investigated the intensity distribution of cosmic radio waves at 1210 Mc/s ($\lambda = 25$ cm). Since the intensity of the radiation at such high frequencies is not very great, they used a modulation method. As a "dummy load" they used a second antenna directed toward a region of the sky not far from the galactic pole, where the brightness temperature is low. The receiver was periodically switched from the antenna directed toward the "cold"

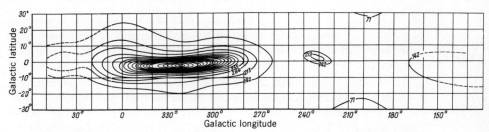

Fig. 27. Isophotes of galactic radio emission at $\lambda = 1.5$ m (Allen and Gum 1950).

part of the sky to the main antenna and back. Figure 28 shows the isophotes at 1210 Mc/s in the region of the galactic center. A sharp maximum at $\alpha = 17^h44^m$, $\delta = -30°$ is caused by a discrete source, Sagittarius A (Sec. 6). After taking into account the influence of this source, the brightness temperature in the region of the galactic center was estimated as 17°K.

At a frequency of 3000 Mc/s ($\lambda = 10$ cm) cosmic radio waves were detected only in the region of the galactic center, where the brightness temperature, according to Piddington and Minnett (1951), is 2.6°K.

Observations of cosmic radio waves at lower frequencies offer considerable interest. This type of investigation is particularly difficult because the resolving power of radio telescopes at long wavelengths is very low. Only during the past few years has the use of crosses, such as the one designed by Mills, enabled us to obtain sufficiently good resolving power even at long wavelengths.

Jansky's first observations at a wavelength of about 15 m had already shown that the brightness temperature of the sky in this range was very high. Shain (1951) published a series of observations

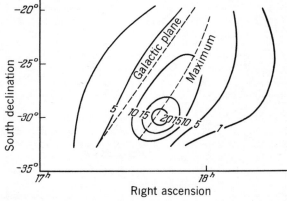

Fig. 28. Isophotes of galactic radio emission at $\lambda = 25$ cm (Piddington and Minnett 1951).

at 18.3 Mc/s (λ = 16.4 m), obtained with a fixed antenna consisting of eight half-wave dipoles pointed toward the zenith. Because of the daily rotation of the celestial sphere, the directional diagram swept a strip of the sky at a constant declination, equal to the latitude of the place at which the observations were taken ($-34°$). The galactic center and the South Galactic Pole are included within this strip.

In 1954 new investigations of the intensity distribution of cosmic radio waves at 18.3 Mc/s were completed by Shain (1954). This time he used an antenna of considerably higher directivity. The half-width of the main lobe of the directional diagram was about 17°. The antenna system consisted of 30 half-wave dipoles, mounted at a height of 0.2λ over the ground, and oriented approximately in a north–south direction. Minimum phase between the different dipole groups could be obtained by inclining the main lobe at angles up to 20° north and south of the zenith. This rendered it possible to investigate a wide strip of sky at declinations $-52° < \delta < -12°$, limited by the daily rotation of the celestial sphere over the zenith at the observing station.

Figure 29 shows the isophotes obtained from Shain's observations. As we can see from the diagram, the intensity distribution at 18.3 Mc/s is approximately the same as that obtained at higher frequencies (see Fig. 26). Notice, in particular, the very high value of the brightness temperature and the relatively small contrast between the "bright" and "dark" regions of the sky.

In the same year, Shain and Higgins (1954) carried out observations of cosmic radio waves at very low frequencies (9.15 Mc/s, λ = 32.8 m). Observations in this frequency range are difficult because

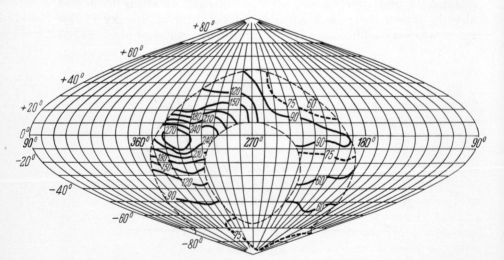

Fig. 29. Isophotes of galactic radio emission at λ = 16.4 m (ν = 18.3 Mc/s), in units of 1000°K (Shain and Higgins 1954).

of ionospheric effects. The antenna was fixed and pointed toward declination $-32°$; the width of the directional diagram was $31° \times 26°$. A curve was obtained giving the dependence of antenna temperature on sidereal time. This curve showed a maximum when the region of the galactic center was incident on the main lobe of the antenna. The brightness temperature in this region was estimated at 10^6 °K. Since the brightness temperature of the sun at these frequencies is of the same order of magnitude as the kinetic temperature of the outer corona, that is, 10^5 °K, the sun should appear as a dark spot against the bright background of cosmic radio emission.

Reber and Ellis (1956) have performed a series of very interesting observations of cosmic radio waves at extremely low frequencies: 2130 kc/s ($\lambda = 140$ m), 1435 kc/s ($\lambda = 210$ m), 900 kc/s ($\lambda = 333$ m), and 520 kc/s ($\lambda = 575$ m). These observations suffer from strong ionospheric interference. However, at the time of minimum solar activity, in some places on the Earth at high magnetic latitudes the ionospheric critical frequency at night reaches sufficiently low values to make the observations possible. Observations carried out in Tasmania will be described below.

For $\nu = 1435$ kc/s and $\nu = 900$ kc/s, Reber and Ellis used an antenna built with simple half-wave dipoles; for $\nu = 2130$ kc/s and $\nu = 520$ kc/s the antennas were also very simple. All the antennas were directed toward the zenith. The bandwidth of the receivers was about 4 to 6 kc/s. Because of the atmospheric interference the time constant of the recording device had to be very small, about 0.01 sec. The cosmic radio waves were recorded during very small intervals of time between atmospheric discharges.

The most complete observations were carried out at 2130 kc/s. During the period of observations the critical frequency fluctuated between 1.6 and 2.1 Mc/s. A very strong correlation was observed between the critical frequency and the level of cosmic radio waves (Fig. 30). In those cases when the critical frequency dropped below 1.6 Mc/s, the intensity of cosmic radio waves no longer increased. An analogous situation was observed at 1435 kc/s. At even lower frequencies, when the critical frequency was higher than that of the cosmic radio waves, observations were possible only during some isolated, peculiar ionospheric disturbances.

Figure 31 shows the intensity of cosmic radio waves as a function of sidereal time, at 2130 kc/s. Only those records obtained when the critical frequency was less than 1.6 Mc/s were used. The maximum intensity was observed when a region of the Milky Way located fairly near the galactic center was over the zenith.

Figure 32 again shows intensity as a function of sidereal time, at 1435 kc/s. According to this graph the intensity remains practically

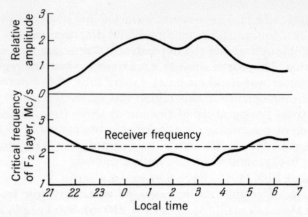

Fig. 30. Cosmic radio-wave intensity ($\nu = 2130$ kc/s) and the critical frequency of the F_2 layer as a function of sidereal time (Reber and Ellis 1956).

constant when going from the region of the galactic center to regions of higher galactic latitude.

Ellis (1957) has carried out an investigation of cosmic radio waves at very low frequencies, namely 10.05, 5.65, 4.4, and 3.8 Mc/s. For the first three of these frequencies he used a half-wave dipole; at 3.8 Mc/s he used a more complicated antenna, consisting of eight half-wave dipoles at a distance of 0.1λ from the ground. The directivity of this latter antenna was $26° \times 50°$, and the passband width of the receiver was 4.2 kc/s.

Figure 33 shows the dependence of the antenna temperature on sidereal time at $\nu = 3.8$ Mc/s. This dependence is approximately the same as that observed at $\nu = 2.13$ Mc/s (Fig. 31). There is a well-defined minimum between 22^h and 3^h, corresponding to the passage of a "cold" region of the sky over the directional diagram of the antenna.

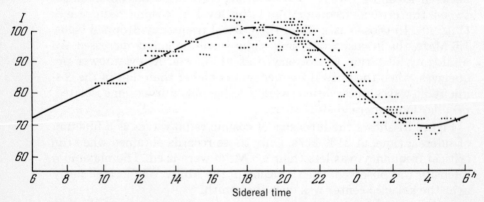

Fig. 31. Cosmic radio-wave intensity ($\nu = 2130$ kc/s) as a function of sidereal time (Reber and Ellis 1956).

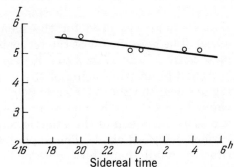

Fig. 32. Cosmic radio-wave intensity ($\nu = 1435$ kc/s) as a function of sidereal time (Reber and Ellis 1956).

All the investigations of the intensity distribution of cosmic radio waves at different frequencies which we have discussed so far were performed with antennas of low resolving power. Recently, surveys of cosmic radio waves have been made with larger antennas, which expose an amount of detail in the intensity-distribution charts that had not been detected before.

McGee, Slee, and Stanley (1955) in Australia investigated the intensity distribution of cosmic radio waves in the region of the galactic center at 400 Mc/s ($\lambda = 75$ cm). They used a reflector of about 25-m diameter, having a beamwidth of 2°. The antenna was pointed toward a region of the sky of chosen declination and remained fixed during the course of the day. Because of the rotation of the celestial sphere, regions of the sky of different right ascension drifted over the antenna. In successive days the antenna was shifted by about $\frac{1}{2}$° to 1° in declination, and in this way the sky between $\delta = -17°$ and $\delta = +49°$ was investigated. With the exception of occasional errors, the records of

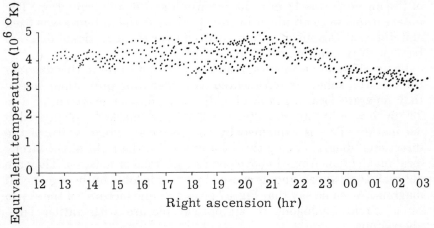

Fig. 33. Antenna temperature as a function of sidereal time at $\nu = 3.8$ Mc/s (Ellis 1957).

repeated observations of the same region were identical. A modulation method was used for the measurement of the cosmic radio waves. The noise factor of the receiver was equal to 11, and therefore its sensitivity was $\Delta T \approx 2$ to 3 K deg.

Absolute measurements of the brightness temperature of "cold" regions near the South Galactic Pole ($\alpha = 4^h30^m$, $\delta = -33°.9$) led to a value $T_b = 10° \pm 5°$K. Figure 34 shows the isophotes of cosmic radio waves in the region of the galactic center as obtained from the observations described above. The isophotes show considerably more quantitative detail than those obtained with radio telescopes of lower resolving power.

Piddington and Trent (1956) have published isophotes of cosmic radio waves at 600 Mc/s. Their antenna was a fixed reflector about 12 m in diameter, which could be set at different declinations; the width of the directional diagram was $3°.3$. Figure 35 shows Piddington and Trent's isophotes, which give the antenna temperature after correction for the width of the main lobe. These observations do not give a direct determination of the brightness of the background, that is, the absolute value of the brightness temperature of the cold regions of the sky. Extrapolating the data obtained by other investigators at lower frequencies, we can estimate that the brightness temperature of the cold sky at 600 Mc/s is between $4°$ and $8°$K or, taking a weighted mean, $5.7°$K.

Among other surveys of cosmic radio waves performed with radio telescopes of relatively high resolving power we may mention that of Kraus and Ko (1955), at a frequency of 250 Mc/s ($\lambda = 1.2$ m). For their observations they used the 96-element helical antenna at The Ohio State University (Fig. 36). The width of the directional diagram of the antenna was $1°$ in right ascension and $8°$ in declination. The observations were obtained in the usual way: the antenna was fixed and different regions of the sky drifted over the directional diagram because of the diurnal rotation.

Figure 37 shows isophotes of antenna temperature. In this diagram the celestial sphere is represented as a Mercator projection. Notice that separate local maxima of radio emission (for example, at $\alpha = 23^h30^m$, $\delta = +58°30'$; $\alpha = 20^h$, $\delta = +40°$) are markedly elongated in declination. This is explained by the relatively great width of the directional diagram along this coordinate. To the right of the strong local maxima mentioned above one can see weaker maxima. They are not real: their appearance is caused by the side lobes of the directional diagram. We could say that Fig. 37 gives a "radio picture" of the sky to some extent analogous to an optical picture with rather large aberrations.

The isophotes in Fig. 37 correspond to an intensity of 3.22 $\times$

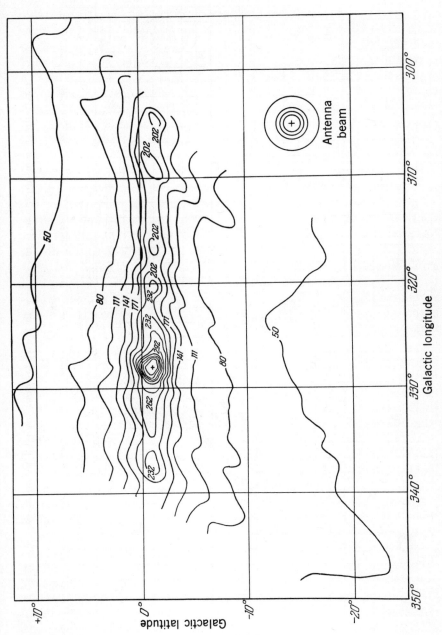

Fig. 34. Isophotes of galactic radio emission at $\lambda = 75$ cm (McGee, Slee, and Stanley 1955).

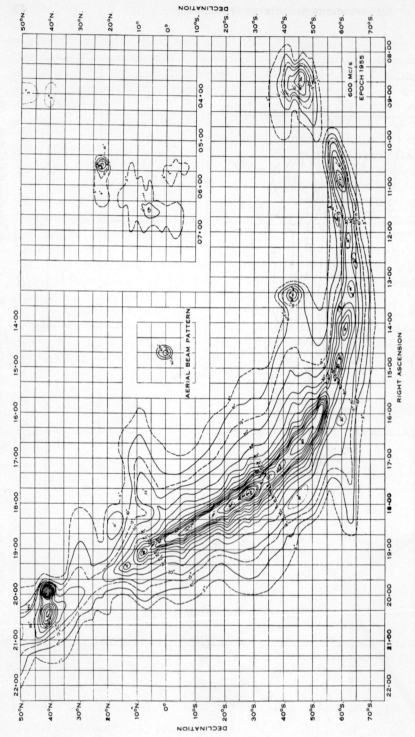

Fig. 35. Isophotes of galactic radio emission at $\lambda = 50$ cm (Piddington and Trent 1956).

10^{-22} w m^{-2} (c/s)$^{-1}$ sterad^{-1}. In their isophotes Kraus and Ko have plotted the intensity corresponding to the measured antenna temperature, multiplied by 1.26 (Piddington and Trent 1956). In order to compute the brightness temperature averaged over the main lobe we have to multiply the antenna temperature by 1.5 (to take the side lobes into account). By doing this we find that the intensity units of Kraus correspond to a brightness temperature $T_b = 20.6°\,\mathrm{K}$.

Hanbury Brown and Hazard (1953a), using the large fixed radio telescope of the University of Manchester (diameter $= 67$ m), investigated the intensity distribution of cosmic radio waves at 158.5 Mc/s ($\lambda = 1.89$ m). From the observed antenna temperature and a knowledge of the directional diagram, they estimated the brightness temperature averaged over the main lobe ($2° \times 2°$). The observations covered the region in the sky $40° < l < 130°$, $|b| \leq 12°$. Figure 38 gives the brightness-temperature isophotes. In comparison with the isophotes of Reber at similar frequencies (Fig. 23), those of Hanbury Brown and Hazard show greater quantitative detail.

A very important investigation of the intensity distribution of cosmic radio waves at 81.5 Mc/s was carried out by Baldwin (1955a) with one of the elements of the Cambridge interferometer. The width of the directional diagram was $1° \times 7°.5$. Figure 39 shows Baldwin's

Fig. 36. The 96-element helical antenna at The Ohio State University.

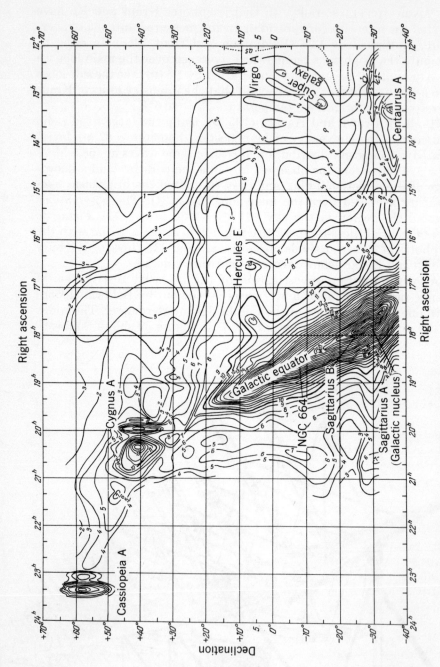

Fig. 37. Isophotes of galactic radio emission at λ = 1.2 m (Kraus and Ko 1955).

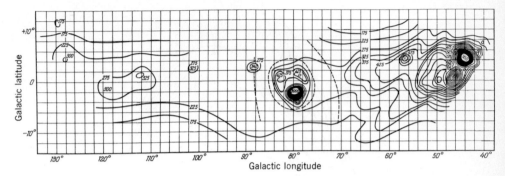

Fig. 38. Isophotes of galactic radio emission at $\lambda = 1.89$ m (Hanbury Brown and Hazard 1953*a*).

isophotes. The relatively low frequency at which the observations were carried out made possible the absolute determination of the antenna temperature for the cold regions of the sky.

Recently, the French radio astronomers Denisse, Leroux, and Steinberg (1955) investigated the intensity distribution at the rather high frequency $\nu = 910$ Mc/s ($\lambda = 33$ cm). For an antenna they used a German radar paraboloid, the Great Würzburg, 7.5 m in diameter. The width of the directional diagram was about $2°.5$; the sensitivity of the receiving equipment, $\Delta T \approx 1.5$ K deg. Figure 40 shows the isophotes of brightness temperature of cosmic radio waves with $\lambda = 33$ cm. Each unit corresponds approximately to 3 K deg.

The surveys of the intensity distribution of cosmic radio waves at different frequencies discussed so far include most of the major investigations performed up to the present time. In the next section we shall discuss the results of these observations.

4. ANALYSIS OF THE RESULTS

A rapid inspection of the isophotes presented in the previous section, obtained by different authors at different frequencies, shows that they all give the same general picture. We should point out, however, that when comparing different sets of isophotes we must keep in mind the very low resolving power of radio telescopes. This is particularly true for the earlier observations. Therefore we can only compare the isophotes obtained with radio telescopes of approximately the same directivity.

We began by presenting isophotes obtained with very low directivity. From these we see that cosmic radio emission is concentrated toward the galactic equator and in particular toward the galactic center. This shows that a considerable fraction of the sources of cosmic radio waves are within our stellar system. However, it is not possible to deduce a

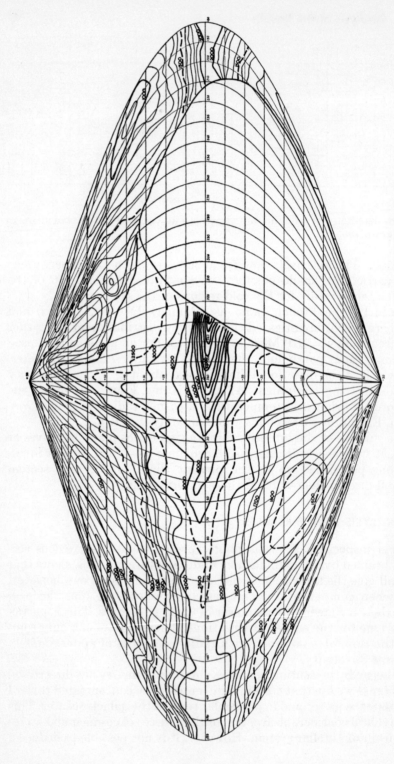

Fig. 39. Isophotes of galactic radio emission at $\lambda = 3.7$ m (Baldwin 1955a).

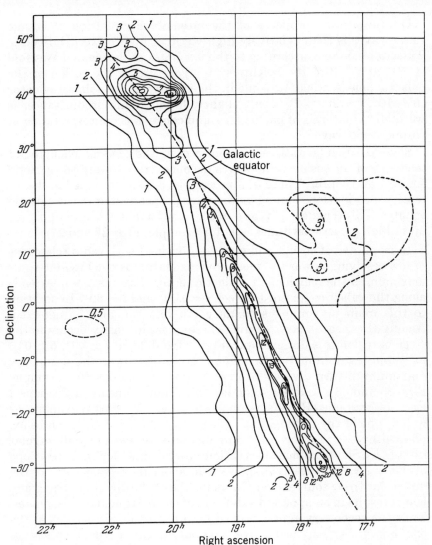

Fig. 40. Isophotes of galactic radio emission at λ = 33 cm (Denisse, Leroux, and Steinberg 1955).

reliable map of the intensity distribution of cosmic radio waves at low galactic latitudes from these early studies. In this region the intensity of the radio waves decreases very rapidly with increasing latitude; thus not even the most important details of the intensity distribution in this region of the sky could be detected with radio telescopes of low resolving power. On the other hand, at intermediate and high galactic latitudes, where changes occur slowly, the early observations have not lost their value.

One important property of the intensity distribution of cosmic radio waves is their relatively high value at high galactic latitudes. Thus, for example, according to the isophotes of Bolton and Westfold (1950), at $\lambda = 3$ m the brightness temperature in the region of the galactic pole is $\approx 600°$K, while in the vicinity of the galactic equator ($b = 10°$, $l = 200°$) T_b is only slightly higher. This circumstance, as we shall see later, is of particular significance in the interpretation of cosmic radio waves.

Several irregularities can be noticed in the general background tendency of the cosmic radio waves to concentrate toward the galactic equator and center. There is, for example, a wide "tongue" in the intensity distribution running more or less perpendicular to the galactic equator. This tongue begins at $b = 0°$, $l = 0°$ and it shows in isophotes up to high galactic latitudes (see, for example, Figs. 26 and 27).

In isophotes of cosmic radio waves obtained with radio telescopes of low resolving power the most obvious feature is the high-intensity maximum in the region of the galactic center. This maximum extends along the galactic equator between $l = 260°$ and $l = 30°$. In addition to this main maximum, secondary maxima can also be detected. Among these secondary maxima the highest intensity occurs in the Cygnus region (see, for example, Figs. 23 and 24, at $l \approx 40°$, $b \approx 0°$); in isophotes obtained with higher resolving power the secondary maximum in turn divides into two close maxima (see, for example, Figs. 25 and 38). Another secondary maximum appears in Cassiopeia (see, for example, Reber's isophotes in Fig. 23). This maximum, as well as one of the maxima in Cygnus ($\alpha = 20^h$, $\delta = +40°$), can be explained by the presence of strong discrete sources of small angular extent (see Sec. 6). When observations are taken with a radio telescope of relatively small resolving power, there occurs a "smearing" of the flux from strong sources over the main lobe of the directional diagram, which results in an apparent extension of the maxima in the isophotes. Thus the secondary emission maximum in Cassiopeia, as well as the one in Cygnus mentioned above, does not appear as a detailed feature in the intensity distribution of cosmic radio waves because of the principal shortcoming of radio telescopes, their very small resolving power, which results in considerable "smearing" of separate bright sources. However the main maximum in the region of the galactic center, and the secondary maximum in Cygnus at $\alpha = 20^h 20^m$, $\delta = +43°$ (see, for example, Fig. 40), clearly appear as very extended sources of radio waves with complex intensity distribution, and also show large-scale structural detail in the spatial distribution of cosmic radio emission.

A similar extended secondary maximum can be distinguished in the isophotes of Fig. 23, in the region of the galactic anticenter ($l \approx 150°$,

$b = 0°$). Its intensity, however, is considerably lower than that of the secondary maximum in Cygnus. Furthermore, this secondary maximum appears distinctly only at relatively high frequencies (see, for example, Reber's isophotes for 480 Mc/s, Fig. 25).

A characteristic feature of all extended secondary intensity maxima of cosmic radio waves is their distribution along the galactic equator. They therefore delineate some coarse structural detail of galactic radio emission.

Another interesting feature is the fact that in the region of the galactic center the maximum intensity of the radio waves does not occur exactly on the galactic equator (at $b = 0°$), but somewhat to the south of it. Table 1 gives the coordinates of the intensity maximum in the region of the galactic center, as obtained by different observers at different frequencies.

In spite of the significant scatter in the data obtained by different authors, the galactic latitude of the main maximum is undoubtedly negative. Highest weight should be given to the last three observations listed in Table 1, since they were obtained with telescopes of higher resolving power. The weighted mean of the galactic latitude at the maximum is $b = -1°.5$. As McGee, Slee, and Stanley (1955) point out, over the entire central region the intensity reaches its maximum value around $b = -1°$.

In 1952 the present author explained this remarkable feature of cosmic radio-wave isophotes in the region of the galactic center by pointing out that the system of galactic coordinates accepted at present ("Ohlsson's elements") contains a small error (Shklovsky 1952a). The actual position of the galactic plane can be better determined from radio observations, which are not influenced by absorption due to interstellar gas. It is found that the fundamental plane defined by the coordinate system accepted today does not coincide with the plane of symmetry of our stellar system, but is displaced from it by an angle of about $1°.5$.

TABLE 1. The position of the galactic equator.

Observer	Frequency (Mc/s)	α	δ	l	b
Reber (1940)	160	17^h50^m	$-25°$	$332°$	$-2°.0$
Reber (1948)	480	53	-24	333	-2.0
Bolton and Westfold (1950)	100	33	-30	326	-1.0
Allen and Gum (1950)	200	30	-35	322	-3.0
Piddington and Minnett (1951)	1210	50	-27.5	330	-2.7
Kraus and Ko (1955)	250	43	-28.8	328	-1.4
McGee, Slee, and Stanley (1955)	400	42	-28.5	328	-1.0
Piddington and Trent (1956)	600			330	-1.3

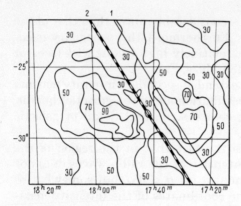

Fig. 41. Isophotes of the galactic-center region in the infrared, at $\lambda = 10\,000$ A: *1*, the position of the galactic equator originally adopted; *2*, the new position (Kalinyak, Krasovsky, and Nikonov 1951).

This conclusion agrees with the results of Kalinyak, Krasovsky, and Nikonov (1951), who observed the galactic nucleus in infrared ($\lambda \approx 10\,000$ A). Figure 41 shows schematically the intensity distribution observed by these authors in the region of the galactic center. Two intensity maxima are detectable. One is a well-known star cloud in Sagittarius, the brightest region of the Milky Way; the other, almost as bright, was detected south of the galactic equator. On common blue photographic plates this second region cannot be observed since it is obscured from us by a cloud of interstellar dust which strongly absorbs the short wavelengths of the visual spectrum. Notice that in Fig. 41 two "luminous bodies" seem to be connected by a dark "bridge." Obviously in this region the absorption is so strong that not even infrared radiation can penetrate. This shows, then, that in this region visual light is absorbed in a layer of dust extending along the middle of the galactic plane. It follows that the "actual" galactic equator should be placed along a dark lane between the two luminous regions seen in the direction of the center, that is, 1°.6 farther south than the presently accepted position (Shklovsky 1953*b*).

Along with their intensity distribution, a very important feature of cosmic radio waves is their spectral distribution. In this respect there arises the question of the variation of the spectral distribution over the sky. Here it is again necessary to take into account the low resolving power of the telescopes used, which produces a "smearing" of the spectral characteristics from regions of relatively small angular extent.

Moxon (1946) made the first attempt to determine the frequency dependence of the intensity of cosmic radio waves, observing their intensity at frequencies $\nu = 40, 90,$ and 200 Mc/s. In the region of the sky where the intensity was a maximum, the antenna temperature decreased according to the relation

$$T_{\max} = \text{const} \cdot \nu^{-2.69}.$$

For the region of minimum intensity the dependence was given by

$$T_{\min} = \text{const} \cdot \nu^{-2.1}.$$

Herbstreit and Johler (1948) made a similar investigation at 25 and 110 Mc/s, and found the relation

$$T = \text{const} \cdot \nu^{-2.41},$$

which applies both to the region of maximum intensity and to "colder" regions.

Piddington (1951) carried out a detailed investigation of the spectral distribution of cosmic radio waves. He analyzed all the isophotes at different frequencies available at the time. Only the observations at 100 and 64 Mc/s (see Figs. 26 and 24) gave the absolute value of the intensity in all regions of the sky investigated. At frequencies $\nu \approx$ 100 Mc/s the weak emission from the "cold" regions of the sky had not yet been detected. Therefore isophotes at $\nu = 200$ Mc/s, for example, gave the *difference* in intensity between a given region of the sky and a "cold" region.

The first observations at 18.3 Mc/s, obtained with a fixed antenna pointed toward the zenith, gave the antenna temperature for a fairly narrow strip of sky which passed over the zenith at the point of observation (Shain 1954). Brightness temperatures derived in this work are very unreliable, since Shain calibrated his observations on the basis of Bolton and Westfold's isophotes for $\lambda = 3$ m. Piddington made an attempt to reduce all results to a standard directional diagram. Moreover, he estimated the brightness temperature of the cold regions of the sky at high frequencies.

The emission spectra of three regions of the sky were investigated: $A\ (l = 300°,\ b = -2°)$; $B\ (l = 180°,\ b = 0°)$; $C\ (l = 200°,\ b = -30°)$. Region A is near the galactic center; region B is on the galactic equator, at a point where the intensity of cosmic radio waves is relatively low; region C is located off the galactic equator, in a relatively cold region of the sky. Table 2 gives the brightness temperature for the

TABLE 2. Brightness temperatures.

ν (Mc/s)	T_A (°K)	T_B (°K)	T_C (°K)
18.3	200 000	75 000	50 000
40	67 000	11 900	8 500
64	21 000	3 100	2 200
90	7 700	−	−
100	6 000	720	490
160	2 180	−	−
200	1 190	120	70
480	145	16.6	−
1 200	17.9	−	−
3 000	2.77	−	−

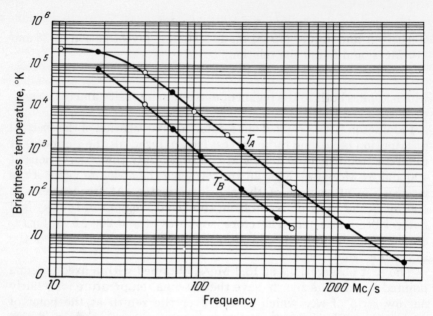

Fig. 42. Brightness temperature as a function of frequency, for two regions of the sky (Piddington 1951).

regions A, B, and C at different frequencies, as computed by Piddington.

Figure 42 shows graphically the dependence of T_A and T_B on frequency. The curve for T_C is not shown, since it practically coincides with that for T_B. At intermediate frequencies the dependence of T_B on frequency follows the law $T_B \propto \nu^{-2.73}$, and (also for intermediate frequencies) T_A is given by $\nu^{-2.51}$. For $\nu > 1000$ Mc/s, $T_{A,B} \propto \nu^{-2}$. For low frequencies, T_A is practically independent of frequency.

Piddington's results need to be reconsidered in view of more recent data. Thus, for example, according to Shain's measurements at $\nu = 18.3$ Mc/s, $T_A = 270\,000°$K (see Fig. 29), and from Shain and Higgins' work (1954) at 9.15 Mc/s, $T_A \approx 10^6$ °K. From these data we can draw the conclusion that for the lowest frequencies the sky's brightness temperature rises steadily with diminishing frequency.

With the help of recent data we are now in the position to describe the spectral distribution of cosmic radio waves in the cold regions of the sky near the galactic pole. According to Bolton and Westfold's (1950) absolute measurements at $\nu = 100$ Mc/s, in a region of the sky around $\alpha = 4^h30^m$, $\delta = -34°$, $T_b \approx 500°$K (see Fig. 26). According to Shain (1954), in the same region of the sky at 18.3 Mc/s, $T_b = 60\,000°$K (see Fig. 29). At 400 Mc/s the brightness temperature of this region is $T_b = 10° \pm 5°$K (McGee, Slee, and Stanley 1955). From these data we conclude that between $\nu = 18.3$ Mc/s and $\nu = 100$ Mc/s,

$T_b \propto \nu^{-2.8}$. In the frequency interval between $\nu = 100$ Mc/s and 400 Mc/s, T_b is also proportional to $\nu^{-2.8}$.

The recent observations by Reber and Ellis (1956) at very low frequencies significantly extend the wavelength range investigated. Figure 43 shows Ellis' results (1957) at low frequencies. It also includes the observations of Bolton and Westfold (1950) at $\nu = 100$ Mc/s, Shain (1951) at $\nu = 18.3$ Mc/s, and Shain and Higgins (1954) at $\nu = 9.19$ Mc/s, as well as the results of Reber and Ellis (1956) for lower frequencies. The black dots correspond to maximum intensity, the white dots to minimum. From the figure it is clear that at 10 Mc/s

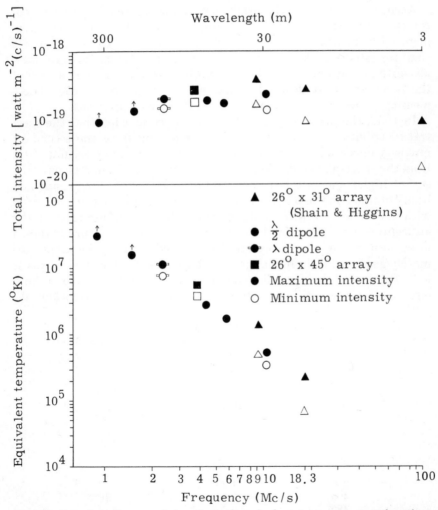

Fig. 43. Cosmic radio-wave intensity and equivalent temperature as functions of frequency (Ellis 1957).

the spectrum of cosmic radio waves is considerably smoother than at higher frequencies, within the limits of observational error. This result has great significance for analyses of the nature of cosmic radio waves.

We can thus adopt for high galactic latitudes

$$T_b \propto \lambda^\alpha, \tag{4-1}$$

where $\alpha = -2.8$ for $\lambda < 30$ m, and $\alpha = -2$ for $\lambda > 30$ m. Table 3 shows the values of T_b in the cold regions of the sky near the galactic poles at different frequencies, computed from Eq. (4-1) by interpolation and extrapolation.

According to Fig. 40, the brightness temperature of cold regions of the sky at $\nu = 910$ Mc/s ($\lambda = 33$ cm) is around $1\,^\circ$K, in good agreement with the extrapolated value shown in Table 3. In the future we shall repeatedly make use of data shown in this table. However, new absolute measurements of the sky brightness in the neighborhood of the poles are most desirable, especially at high frequencies. These measurements, of course, demand very sensitive apparatus.

In 1952, the present author reached the conclusion that the spectral distribution of cosmic radio waves arising from the secondary intensity maxima in Cygnus and at the anticenter differed significantly from the spectral distribution of radiation received from regions of the sky at higher galactic latitudes (Shklovsky 1952a). This result had been obtained on the basis of isophotes available at the time, observed with telescopes of low resolving power. It had been found that in the anticenter and Cygnus regions the brightness temperature obtained after subtracting the background depended on frequency approximately as ν^{-2}, while at higher galactic latitudes the dependence was as $\nu^{-2.8}$. Because of their peculiar spectrum, the secondary maxima stand out considerably more strongly at high frequencies than at low. We

TABLE 3. Temperatures near the galactic pole.

ν (Mc/s)	λ (m)	T_b ($^\circ$K)
9.15	32.8	425 000
18.3	16.4	60 000
64	4.7	1 700
81	3.7	900
100	3	500
160	1.87	128
200	1.50	72
250	1.20	37
400	0.75	10
480	0.625	6
910	0.33	1
1 420	0.21	0.3

can now study these effects on the basis of higher-quality isophotes at $\lambda\lambda$ 33, 120, and 189 cm (Figs. 40, 37, 38).

According to Fig. 38, at $\lambda = 1.89$ m the brightness temperature in the region of the secondary maximum in Cygnus ($l = 48°$, $b = 0°$) is $T_b = 1070°$K, corrected for a 2° main lobe. The brightness temperature of the sky surrounding this maximum is about 200°K, which means that the brightness temperature of the secondary maximum itself is 870°K. At a frequency $\nu = 910$ Mc/s the brightness temperature corrected for a 2°.5 main lobe is 30°K, by Fig. 40. The background in this region does not exceed 3°K, and hence the brightness temperature of the Cygnus maximum at $\lambda = 33$ cm is close to 30°K.

At $\lambda = 1.20$ m, the brightness temperature of the region in question (after subtraction of the background) is around 240°K. Notice, however, that these observations were obtained with a directional diagram of 8° $\times$ 1° (see Sec. 3), and hence the brightness temperature must be corrected in declination. If we make this correction, the brightness temperature turns out to be about 320°K. It is easy to see that the brightness temperature of the emission from the secondary maximum in Cygnus at frequencies of 910, 250, and 158 Mc/s depends on frequency as ν^{-2}, as we had set out to prove.

Piddington and Minnett (1952) observed in the Cygnus region, at wavelength $\lambda = 25$ cm, the extended source which they called "Cygnus X." The position of this source coincides with the position of the secondary maximum observed before. They also stress that the spectrum of the source Cygnus X differs significantly from the spectrum of the surrounding region of the sky.

Table 4 shows the brightness temperature of the secondary maximum in the anticenter region at different wavelengths, after subtracting the brightness temperature of the surrounding background (Shklovsky 1952a). The fourth column shows the temperature computed on the assumption that $T_b \propto \nu^{-2}$. Comparison with the observed temperature shows that this assumption holds quite accurately. As will be shown in Sec. 9, such a spectral distribution of the radio intensity from the secondary maxima allows us to make a reliable inference about their nature, and to understand the emission mechanism.

TABLE 4. Brightness temperatures of a secondary maximum.

ν (Mc/s)	λ (m)	T_b (°K)	
		Obs.	Theor.
480	0.625	19	16
200	1.50	89	89
100	3.00	375	356
64	4.70	850	860

5. SPATIAL DISTRIBUTION OF THE COMPONENTS OF RADIO EMISSION

The peculiarities in the spectral distribution of the radiation from the secondary maxima in Cygnus and at the anticenter force us to conclude that their mechanism of radio emission is substantially different from that of the sources observed at intermediate and high latitudes. That is, the observed cosmic radio waves consist of at least two components, whose nature and spatial distribution are quite distinct. Since the secondary maxima of cosmic radio waves—which have a very peculiar spectrum and, we may assume, a peculiar nature as well—are distributed along the galactic equator, the sources of the radiation must be identified with objects lying in a thin stratum near the galactic plane (Baade's galactic population I). As the frequency increases, the brightness temperature of the radiation from the sources on the galactic plane diminishes considerably more slowly than that from sources at intermediate and high latitudes. Generally speaking, therefore, at relatively high frequencies this *flat component* may play a dominating role at low galactic latitudes.

That component of cosmic radio waves which is observed at intermediate and high latitudes could be due to sources found within our stellar system, or to sources distributed outside our Galaxy. Westerhout and Oort (1951) presented a model for the spatial distribution of sources of cosmic radio waves, according to which the sources occurred both within and outside the Galaxy. The galactic sources were distributed in the same way as common stars (or, more exactly, according to the model of galactic mass distribution also proposed by Oort). Superimposed on the radiation from the galactic sources is the radiation from the so-called *isotropic component* of cosmic radio waves, caused by the sources distributed outside the Galaxy. This component accounts for practically all the intensity observed at the galactic poles. Considering the radio flux from the whole sky, the radiation from the isotropic component is a few times as great as that from the galactic component. On the basis of the difference in distribution between the galactic sources and the isotropic component, Oort and Westerhout were able to explain satisfactorily the isophotes obtained by Bolton and Westfold (1950) at $\lambda = 3$ m. We should point out, however, that an exact correspondence should not be expected.

In 1952 the present writer had already derived a spatial distribution of sources of cosmic radio waves which differs significantly from the one proposed the year before by Oort and Westerhout (Shklovsky 1952a). The very concept of a distribution "like the most common stars" appeared to us highly artificial. It is well known that different objects in our stellar system have significantly different spatial distributions. Thus, for example, globular clusters form a gigantic spherical corona around the center of the Galaxy. Their concentration to the galactic center is very pronounced, but there is practically no concentration to the galactic plane. On the other hand, the clouds of interstellar gas move only in a narrow layer 100 to 200 pc in thickness, near the galactic plane.

Analysis of isophotes of cosmic radio waves first of all forces us to conclude that a significant part of the intensity of this radiation in the vicinity of the galactic poles is due to sources occurring within our stellar system. This is particularly clear in Fig. 44, which shows the isophotes at the galactic "polar caps" obtained by Bolton and Westfold (1950) at $\lambda = 3$ m. It is easy to see that the intensity of cosmic radio waves diminishes from the poles toward the anticenter (if we subtract the secondary maximum in the anticenter region, which is due to a source of the same type as those that form the flat component). At the same time we also see from Fig. 44 that the intensity of the emission increases in the direction of the center, in agreement with previous isophotes. These two properties of the distribution are typical of a system of objects with spherical components. If the greater share of the emission in the region of the galactic poles were due to sources located outside our Galaxy, the intensity distribution in galactic latitude would be different from that observed. In such a case the isophotes for the region surrounding the galactic center would be circular, rather than elongated ellipses along the galactic equator (see, for example, Fig. 27, Sec. 3). The globular clusters, for example, have such a distribution; none are observed in the half of the celestial sphere opposite to the galactic center.

If we accept now that all the secondary maxima observed in the

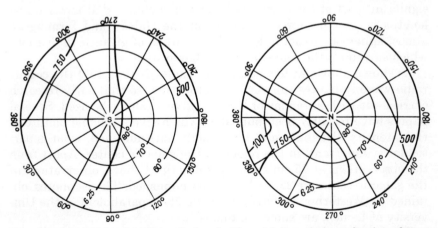

Fig. 44. Isophotes of radio emission in the vicinity of the galactic poles at $\lambda = 3$ m (Bolton and Westfold 1950).

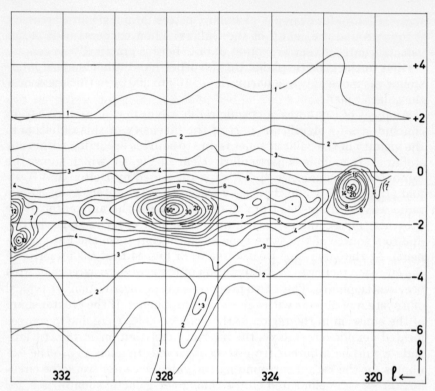

Fig. 45. Isophotes of radio emission in the vicinity of the galactic center at
$\lambda = 22$ cm (Westerhout 1957b).

intensity distribution of cosmic radio waves are due to sources belonging to the flat component, we reach the basic conclusion that a
significant part of the cosmic radio emission observed at meter wavelengths is due to sources localized in the Galaxy and forming an
almost spherical, structureless halo or corona of gigantic dimensions
which surrounds the center of our stellar system.

As shown above, at low galactic latitudes ($|b| < 15°$) the cosmic
radio-wave isophotes run very close together, so that early research
carried out with antennas of low directivity cannot be used. Several
papers have recently been published which give a reliable picture of
the intensity distribution at low galactic latitudes and, what is particularly important, in the region of the galactic center. Apart from
the observations at 75 cm and 50 cm described above, observations of
the galactic center at 22 cm have been published. The isophotes, obtained by Westerhout (1957b) with the 25-m paraboloid of the University of Leiden, are shown in Fig. 45.

The isophotes exhibit a very high intensity maximum centered at
$l = 327°.7$, $b = -1°.47$, practically coincident with the galactic center. This sharp maximum is none other than the discrete source

Sagittarius A, discovered by Piddington and Minnett. A comparison
of Figs. 28 (Sec. 3) and 45 shows the tremendous improvement in the
technique of radio observations over the last five or seven years.

For a long time the question had been open whether Sagittarius A
constituted a "radio nucleus" of our Galaxy, or whether it had an entirely different nature and only happened to be projected on the
galactic center. This seemed a very difficult question to answer (see
Sec. 6). Looking ahead, let us say that today we can consider it definitely established that Sagittarius A is indeed the galactic radio
nucleus.

From the isophotes shown in Fig. 45 it follows that the brightness
temperature in the region of the galactic nucleus ($l = 327.7, b = -1°.47$)
is approximately 25 times as great as that for the same value of l but
for $b = +1°$. The extent of the galactic radio nucleus between points
where the brightness is half the maximum is $0°.83 \times 0°.64$, and between points where the brightness decreases by a factor of ten it is
$2°.60 \times 1°.44$.

Figure 46 shows the isophotes of the intensity of cosmic radio

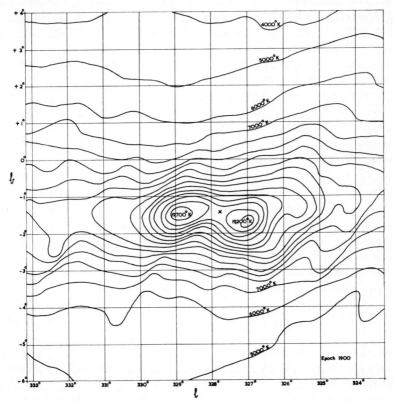

Fig. 46. Isophotes of radio emission in the vicinity of the galactic center at
$\lambda = 3.5$ m (Mills 1956).

waves at 3.5 m in the region of the galactic center as obtained by Mills (1956). The resolving power of the antenna used by Mills was 50'. In general, the intensity distribution at 22 cm and at 3.5 m is quite similar. In both cases an enormous concentration toward the galactic nucelus is clearly observed. However, the isophotes for 3.5 m show one important peculiarity: a well-defined minimum in the immediate neighborhood of the galactic nucleus. As we shall see later, this is a case of *absorption* of cosmic radio waves by the interstellar medium. This absorption is particularly significant at long wavelengths.

From Figs. 45 and 46 we can see that at a distance of 1° or 2° from the nucleus the isophotes already run parallel to the galactic equator. The width of the bright band of radio emission along the galactic equator between isophotes whose intensity is half the maximum is about 7°. This bright band extends over the entire galactic equator; it can be traced, for example, in Baldwin's isophotes for $\lambda = 3.7$ m (Fig. 39, Sec. 3).

This important detail in the intensity distribution of cosmic radio waves escaped detection during the first investigations, carried out with radio telescopes of low resolving power. Thus, for example, no bright lane along the galactic equator is seen in the isophotes shown in Figs. 22 and 26. This is easily understood: a structural detail of such small angular dimension could not of course be detected by a radio telescope of low directivity.

Scheuer and Ryle (1953) made a successful attempt to observe this bright lane predicted theoretically, by an interferometric method. For this purpose they built a radio interferometer of variable spacing. The basic problem was the necessity of high resolving power in one coordinate, namely the galactic latitude b, while high resolution in galactic longitude l was not required.

Let the two-antenna interferometer have a spacing equal to n wavelengths, with the axis pointing to the galactic pole. Also let both antennas of the interferometer be directed toward a point on the galactic equator of longitude l_0. The directional diagram of each antenna is given by the function $A(l, b)$. If the brightness temperature of the region of the sky under investigation is $T(l, b)$, then the power absorbed by each antenna from the region of the sky centered at (l, b), within a solid angle $d\Omega$ is given by the relation

$$\delta T_a = \frac{T(l, b)d\Omega \cdot A(l, b)}{\int\int A(l, b)\cos b\, db dl} = \frac{G}{4\pi}T(l, b)d\Omega \cdot A(l, b), \qquad (5\text{-}1)$$

where G is the antenna gain.

If the two antennas are connected in such a way as to form a swept-lobe interferometer, the deflection in the recording device due to radiation emitted by an element $d\Omega$ of the sky will be given by the formula

$$C\delta T_a = C\frac{G}{4\pi} T(l, b)A(l, b)d\Omega \cdot \cos (2\pi n \sin b), \tag{5-2}$$

where the constant C depends on the sensitivity of the recording apparatus. The deflection $d(n)$ due to radiation from the whole sky is given by

$$d(n) = C\frac{G}{4\pi} \int T(l, b)A(l, b) \cos (2\pi n \sin b)d\Omega. \tag{5-3}$$

We make the following simplifying assumptions: (1) the variation of $T(l, b)$ with longitude is linear within the directional diagram; (2) $A(l, b)$ can be written as a product $L(l)B(b)$, where $L(l)$ is symmetric with respect to l_0. Then Eq. (5-3) can be rewritten as

$$d(n) = C\frac{G}{4\pi} \int_{-\pi}^{\pi} L(l)dl \int_{-\frac{\pi}{2}}^{\frac{\pi}{2}} T(l_0, b)B(b) \cos (2\pi n \sin b) \cos b \, db$$

$$= C\frac{G}{4\pi} \int_{-\pi}^{\pi} L(l)dl \int_{-\infty}^{\infty} X(x) \cos (2\pi n x)dx, \tag{5-4}$$

where

$$x = \sin b, \qquad X(x) = \begin{cases} T(l_0, b)B(b) & \text{for } |x| \leq 1, \\ 0 & \text{for } |x| > 1. \end{cases}$$

By using different spacing we can observe $d(n)$ as a function of the length n of the baseline. If the functions $A(l, b)$ and $T(l, b)$ are symmetric functions of b, then $X(x)$ would be an even function of x. By Eq. (5-4) we can write the relation

$$T(l_0, b)B(b) = X(x) = 8\pi \left[CG \int_{-\pi}^{\pi} L(l)dl \right]^{-1} \int_0^{\infty} d(n) \cos (2\pi n x)dx. \tag{5-5}$$

Equation (5-5) allows us, in principle, to compute the unknown distribution of brightness temperature if $d(n)$ is known. In our case we are interested only in the distribution of brightness temperature with galactic latitude in a small region around $b = 0°$, so that we may set $x = \sin b = b$, $B(b) = 1$.

Observations of the predicted bright lane along the galactic equator were carried out at frequencies of 210 and 81.5 Mc/s ($\lambda = 1.43$ and 3.70 m). The maximum spacing was 60λ (that is, $n = 60$). Figures 47 and 48 depict the results of the observations in the form of the dependence on spacing of the amplitude of the tracings (expressed in arbitrary units).

By means of these curves and Eq. (5-5) it was possible to determine the unknown distribution of brightness temperature with galactic latitude. The results are shown in Figs. 49, 50, and 51.

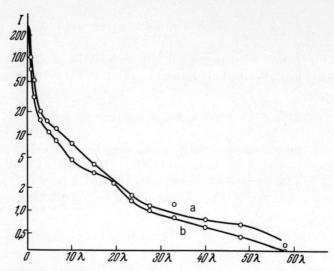

Fig. 47. Cosmic radio-wave intensity as a function of interferometer spacing at $\lambda = 3.7$ m ($\nu = 81.5$ Mc/s): curve a, for $l = 353°$; curve b, for $l = 13°$ (Scheuer and Ryle 1953).

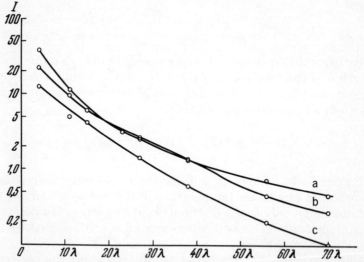

Fig. 48. Cosmic radio-wave intensity as a function of interferometer spacing at $\lambda = 1.43$ m ($\nu = 210$ Mc/s): curve a, for $l = 338°$; curve b, for $l = 358°$; curve c, for $l = 16°$ (Scheuer and Ryle 1953).

The dotted lines in Figs. 49 and 50 show the intensity distribution according to the isophotes of Bolton and Westfold (1950), at 100 Mc/s (Fig. 26, Sec. 3). The absolute value of the brightness temperature was obtained from the observations at 81.5 Mc/s. At 210 Mc/s the observational results give only the relative distribution of brightness

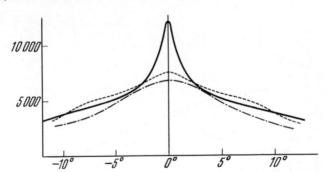

Fig. 49. Brightness temperature as a function of galactic latitude at $l = 353°$ and $\lambda = 3.7$ m (Scheuer and Ryle 1953).

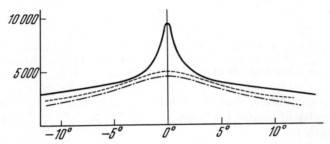

Fig. 50. Brightness temperature as a function of galactic latitude at $l = 13°$ and $\lambda = 3.7$ m (Scheuer and Ryle 1953).

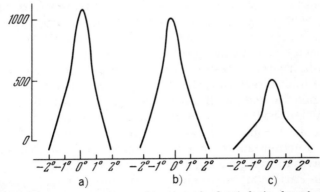

Fig. 51. Brightness temperature as a function of galactic latitude at $\lambda = 1.43$ m: curve a, $l = 338°$; curve b, $l = 358°$; curve c, $l = 16°$ (Scheuer and Ryle 1953).

temperature; the temperature at this frequency was obtained by an indirect method and is not sufficiently reliable.

Figures 49, 50, and 51 clearly reveal a bright emission lane extending along the galactic equator. The width of this layer is about 2°. Figure 52 gives the distribution of excess brightness temperature as a function of galactic longitude; the crosses express the probable error

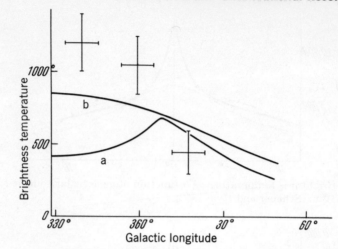

Fig. 52. Excess brightness temperature as a function of galactic longitude at λ = 1.43 m (Scheuer and Ryle 1953). The crosses express the observational error; curves *a* and *b* are theoretical distributions.

of the observations. From this figure we see that in the bright lane the intensity of cosmic radio waves is markedly concentrated toward the galactic center.

Figure 53 presents the brightness temperature distribution for three "sections" through the Milky Way, at declinations δ = +2°, −11°, and −29°.5 (this last "section" contains the region of the galactic center), as obtained by Mills (1955) with his cross-shaped antenna. These observations clearly demonstrate the existence of a bright band extending along the galactic equator.

The observational results described above, obtained with an antenna of high directivity, show that sources of cosmic radio waves form a galactic *disk,* with fairly strong concentration to the center. Since the half-width of the bright band changes from about 7° or 10° to 2° or 3°, its linear thickness is of the order of a few hundred parsecs. Further, as we have emphasized, other sources of galactic radio waves form a gigantic, almost spherical "halo."

At the Paris symposium on radio astronomy held in August 1958, Mills presented new observational results obtained with a cross at 3.5 m. He constructed isophotes for a strip of sky 10° wide along the galactic equator, between *l* = 13° and *l* = 223° (the part of these isophotes corresponding to the vicinity of the galactic center had been published earlier, and is presented in Fig. 46). By analyzing these isophotes Mills found that the nonthermal sources in the disk possess spiral structure; the width of these arms is several hundred parsecs. Basically they coincide with the spiral arms detected by other methods (see Sec. 15). (Let us note that the "usual" spiral structure is due

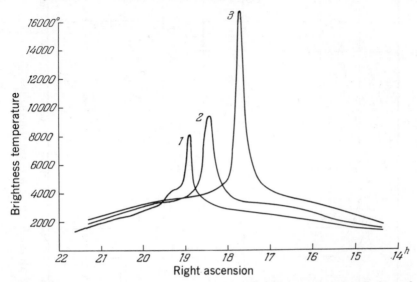

Fig. 53. Distribution of brightness temperature for three "sections" through the Galaxy: *1*, $\delta = +2°$; *2*, $\delta = -11°$; *3*, $\delta = -29°.5$ (Mills 1955).

to the interstellar gas and associated hot stars, which form a very compact system.) The thickness of the spiral arms of nonthermal emission considerably exceeds that of the "gaseous" arms. The sun appears to be within a "radio spiral."

The spatial distribution of sources in the "disk" is quite similar to the distribution of stars. Thus, for example, the distribution of radio sources near the plane of symmetry of the Galaxy resembles the model proposed by Westerhout and Oort (1951). However, at greater distances from the galactic plane their model, based on the assumption of an "isotropic component" of extragalactic origin, is untenable.

It is very important to understand the relation between the distribution of radio sources in the disk and in the halo. Is there a continuous transition between the two distributions? Are the sources in disk and halo of the same or of a different nature? The latter question is fundamental.

We shall attempt to answer the former question first. Blythe (1957*b*) has recently published a series of isophotes obtained at Cambridge at 38 Mc/s ($\lambda = 7.9$ m), with their new interferometer. The width of the main lobe (for half-power points) was 2°.2 in the east–west direction; in the north–south direction the width varies with declination: 2°.3 for $\delta = 52°$ and 7°.4 for $\delta = -20°$. The survey covers the sky north of $\delta = -20°$, except that several regions surrounding strong sources were excluded. The isophotes are reproduced in Fig. 54. The inhomogeneous character of the intensity distribution, even at high galactic latitudes, is very striking.

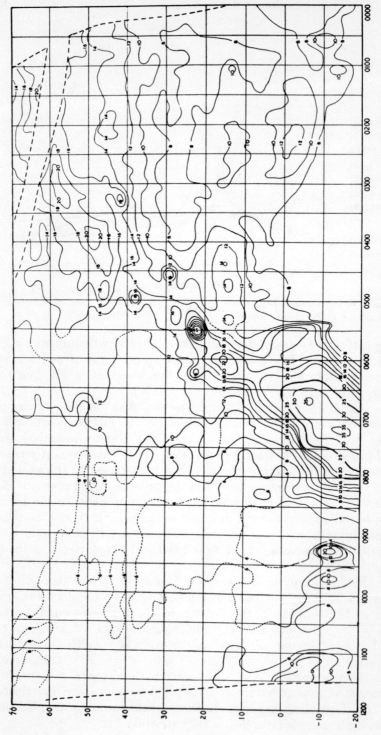

Fig. 54a. Isophotes of galactic radio emission at $\lambda = 7.9$ m ($\nu = 38$ Mc/s), for $0^h \leq \alpha \leq 12^h$ (Blythe 1957b).

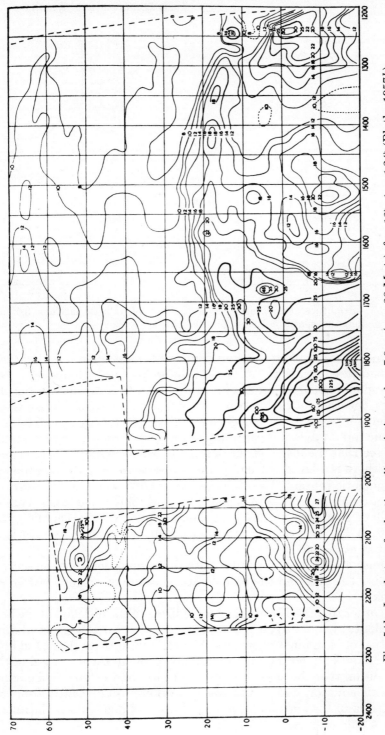

Fig. 54b. Isophotes of galactic radio emission at $\lambda = 7.9$ m ($\nu = 38$ Mc/s), for $12^h \leq \alpha \leq 24^h$ (Blythe 1957b).

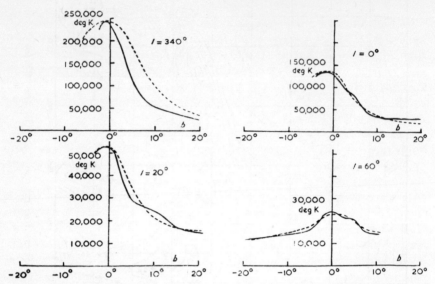

Fig. 55. Brightness temperature as a function of galactic latitude for four galactic longitudes, at $\lambda = 7.9$ m (Blythe 1957b).

From these isophotes we can deduce the dependence of brightness temperature on galactic latitude at different galactic longitudes. The corresponding curves are shown in Fig. 55, where the dotted line denotes the theoretical distribution given by the Oort–Westerhout model. We see that this model closely fits the intensity distribution of cosmic radio waves at low galactic latitudes. From the curves we can also see that the disk distribution passes smoothly into the halo, and that there is a large concentration of disk sources in the direction of the galactic center. The brightness temperature of the sky near the galactic pole at $\lambda = 7.9$ m is about 7000°K, only some three times as small as that near the galactic equator (at $l = 60°$). On the other hand, at $l = 340°$, $b = 0°$, the sky brightness is about ten times as great as at $l = 60°$, $b = 0°$.

Thus, the latest data substantiate the assumption that the sources forming the disk and the halo comprise a unified system within our Galaxy.

In order to establish whether the sources in the disk and in the halo are identical in nature, it is necessary to investigate the spectral composition of both kinds of sources. A number of authors (Ryle, Piddington, Shklovsky), on the basis of data available to them, arrived at the conclusion that the spectrum of the bright band extending along the galactic equator, due to disk sources, differed significantly from the spectrum of the emission at intermediate and high galactic latitudes, due to halo sources.

It was concluded that the disk spectrum is smoother than the halo spectrum. This gave basis to the assumption that the radiation from disk sources was thermal radiation from the interstellar gas (see, in this respect, Sec. 9), while the radiation from halo sources had a non-equilibrium character. At the present time it has been shown beyond doubt that at least the main fraction of the radiation from disk sources cannot be thermal.

Direct analysis of recently obtained isophotes of cosmic radio emission from the disk (in particular, from the region of the galactic center) indicates that to within observational error the spectrum of the radiation from both disk and halo sources is identical.

Table 5 provides values of the brightness temperature for a region near $l = 322°$, $b = +1°$, at several wavelengths. Since no discrete sources of thermal radio emission (Sec. 10) are present in this region, we can assume that the observed spectrum is due to disk sources. The second column shows the value of T_b taken from the isophotes (the temperature of the cold sky has been added for $\lambda = 50$ cm and 75 cm). It is easily seen that $T_b \propto \lambda^\alpha$, where $\alpha = 2.7$. The computed values of T_b with $\alpha = 2.7$ are given in the third column. The measurement of Piddington and Trent at $\lambda = 50$ cm is badly astray, probably because of a systematic observational error. As for the entry at 350 cm, Dr. Mills has kindly informed us that his measurement of the brightness temperature shown in Fig. 46 is too low by a factor of about 1.5. The corrected value of T_b is given in parentheses. The value $\alpha = 2.7$ practically coincides with the value of 2.8 which we had derived for the spectrum of the cold region of the sky at high galactic latitudes. Absolute measurements in radio astronomy are not, in general, highly accurate, and therefore the error in the computed value of α may be as much as 10 percent.

Recently Westerhout (1958), from an analysis of his own isophotes at $\nu = 1390$ Mc/s and those of Mills at $\nu = 85$ Mc/s, derived for α the most probable value of 2.7.

In any case, from Table 5 it follows that for most of the radio spec-

TABLE 5. Observed and computed brightness temperatures.

λ (cm)	$T_b (°K)$	
	Obs.	$\alpha = 2.7$
22	6.5	6.5
33	20	19
50	115	60
75	180	180
350	7 000 (11 500)	11 300

trum α is considerably greater than 2, which would be the value corresponding to thermal emission from an ionized gas.

Adgie and Smith (1956) in Cambridge made an interesting study of the spectral composition of the different components of galactic radio waves. They utilized three antennas, working at 38, 81.5, and 175 Mc/s, all with the same directional diagram (25° in azimuth, 70° in altitude). They observed the incident radiation as a function of sidereal time, leaving the antenna stationary (see Fig. 56). Because of the broad directional diagram, at any given moment the incident radiation comes from both the spherical and the disk components. At 7^h local sidereal time, when the galactic equator drifted over the main lobe of the antenna, the incident radiation came chiefly from the disk. From Fig. 56 we see that the curves are practically identical; only near 7^h does the intensity at higher frequencies increase significantly. Comparing curves for different frequencies for $t = 0^h$ and $t = 7^h$, we can determine the difference $\alpha_1 - \alpha_2$ between the spectral exponents for the halo and for the disk. At 175 Mc/s and also at 38 Mc/s we find $\alpha_1 - \alpha_2 = 0.06$, a very small quantity. We can conclude, then, that to an accuracy of 0.1 the spectral indices of the sources belonging to the halo and to the disk are equal. Let us not forget, however, that disk sources, particularly those very close to the equator, have a smoother

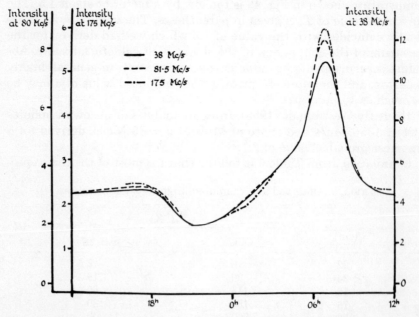

Fig. 56. Galactic radio-wave intensity as a function of sidereal time at three frequencies (Adgie and Smith 1956). The unit of intensity is 10^{-21} w m^{-2} (c/s)$^{-1}$.

spectrum than halo sources. It is necessary to obtain independent confirmation of this result, so critical for the theory (see Sec. 23).

Adgie and Smith derive a spectral index $\alpha = 2.5$, that is, 0.3 smaller than the one we computed from different data. The reason for this discrepancy lies in the large errors involved in absolute radio-astronomical measurements.

Let us turn now to the question of the fundamental properties of the sources forming the spherical component of the galactic radio emission, or *halo*. First of all we shall reanalyze those facts which lead us to postulate the presence of such a peculiar intensity distribution in our stellar system. Let us look, for example, at the interferometric observations of Scheuer and Ryle (1953) at $\lambda = 3.7$ m (Figs. 49 and 50), as well as those of Mills (1955) (Fig. 53), with higher resolving power. According to Bolton and Westfold (1950), the sky's brightness temperature near the galactic poles at $\lambda = 3$ m is in the neighborhood of $600°$K. As we have already emphasized, the greater part of the intensity observed in this region of the sky is due to sources belonging to the spherical component. We estimate that extragalactic sources give, at this wavelength, a brightness temperature $T_b \approx 150° \pm 50°$K (see Sec. 26). It follows, then, that the brightness temperature of the spherical component in the region of the galactic poles is $T_b \approx 450°$K, at $\lambda = 3$ m. Since $T_b \propto \lambda^{2.8}$ we find that at $\lambda = 3.5$, $T_b \approx 700°$K; and at $\lambda = 3.7$ m, $T_b \approx 800°$K. According to Fig. 53, at $\lambda = 3.5$ m the brightness temperature near the galactic center is $\approx 4000°$K. From the curves shown in this figure it is clear that the intensity of the spherical component of the radiation slowly, but unmistakably, diminishes as we move away from the galactic center. Thus, for example, at $\delta = -29°.5$, $\alpha = 14^h20^m$ ($l = 294°$, $b = +27°$), the brightness temperature of the spherical component has decreased by about 2000 K deg, to half the maximum value.

Obviously, such change in the intensity of cosmic radio waves is in disagreement with the assumption of Oort and Westerhout as to the existence of an extragalactic isotropic component of very high intensity. If the Oort–Westerhout model were correct, the angular distance between two points in the sky whose brightness temperature differs by a factor of two should be $17°$, while according to Mills's observations the separation between two such points reaches almost $60°$. In the isophotes of Scheuer and Ryle it is clearly seen that the brightness temperature of the sky for $|b| > 2°$ decreases slowly and smoothly with distance from the galactic equator. For $|b| > 10°$ the theoretical curve of Oort and Westerhout begins to fall with increasing b appreciably faster than the observed curve. For high values of b the discrepancy between these two curves should be even more striking.

We shall now examine the model of the distribution of halo sources proposed by us in 1952. The disk sources were excluded from the analysis. First we note that the halo sources show a very small degree of concentration to the galactic center, if we exclude a small region near the center containing Sagittarius A. Thus, for example, at $\lambda = 3.5$ m the ratio of the brightness temperature of the spherical component in a region around the galactic center to that at the poles is 5.3. The ratio of the brightness at the poles to that at the anticenter is 1.25. Notice, however, that this last number was obtained from an analysis of Bolton and Westfold's old observations, made with an instrument of low resolving power. It would be very important to make a detailed study of the intensity variation as we move from the galactic poles toward the anticenter by means of a cross-shaped antenna of high resolving power. If we accept the values given by Bolton and Westfold, the intensity of the spherical component in a region near the center is 6.6 times as great as that in the direction of the anticenter.

Let us assume, as is done in stellar astronomy, that the distribution of the sources in the galactic plane which belong to the spherical component is given by the formula

$$n(R) = n(0)e^{-\alpha R}, \tag{5-6}$$

where R is measured in kiloparsecs. Then the intensity of the radio emission in any direction in the galactic plane will be proportional to $\int n(R)\, ds$. The intensity of the emission in the direction of the galactic center will be proportional to

$$\int_0^{R_\odot} n(R)dR + \int_0^\infty n(R)dR,$$

where $R_\odot$ is the distance from the sun to the galactic center (the second integral gives the contribution of the sources beyond the center). In the direction of the anticenter the intensity will be proportional to

$$\int_{R_\odot}^\infty n(R)dR.$$

The condition that the intensity of the spherical component toward the center be 6.6 times that toward the anticenter can be written as follows:

$$\int_0^\infty e^{-\alpha R}dR + \int_0^{R_\odot} e^{-\alpha R}dR = 6.6 \int_{R_\odot}^\infty e^{-\alpha R}dR, \tag{5-7}$$

whence
$$e^{-\alpha R} = 2/(6.6 + 1) = 0.26.$$

Assuming $R_\odot = 8.2$ kpc, it follows that $\alpha = 0.16$ kpc^{-1}. In stellar astronomy, to describe the density gradient of any cosmic object in the galactic plane one usually introduces the quantity

$$m = d(\log n)/dR.$$

For the sources belonging to the spherical component of galactic radio emission, $d(\log n)/dR = -0.07$, considerably less even than for subsystems which are as weakly concentrated to the galactic center as the long-period Cepheids, for which $d(\log n)/dR = -0.11$. For typical members of the spherical component of our stellar system (globular clusters, subdwarfs, RR Lyrae stars), $m \approx -0.27$. The space density of these objects is approximately one hundred times as great in the vicinity of the galactic center as in the solar neighborhood. On the other hand, the space density of radio sources comprising the spherical component is only 3.7 times as great. Although the brightness temperature of the spherical component near the anticenter is not known with much accuracy, we can confidently say that the space density of sources in the spherical component near the galactic center does not exceed that in the solar neighborhood by more than 2 to 4 times. Let us emphasize once more that we are considering halo sources only; disk sources dominate in the nucleus and in the galactic plane.

Such a spatial distribution is a very peculiar one and is not shared by any objects identified optically. Objects belonging to spherical subsystems always exhibit strong concentration to the galactic center; for objects belonging to flat subsystems this concentration is much weaker, but even so it still far exceeds that shown by the radio sources belonging to the spherical component. The distribution of these sources shows simultaneously a very small concentration to the galactic plane *and* practically no concentration to the center.

The sources comprising the spherical component of galactic radio emission form a gigantic corona around the galactic center, 15 to 20 kpc in radius. Surfaces of equal number density for sources of the spherical component can be obtained empirically from observations of the intensity of the radio emission in different regions of the sky. Figure 57 shows a cross section of these surfaces that passes through the center of the Galaxy and through the sun, and that is perpendicular to the galactic plane. At a relatively small distance from the galactic center the sections are ellipses of eccentricity ≈ 0.9; but at the periphery of our stellar system the sections are almost circular. The density of sources has dropped by a factor of 25 at 20 kpc from the center, and by a factor of 100 at 28 kpc. Of course the computed distribution of sources is highly schematic, serving only as a first rough approximation to reality. In particular, it seems artificial that the distribution of halo sources should be different from that of disk sources.

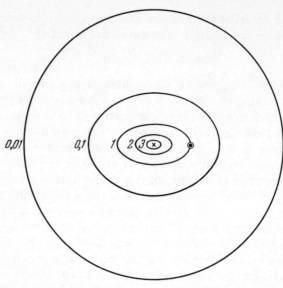

Fig. 57. Distribution of the sources of the spherical component of galactic radio emission.

Since according to the latest data the nature of both types of sources is identical, their distribution should also be the same. Although, as shown above, our model is not perfect, it reveals the main features of the spatial distribution of sources responsible for the spherical component of galactic radio waves.

Baldwin (1955b), on the basis of isophotes which he obtained at $\lambda = 3.7$ m, corroborated our assumption as to the existence of a spherical component of sources of galactic radio waves. He constructed a simple model of the spatial distribution of sources which represented the isophotes quite satisfactorily. In the zone $|b| < 30°$ the distribution of sources also agreed with the Oort–Westerhout model. However, for higher galactic latitudes the greatest contribution comes from sources belonging to the spherical component, or halo. Baldwin's first model consisted of a spherical system of constant density and radius R. If the Galaxy is transparent at $\lambda = 3.7$ m, the intensity of any portion of the sky due to sources in the spherical system should be proportional to the length of a segment extending from the sun to the intersection with the edge of the spherical system. By varying R and the emission coefficient ϵ_ν per unit volume, it was possible to show that the best agreement with the observed isophotes was obtained with $R = 16$ kpc. This is illustrated in Fig. 58.

With this value of R a theoretical intensity distribution was constructed for galactic latitudes 30°, 40°, 60°, and 70°, and compared with the observations. The results of this comparison are shown in

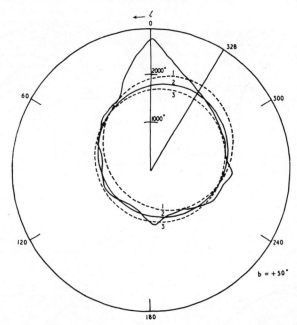

Fig. 58. Brightness temperature as a function of galactic longitude at $b = +50°$. The three circles are the calculated distributions for models having values of R of (1) 12 kpc, (2) 16 kpc, (3) 20 kpc (Baldwin 1955b).

Fig. 59. In this way it was found that $\epsilon_\nu = 1.8 \times 10^8$ w $(c/s)^{-1}$ sterad^{-1} pc^{-3}.

For all longitudes except $l = 0°$, the agreement with observation was not bad. In that region there is a well-known irregularity in the isophotes, the "tongue" to which we referred earlier in this section. At $b = 30°$ the influence of the disk sources becomes apparent (the Oort–Westerhout distribution).

The contribution of extragalactic sources to the observed intensity is relatively small. The brightness temperature of the extragalactic radiation at $\lambda = 3.7$ m should be less than $500°$K; otherwise the theoretical isophotes would differ sharply from those observed. This is illustrated in Fig. 60 for $b = 40°$. Notice that at the galactic pole $T_b = 1100°$K.

Besides the model described above, Baldwin also examined an ellipsoidal model with constant ϵ_ν. In all cases he found that the ratio of the axes of the ellipsoid of revolution was not less than 0.5, if the extragalactic brightness temperature satisfies $T_b < 500°$K.

Finally Baldwin chose a model consisting of three components: (a) an ellipsoid of revolution whose major semiaxis is between 11 and 14.5 kpc, and with a ratio of minor to major semiaxes of about 0.5 or 0.6; (b) a relatively thick disk (corresponding to the Oort–Westerhout

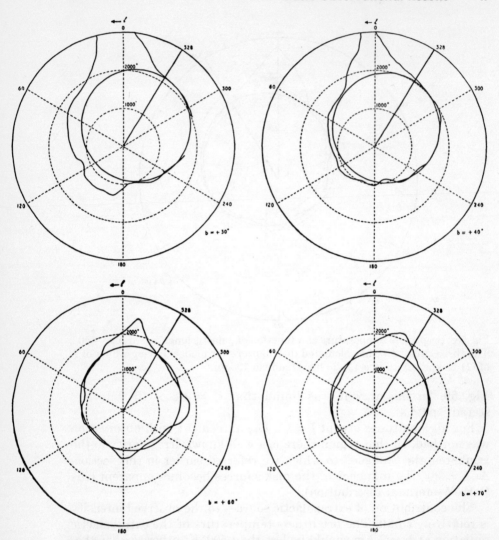

Fig. 59. Comparison of the observed distribution of brightness temperature at four galactic latitudes with the predicted curves for a spherical model having $R = 16$ kpc and $\epsilon_\nu = 1.8 \times 18^8$ w $(c/s)^{-1}$ sterad^{-1} pc^{-3} (Baldwin 1955b).

distribution), the galactic radiation in the zone $|b| < 30°$; (c) a flat component in the galactic plane of thickness about 150 pc. The radiation per unit volume of component a is constant. Component c is responsible for the structural detail shown by the isophotes along the galactic equator. For example, to this component belong the secondary maxima in Cygnus and at the anticenter. The spectrum of component c, as shown above, differs strongly from the spectra of components a and b. For not too low frequencies, $T_b \propto \lambda^2$. As we shall see later, the

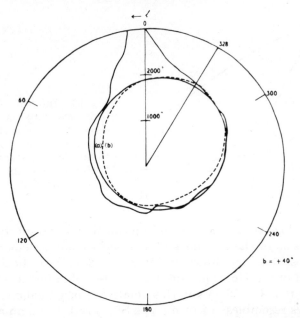

Fig. 60. Computed curves for models having different contributions from extragalactic sources: (a) $\leq 500°$K, (b) $700°$K (Baldwin 1955b).

role of component c in the general picture of cosmic radio waves is relatively unimportant.

We have pointed out that components a and b are identical in nature; therefore to separate them into two different distributions is an idealization of the problem, although a justifiable one during the first stages of the investigation.

Baldwin's inference that ϵ_ν is constant for the whole spherical component is very significant. As he himself points out, that part of the spherical component inside a radius $R_1 = 4$ kpc cannot be easily distinguished from the disk or from the Oort–Westerhout system. On the other hand, the distribution of spherical sources at a distance of more than 10 kpc from the center is uncertain. Thus from Baldwin's model we can only assume that the space density of halo sources changes insignificantly in the interval from $R_1 = 4$ kpc to $R_2 = 10$ kpc. A similar result follows from our schematic distribution shown in Fig. 57: in the interval $4 < R < 10$ kpc, ϵ_ν changes by less than a factor of two.

The concentration of sources (or ϵ_ν) undoubtedly changes at large distances from the galactic center. In the neighborhood of the galactic plane and, particularly, at the center, ϵ_ν probably is considerably greater than Baldwin's value of 1.8×10^8 w (c/s)$^{-1}$ sterad^{-1} pc^{-3}.

In 1952 the author used the intensity distribution of radio emission

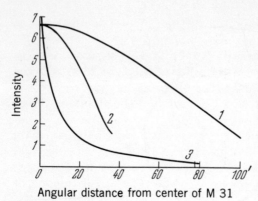

Angular distance from center of M 31

Fig. 61. Intensity distribution (arbitrary units) in M 31:*1,* observed distribution of radio emission; *2,* radio emission according to the Oort–Westerhout model; *3,* observed distribution of optical emission (Baldwin 1954*b*).

in the spiral galaxy M 31, the Andromeda Nebula, to substantiate his arguments as to the existence of a spherical component of galactic radio waves. Hanbury Brown and Hazard (1951), with the 218-ft fixed paraboloid of the University of Manchester, obtained a set of rough isophotes at $\lambda = 1.59$ m for this relatively large galaxy, which in many ways resembles our own (type Sb spiral). From an analysis of these isophotes we can draw the conclusion that the sources of radio waves in M 31 form a spherical system. Since both galaxies—M 31 and our own—are very similar in their basic morphological features, it might be assumed that in our stellar system there are also sources of radio waves forming a spherical system. An independent argument based on the analysis of isophotes of galactic radio waves fully confirms this result.

We must bear in mind, however, that the observations of the intensity distribution in M 31 were obtained with an antenna of relatively low resolving power. It therefore seemed advisable to repeat them using more exact methods.

The problem was solved by Baldwin (1954*b*, 1955*c*) at Cambridge. Interferometric observations were carried out using four different spacings, at a wavelength of 3.7 m; the spacings varied between 6 and 18 wavelengths. The percentage modulation of the interferometric pattern was measured as a function of the length and the direction of the baseline. In Fig. 61, curve *1* gives the intensity distribution of radio waves in M 31 in the direction in which the optical extent is shortest. This distribution was obtained with the baseline approximately east–west. Curve *2* gives the theoretical distribution according to the Oort–Westerhout model. Finally, curve *3* gives the distribution of optical brightness of M 31 in the direction in question.

From Fig. 61 we can see that the sources of radio waves extend to huge distances from the plane of symmetry of the galaxy, forming a spherical system. The effective extent of this spherical corona of

M 31 in the direction perpendicular to the plane of symmetry is at least 10 kpc.

Seeger, Westerhout, and Conway (1957) have investigated the distribution of brightness temperature in M 31 with the 25-m reflector at Dwingeloo in The Netherlands. The observations were carried out at a rather short wavelength ($\lambda = 75$ cm), and the resolving power was $2°.2 \times 1°.8$.

Figure 62 shows the isophotes for M 31. The cross-hatched region has been observed optically. This galaxy has a well-defined ellipsoidal appearance, the ratio of the lengths of its axes being 0.44. The angular extent of M 31 is approximately $10° \times 6°$; the angular extent of the "half-brightness" isophotes, $5°.9 \times 2°.6$. Taking into account the position of the plane of symmetry of M 31 with respect to the plane of the figure we find that this galaxy, at radio wavelengths, is an ellipsoid of revolution with axial ratio 0.4. If the distance to M 31 is 500 kpc, the extent of the region containing radio sources is fully 100 kpc. This value is considerably larger than the optical extent of M 31. Above the plane of symmetry the radio sources extend for several tens of kiloparsecs!

From the isophotes it is quite clear that the density of radio sources in the halo, which determine the magnitude of ϵ_ν, decreases regularly from the center to the periphery. Hence the assumption of a constant

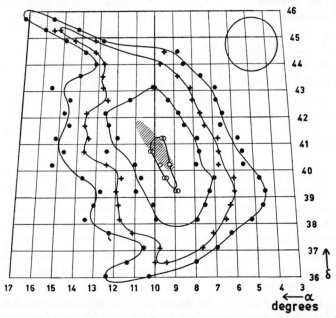

Fig. 62. Isophotes of radio emission from M 31 at $\lambda = 75$ cm; contour interval, $0.8°$K (Seeger, Westerhout, and Conway 1957).

ϵ_ν, which forms the basis of Baldwin's model, appears to be erroneous. We must keep this well in mind, since the presence of a gradient in ϵ_ν has great significance in understanding the nature of the halo (see Sec. 23).

There are two large irregularities present in M 31. These structural details have, probably, the same nature as the bright "tongue" of radio waves in our Galaxy, at $l = 0°$.

The confirmation of the existence of a spheroidal distribution of sources of radio waves in M 31 has forced Western investigators to relinquish the Oort–Westerhout model and, to a greater or lesser degree, to accept our point of view with respect to the distribution of sources of radio waves in the Galaxy, as discussed in detail above. The existence of a very extended, almost spherical system of radio-wave sources, forming a "radio corona" to our stellar system, can now be considered a well-established fact. This, as we shall see later, has great significance in many physical and astrophysical problems, and especially in the problem of the origin of cosmic rays.

6. DISCRETE SOURCES

Hey, Parsons, and Phillips (1946), while investigating the intensity distribution of cosmic radio waves at $\lambda = 4.7$ m, detected a fluctuation in the intensity when the antenna was pointed toward a region in Cygnus. According to their measurements, the approximate coordinates of this region were $\alpha = 20^h$, $\delta = +43°$. The observed relative change in the intensity of the radio waves reached 15 percent. In a first analysis of these observations, they arrived at the conclusion that the angular extent of the radio source did not exceed $2°$.

The discovery by Hey, Parsons, and Phillips of a discrete radio source of varying intensity opened a new epoch in radio astronomy. Today, 12 years later, it is very clear that it was an outstanding development, not only for radio astronomy but for astronomy in general.

Two investigations were published in 1948 that confirmed and improved the conclusions of the group mentioned above. In Australia Bolton and Stanley (1948), using a sea interferometer, were able to determine more precisely the coordinates of the discrete radio source. They established that the fluctuations came from a region of the sky whose angular dimensions were less than $8'$. The coordinates of the source were given as $\alpha = 19^h58^m47^s \pm 10^s$, $\delta = +41°41' \pm 7'$. According to the interferometric observations by Bolton and Stanley at frequencies of 60, 85, 100, and 200 Mc/s, the radiation from this source has two components, one of constant and the other of variable intensity. The intensity of the variable component decreases with increasing frequency; at 200 Mc/s the variable component could not be detected. Measurements of the intensity of the variable component at different frequencies are strongly correlated. No peculiar optical object was observed in the region occupied by the radio source.

In the same year Ryle and Smith (1948), with a two-antenna interferometer operating at 81.5 Mc/s with 500-m spacing, also measured the coordinates of the variable radio source in Cygnus. They found that the spectral flux density (which, for brevity, we shall in the future simply call "flux") was 1.4×10^{-22} w m^{-2} (c/s)$^{-1}$, but that individual "bursts" of 2.0×10^{-22} w m^{-2} (c/s)$^{-1}$ occurred. The duration of such bursts was of the order of tens of seconds. Neither the constant nor the variable component of the radiation was found to be polarized.

The fact that the intensity of the source varied rapidly and randomly with time led to the idea that its radiation could not arise from a very extended region (for example, in the interstellar gas). Indeed, in such a case the fluctuations would be averaged along the line of sight, and the flux emitted would be practically constant. Besides, Hey, Parsons, and Phillips drew attention to the fact that the character of the intensity fluctuations of the Cygnus source was reminiscent of the intensity fluctuations in the radio emission from the disturbed sun. Ryle and Smith, from the fact that the duration of a fluctuation did not exceed 20 sec, drew the reasonable conclusion that the linear dimensions of the radio source cannot significantly exceed the distance which light travels in that length of time—that is, about 6×10^6 km, or of the order of a giant-star diameter.

Thus, the fact that the discrete source of radio waves fluctuated had immediately led to far-reaching inferences as to its character. As we shall explain in due course, however, the opinions then expressed were only a misinterpretation. Nature has turned out to be far richer and more diversified than it had seemed to be during the early development of radio astronomy. The intensity fluctuations, which the first investigators had considered the most outstanding characteristic of the Cygnus radio source, have since proved to have no relation to the source.

During the investigation of the Cygnus source, Ryle and Smith (1948) detected a new and considerably more powerful radio source in Cassiopeia, with coordinates $\alpha = 23^h17^m$, $\delta = +58°$. The intensity fluctuations of this source seemed to be considerably smaller than those of the Cygnus source. The flux emitted was found to be 2.3×10^{-22} w m^{-2} (c/s)$^{-1}$, but the bursts during the fluctuations reached no more than 0.6×10^{-22} w m^{-2} (c/s)$^{-1}$.

Bolton, Stanley, and Slee (1949; Bolton 1948) detected by means of a sea interferometer four more discrete sources (all considerably weaker than those mentioned above), located in Taurus, Coma Berenices [accurate determination of its coordinates places this source in Virgo], Hercules, and Centaurus. The following year Bolton and Stanley (1949) derived fairly accurate coordinates for these sources. The one in Taurus was located at $\alpha = 5^h31^m \pm 30^s$, $\delta = +22°1' \pm 7'$. To within observational error, these coordinates correspond to the famous Crab Nebula (NGC 1952), the remnants of the galactic supernova of A.D. 1054. This was the first identification of a discrete source of cosmic radio waves with an optical object. Further developments in radio astronomy showed that this identification has fundamental significance for many important problems of modern science.

The Virgo source has the coordinates $\alpha = 12^h28^m \pm 37^s$, $\delta = +12°41' \pm 10'$; the Centaurus source, $\alpha = 13^h22^m20^s \pm 10^s$, $\delta = -42°37' \pm 8'$. At the positions of these sources there are two peculiar external galaxies, M 87 (NGC 4486) and NGC 5128.

The fact that two quite intense discrete sources are identified with peculiar galaxies offers considerable interest. Yet for several years no special significance was attached to this fact. This was the *Sturm und Drang* of radio astronomy, a period when discoveries occurred in such rapid succession that it was difficult to evaluate their relative significance.

We should notice the huge value of the flux emitted by the most powerful discrete sources. The strongest source in Cassiopeia emits, at 81.5 Mc/s, a flux of 2.3×10^{-22} w m^{-2} (c/s)$^{-1}$. For comparison, let us compute the thermal flux emitted by the quiet sun:

$$F_\odot = 2kT\Omega\lambda^{-2},$$

where $T \approx 10^6$ °K and $\Omega = 1.35 \times 10^{-4}$ sterad (since the effective diameter of the sun at this frequency is 1.4 times as great as the optical diameter). Introducing these values into the equation above we find $F_\odot = 2.76 \times 10^{-19}$ cgs, or 2.76×10^{-22} w m^{-2} (c/s)$^{-1}$, about the same as for the Cassiopeia source. The flux from the Cygnus source is only twice as small as that from the quiet sun! On the other hand, in the optical range the flux from the sun is many orders of magnitude greater than that from all other cosmical objects combined. Thus if the sun appears to be the dominating source of radiation at optical frequencies, in the radio range it plays a much more modest role. For example, in Sec. 4 we have seen that at a frequency of 9.15 Mc/s the sun's surface brightness is lower than that of the galactic nucleus, and hence it is observed as a dark spot against a bright background of cosmic radio waves. The relative brightness and flux of cosmic objects in the radio range are entirely different from those observed optically.

After the first observations had revealed the strongest discrete sources of cosmic radio waves, further investigations detected ever weaker sources. Stanley and Slee (1950) in Australia, using a sea radio interferometer, were able to prepare a catalogue of 18 sources, of which 13 were new. The most important contribution of this work was an investigation of the spectrum of the emission from a few of the most intense sources. For the Cygnus and Taurus sources observations were carried out at 40, 60, 85, 100, and 160 Mc/s; for the Virgo source, at 60, 85, 100, and 160 Mc/s; and for the Centaurus source, at 60, 100, and 160 Mc/s.

Figure 63 shows these first results of the investigation of the spectrum of these sources. Although the observations were not very accurate, they were successful in revealing the real differences in the various spectra. While for the Cygnus, Virgo, and Centaurus sources the flux increases rapidly as frequency decreases, for the Taurus source (identified with the Crab Nebula) the flux remains practically constant throughout the range of frequencies investigated.

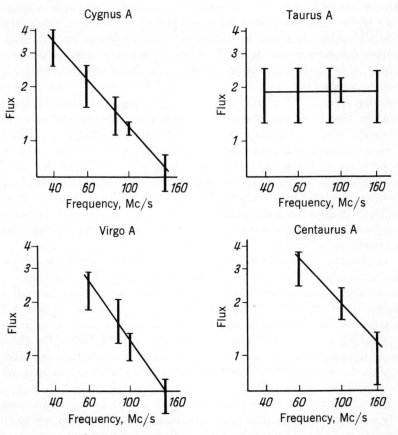

Fig. 63. Spectral composition of the radio emission received from four intense discrete sources (Stanley and Slee 1950).

The Australian investigators made a detailed study of the fluctuations in the intensity of the emission from the Cygnus region. They discovered that the amplitude of the fluctuations (with respect to the mean value of the flux) increased with decreasing frequency. Although there was a general correlation among simultaneous fluctuations at different frequencies, a detailed correlation could not be established.

To investigate this question further, observations of the intensity fluctuations of the Cygnus source were carried out simultaneously at two stations, one in Australia and the other in New Zealand. No correlation between the two stations was shown by fluctuations at a given frequency. On the other hand, when these two stations observed the fluctuations in the intensity of radiation from the disturbed sun, the correlation was fully 90 percent. This fact forces us to conclude that the fluctuations in the Cygnus source (and, for that matter, in the other sources) are not actual variations in the emitted intensity, but are caused by the Earth's atmosphere (more exactly, by the ionosphere). This basic conclusion was later confirmed by further observations which showed that the fluctuations in the Cygnus and Centaurus sources followed an annual cycle. Further, in every instance the fluctuations were particularly noticeable when the altitude of the source was less than 2°.

Smith (1950), in England, confirmed the suggestion that the fluctuations in the emission of discrete sources are purely an ionospheric phenomenon, analogous to the optical effect of stellar scintillation.

Smith observed the intensity fluctuations of the powerful sources in Cygnus and Cassiopeia at wavelengths of 3.7 and 6.7 m. He discovered that high correlation between the fluctuations at the two wavelengths obtained only as long as the two antennas used were sufficiently close. When the antenna spacing exceeded 20 km the correlation vanished. Hence the observed intensity fluctuations must originate in the ionosphere.

Little and Lovell (1950) arrived at the same conclusion at about the same time. They investigated the intensity fluctuations of the same two sources at $\lambda = 3.7$ m from two stations 210 km apart. It was found that the flux from the sources at times remained constant, but at times varied simultaneously at both stations. In this latter case no correlation was observed between the stations. By shortening the distance between the two antennas to 3.9 km, a significant correlation was observed (0.4 to 0.95), and the correlation became complete when the distance had decreased to 100 m.

The observations listed above decisively established the ionospheric origin of the observed intensity fluctuations in the radio waves from discrete sources. The study of these fluctuations has become an important method for investigating the microstructure and motions of

ionospheric clouds. Such problems enter into the realm of the geo-
physical sciences, however, and as such fall beyond the scope of this
book.

Ryle and Elsmore (1951) published the results of a year and a half
of systematic observations whose aim was to detect possible variations in the radio flux from discrete sources. For this purpose they used
a fixed meridian interferometer, at $\lambda = 3.7$ m. The two antennas were
of greater dimension in an east–west direction than in a north–south
direction, so that the directional diagram of the interferometer
formed a narrow strip extending along the meridian from $\delta = \pm 10°$
to $\delta = +80°$ (Fig. 64). By means of such a system it was possible to
observe almost all the discrete sources in the northern hemisphere of
the sky. From the results of the observations a catalogue was assem-
bled (Ryle, Smith, and Elsmore 1950) in which are tabulated the co-

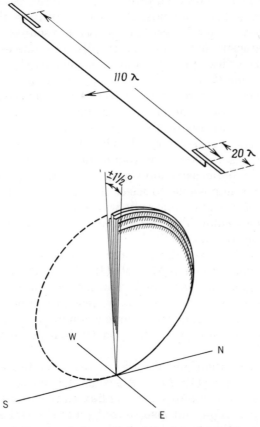

Fig. 64. Directional diagram (schematic) of the interferometer used by Ryle
and Elsmore (1951).

ordinates and intensity of approximately 50 sources in the northern sky—a valuable complement to the Australian survey.

On the basis of their observations Ryle and Elsmore were able to conclude that, after taking into account the ionospheric fluctuations, not a single discrete source showed a true intensity oscillation exceeding 10 percent of the mean value of the flux. For the strong sources in Cygnus and Cassiopeia one could even assert that the true intensity variations did not exceed 5 percent.

It is most remarkable that the flux emitted by discrete sources is practically constant. From optical astronomy we know that the flux from variable stars changes by large amounts, with periods anywhere from a few hours to thousands of days.

We have seen that laborious investigations have merely served to demonstrate that the radio flux from discrete sources is to a high degree constant. And yet these sources were discovered, one might say, just because their observed intensity fluctuated rapidly! It required 4 years to establish the ionospheric nature of these fluctuations, and the constancy in "brightness" of the discrete sources. Meanwhile, the apparent intensity fluctuations of the discrete sources had led theoreticians to try to interpret the sources along entirely incorrect lines. It was at this time that several authors proposed and elaborated on hypotheses that the discrete sources were a special type of star, emitting radio waves at a rate much greater than the sun, and with swiftly and randomly varying intensity (we shall refer briefly to these hypotheses in Sec. 11). By now it has become perfectly clear that investigations in this direction lead to a dead end. But we must remember that early in radio astronomy's agitated development it was very difficult, if not impossible, to select the right line of theoretical attack toward the nature of cosmic radio waves in general, and of the discrete sources in particular. The "radio universe" seemed very peculiar and different from the optical.

At that period more data concerning discrete sources were being gathered. Piddington and Minnett (1951), while investigating the region of the galactic center at 25 cm, detected a new source of radio waves. The emitted flux at this wavelength appeared to be 2.6×10^{-23} w m^{-2} (c/s)$^{-1}$; thus this source is among the strongest. We shall often have occasion to touch upon the nature of this remarkable source.

Further investigations of radio waves from discrete sources were carried out by Mills (1952a) in Australia. His purpose was not only to determine the coordinates and the flux emitted by the sources with maximum precision, but also to make a statistical analysis of their distribution in the sky and to classify them.

The observations were made at $\lambda \approx 3$ m with an interferometer consisting of three antenna arrays in an east–west line and two receivers so connected that one could record simultaneously the signal received by two pairs of arrays. Thus Mills's arrangement was in fact equivalent to two two-antenna interferometers of different spacing, operating simultaneously. Maximum spacing between antennas (250 m) yielded a directional diagram with a main lobe of 40' and a relatively small spacing of 3°. Each antenna array comprised 24 half-wave dipoles and had a directional diagram of 24° × 14°. Simultaneous observations with two interferometers of different spacing permitted rough estimates of the angular extent of the source, if in excess of 10'.

The observational procedure consisted in setting the three-antenna interferometer at a given declination. Because of diurnal rotation it was possible to observe all the sources that drifted over the directional diagram within the declination zone under investigation. Observations at a given declination were carried on for 48 hr, after which the declination of the antennas was shifted by 10°; the measurements were then repeated in the same fashion. The entire set of observations covered the vast declination zone $+50° > \delta > -90°$. Right ascensions and declinations of the sources were derived by the method described in Sec. 2. The error in right ascension of the strong sources was 2^m; in declination, 20'. The coordinates of the weak sources were determined with far less accuracy; here the error approached 2°. The flux from the relatively powerful sources was determined to 10 percent accuracy, and from the weaker ones, to 30 percent accuracy.

From the results of these observations Mills compiled a catalogue containing 77 sources of radio waves. At least three sources—one in Vela ($\alpha = 8^h 35^m$, $\delta = -42°$), one in the vicinity of the galactic center, and one in Centaurus—appeared to be extended. When the observations were carried out with an interferometer of large spacing their flux was ten times as small as when observed with small spacing. Mills deduced that the angular extent of these sources was of the order of 35'. The second source mentioned seemed to be the same as one that had been observed by Piddington and Minnett at 1210 Mc/s. The source in Vela may be responsible for the small secondary maximum in Allen and Gum's isophotes (Fig. 27, Sec. 3). The intensity of the source in Centaurus was half as great when observed with large spacing as when observed with small spacing. Hence it was possible to conclude that this source (one of the most intense) has an angular extent of the order of 20'. It is notable that this is of the same order as the angular extent of the galaxy NGC 5128, with which this source has been identified.

From a statistical analysis of the data provided by the observations, Mills arrives at some interesting conclusions about the distribution of sources over the celestial sphere and in depth; we shall return to this point in Sec. 8.

As a result of the investigations by the Australian and English radio astronomers, the number of discrete sources of cosmic radio waves discovered by 1952 was close to one hundred. It became urgent to decide on some system to designate both old and newly discovered sources which, as a rule, were not identified with any optical object. Several notations were proposed. One of the systems was to write the name of the constellation in which the source is located, followed by a capital Roman letter: A, B, C, ... The letter A would be assigned to the strongest source in a given constellation, the letter B to the second strongest, and so on. Thus, the very first discrete source discovered was called "Cygnus A"; the strongest source, "Cassiopeia A"; the one identified with the Crab Nebula, "Taurus A"; and the like. This designation for the strongest sources has been retained to the present time.

In another system the source Cygnus A would be called "19 + 4A." This code means that the right ascension of the source lies between 19^h and 20^h, and that the declination is positive and lies between 40° and 50°; "A" means that within the specified intervals of α and δ this is the strongest source. This system was used mainly by Mills. A variation of this nomenclature consists in replacing the + or − sign by the letter N or S; this last system was adopted by the International Astronomical Union in 1955.

Bolton, Stanley, and Slee (1954) have published a large catalogue of discrete sources. The observations were carried out with a sea interferometer at a frequency of 100 Mc/s; the width of the lobe was about 1°. The observations covered the zone of sky $+50° > \delta > -50°$, in which 104 discrete sources were detected. The authors of these investigations compared their catalogue with other catalogues; they found good agreement among the positions and fluxes reported for the stronger sources in the different surveys, and even for the weaker ones if isolated. The agreement among the various catalogues breaks down for weak sources located too close to each other, or to stronger sources, because of the overlapping of the interference patterns of the different sources. In this respect the sea interferometer has a definite advantage over a pair of antennas, since the interference pattern of an observed weak source will not be perturbed by the interference pattern of any nearby stronger source which has not yet risen above the horizon (we have already discussed this point in Sec. 2).

By comparison with other catalogues, Bolton, Stanley, and Slee concluded that, of the 104 sources they observed, 73 were real, while the reality of the rest was probable but not certain.

Hanbury Brown and Hazard (1953b) applied an entirely different method in the study of discrete sources. Instead of the usual interferometric method they employed the large 218-ft fixed reflector of the University of Manchester. The investigations were carried out using the main lobe of the antenna at $\lambda = 1.89$ m. When a discrete source passed through the main lobe because of diurnal rotation, the antenna temperature increased and the continuous record showed a maximum (Fig. 65). From the time at which the maximum was recorded it was possible to compute the right ascension of the source. The declination was computed from the angle between the main lobe and the horizon. The flux emitted by the source was derived from the excess of the antenna temperature over the background temperature of cosmic radio waves, whose absolute intensity was assumed known.

In this way it was possible to determine the coordinates and flux of 23 sources in the region of the sky accessible to the fixed radio telescope at Jodrell Bank. This region lies within the limits $40° < l < 140°$, $|b| < 15°$. The weakest source observed had a flux $F \approx 5$ to 7×10^{-26} w m^{-2} (c/s)$^{-1}$.

Kraus, Ko, and Matt (1954) used the same method to observe discrete sources with the helical antenna of The Ohio State University, at 1.20 m. Figure 66 shows one of the records obtained with the antenna at a fixed declination of about $-45°$. The large maximum is caused by the passage of the source Centaurus A through the main lobe of the directional diagram. For the investigation of discrete sources they used a 48-helix antenna; the width of the main lobe was $1°.2$ in right ascension, $17°$ in declination.

In all, 207 discrete sources were detected over a region comprising two-thirds of the sky. The error in the computed coordinates was, of course, rather large (up to $5°$). The weakest flux recorded was $< 10^{-24}$ w m^{-2} (c/s)$^{-1}$. The error in the measured flux was 20 percent for strong sources, up to 100 percent for weaker ones. Forty-seven sources in Mills's catalogue lay within the region of sky observed by Kraus, Ko, and Matt. In 39 cases, the positions agreed, to within ob-

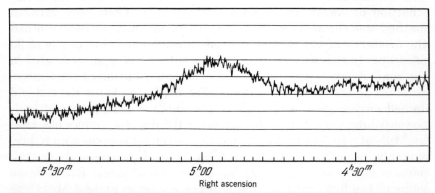

Fig. 65. Record produced by a discrete source observed with the 218-ft radio telescope (Hanbury Brown and Hazard 1953b).

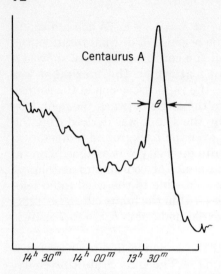

Fig. 66. Record produced by a discrete source observed with the helical antenna of The Ohio State University (Kraus, Ko, and Matt 1954).

servational error, with the positions listed in the catalogue. Of the 22 sources in Ryle and Elsmore's catalogue which were located in the zone of sky swept by the helical antennas, the coordinates agreed in 11 cases. Considering the differences in the methods used by the different investigators, such agreement is not disappointing. It must also be kept in mind that the observations were carried out at considerably different frequencies, and the unequal spectral distribution of the emission from different sources may influence the results.

In this connection, investigations of discrete sources at extremely low and high frequencies offer special interest. At 18.3 Mc/s we have the work of Shain and Higgins (1954), which up to the present time remains the only large-scale survey of discrete sources at very low frequencies. For this investigation Shain and Higgins used the same antenna that had been employed in Shain's study of the intensity distribution of cosmic radio waves (see Sec. 3). Their observational method was the same as that just discussed; that is, they recorded the discrete sources as they drifted through the beam of the radio telescope. In this way they detected a total of 37 sources located in the zone of sky $-12° > \delta > -52°$. The distribution of the sources observed at 18.3 Mc/s is charted in Fig. 67. In the figure the full and dashed curves show the limits of the antenna maximum and half-power points; the white circles represent the discrete sources observed by Mills at 110 Mc/s; the dashed lines encircle sources which, presumably, were observed at both frequencies. In 23 cases the coordinates of the sources agreed within observational error. The absolute value of the flux emitted by the discrete sources at 18.3 Mc/s was very high. Thus, for example, the flux from Centaurus A at this frequency was 10^{-22} w m^{-2} (c/s)$^{-1}$, while at 101 Mc/s it was only

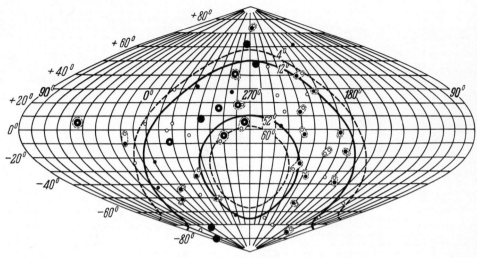

Fig. 67. A chart of discrete sources (Shain and Higgins 1954).

2.4×10^{-23} w m^{-2} (c/s)$^{-1}$. It would be highly desirable to observe the discrete sources at even lower frequencies. For this purpose cross-shaped antennas are required.

During the last few years several investigations of discrete sources at high frequencies have been published. Hagen, McClain, and Hepburn (1954) observed 20 discrete sources at $\nu = 1420$ Mc/s ($\lambda = 21$ cm) with the 50-ft precision radio telescope of the Naval Research Laboratory at Washington. At this frequency the width of the directional diagram was about 1°, which allowed reliable determination of the coordinates of the sources to an accuracy of 15′. Somewhat earlier, and with the same radio telescope, Haddock, Mayer, and Sloanaker (1954a, b) investigated discrete sources at an even higher frequency, $\nu = 3000$ Mc/s ($\lambda = 9.4$ cm). The width of the directional diagram was 24′ × 27′, and the antenna temperature could be measured to an accuracy of 2 K deg.

Table 6 lists the results of the American investigators for the high-

TABLE 6. Observed fluxes.

Source	F_ν [10^{-25} w m^{-2} (c/s)$^{-1}$]		$\alpha(1950)$	$\delta(1950)$
	$\lambda = 21$ cm	$\lambda = 9.4$ cm		
Cassiopeia A	266	150	23^{h}21^{m}3	$+58°32'$
Sagittarius A	170	48	17 42.5	$-29\ 01$
Cygnus A	146	70	19 57.8	$+40\ 34$
Taurus A	106	80	5 31.4	$+21\ 54$
Virgo A	30	14	12 28.2	$+12\ 37.5$

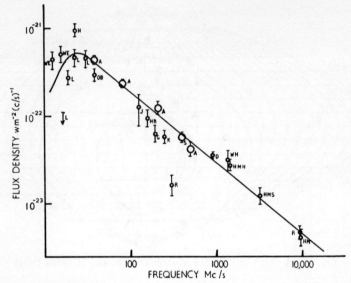

Fig. 68. The spectrum of Cassiopeia A (Whitfield 1957).

frequency radio waves from a few of the discrete sources we have discussed before. Data on other sources will be presented in Sec. 10. It is interesting to note that at $\lambda = 9.4$ cm, the source Taurus A becomes stronger than Cygnus A. This is evidence for the difference between the spectra of these two sources: at a wavelength of 3 m the flux from Cygnus A is almost seven times as great as that from Taurus A.

During 1953 and 1954 several discrete sources were observed in the U.S.S.R. at a wavelength of 3.2 cm, with a 4-m reflector (Kaidanovsky, Kardashev, and Shklovsky 1955; Plechkov and Razin 1956). According to Plechkov and Razin the fluxes received from Cassiopeia A, Cygnus A, and Taurus A equal, respectively, 6.6×10^{-24}, 4.5×10^{-24}, and 6.0×10^{-24} w m^{-2} (c/s)$^{-1}$. Similar results were obtained by Kaidanovsky *et al.*

Some interesting anomalies have recently been observed in the long-wavelength region of the spectra of discrete sources. Hey and Hughes (1954) observed the radiation from the strongest sources in Cygnus and Cassiopeia at a frequency of 22.6 Mc/s ($\lambda = 13.3$ m). They found that the intensity ratio of the Cygnus to the Cassiopeia source is smaller at 13.3 m than at shorter wavelengths.

Lamden and Lovell (1956) have investigated in detail the low-frequency portion of the spectra of the strongest sources in Cygnus and Cassiopeia. The observations were carried out at 16.5, 19, 22.6, and 30 Mc/s. Beginning at 22.6 Mc/s they observed a strong "break" in the spectra of Cygnus A and, particularly, of Cassiopeia A. If, for example, the ratio of the fluxes from the two sources at 22.6 Mc/s is 0.5, then at 19 Mc/s it is 0.7, and at 16.5 Mc/s it exceeds unity.

These results were not confirmed by the work of Wells (1956), who also investigated the low-frequency spectra of the strongest discrete sources. The observations were made at 26.75 Mc/s and 18.5 Mc/s, and Wells also detected a low-frequency break in the spectra. However, the ratio of the fluxes of the Cygnus and Cassiopeia sources was almost the same at both frequencies.

Recently Whitfield (1957) carried out a thorough analysis of the spectra of a large number of discrete sources on the basis of all the available measurements at different frequencies. To begin with, he constructed a spectrum of Cassiopeia A, which is the strongest and best-studied source (see Fig. 68). The large dots correspond to observations which were carefully calibrated, and for which the antenna gain was taken into account. Naturally, these observations carry the greatest weight.

Later Whitfield used this spectrum of Cassiopeia A as a standard, to which he tied the spectra of other sources. Since the relative measurements were made with considerably more accuracy than the absolute ones, Whitfield's spectra seem to be the most reliable available.

Figure 69 shows the spectra of three other strong sources obtained

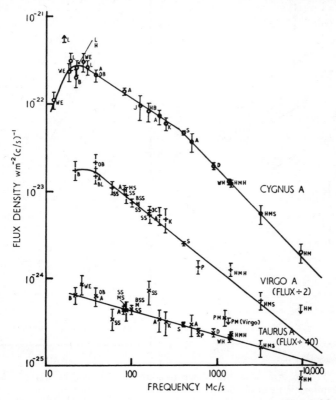

Fig. 69. The spectra of Cygnus A, Virgo A, and Taurus A (Whitfield 1957).

by comparison with Cassiopeia A. Cygnus A shows a large drop at low frequencies; also there is a characteristic break in the spectra around 350 Mc/s. The spectrum of Taurus A slopes more gently than those of other sources, although it is definitely not horizontal, contrary to our earlier assumption (see Fig. 63).

Figure 70 shows the spectra of several sources, as derived by Whitfield. In certain cases there are optically observed objects that have been identified with them (see Sec. 8). Since the spectra appear as straight lines on a logarithmic scale they must be of the form $F_\nu \propto \nu^x$, where x is called the *spectral index*. This quantity is very significant for the theoretical interpretation of the radiation from discrete sources (see Secs. 13, 20, 21). The spectral index varies for different sources between rather wide limits. The largest value is $x = -0.28$, observed

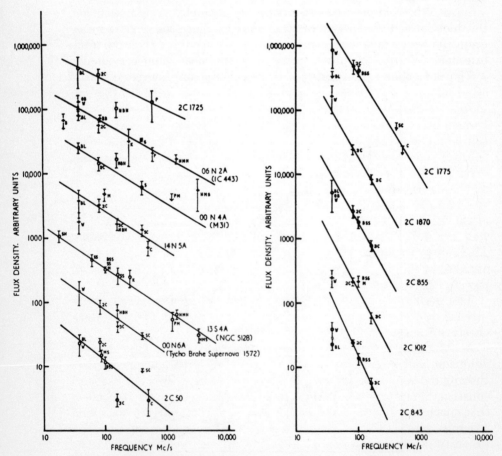

Fig. 70. The spectra of several discrete sources, arranged in order of decreasing spectral index (Whitfield 1957).

for Taurus A; the smallest value is $x = -2.2$. The majority of the sources have spectral indices in the range $-0.6 > x > -1.2$.

7. ANGULAR EXTENT OF THE DISCRETE SOURCES

In the preceding section we have already quoted the results of Mills (1952a), who found, from observations with an interferometer at two different spacings, that the angular extent of the sources Sagittarius A, Vela A, and Centaurus A is of the order of 20′ to 30′. The question of the angular extent of the strongest discrete sources in Cygnus and Cassiopeia remained unsettled, however.

This question has great significance since it is intimately connected with the problem of their nature. During the first years after the discovery of the discrete sources, particularly before the reason for their intensity fluctuations was cleared up, the most popular idea was that the majority of the discrete sources were of stellar nature. In this case we would expect their angular extent to be of the order of hundredths of a second of arc or even less. Radio telescopes, with their very low resolving power, obviously could not detect such a small dimension. On the other hand, shortly after the discovery of one of the strongest discrete sources (Taurus A), an identification was made with the Crab Nebula. Immediately the question was raised: with which object should this source be identified, with the nebula itself, or with the stellar remnants of the supernova responsible for it? In the first case the angular extent of the Taurus A source would be of the order of 5′; in the second case, vanishingly small.

All efforts undertaken before 1950 to determine the angular extent of the most intense discrete sources—Cygnus A and Cassiopeia A (as well as a number of other sources)—invariably gave negative results. These measurements only allowed an upper limit to be placed on the angular extent of these sources, which depended on the resolving power of the equipment used.

The problem of a reliable determination of the angular extent of intense discrete sources was solved almost simultaneously at Manchester, Cambridge, and Sydney.

At Manchester the problem was taken up by Hanbury Brown, Jennison, and Das Gupta (1952) in 1950. Since it was believed that the angular diameters would be negligibly small, it was necessary to construct an interferometer of very high resolving power. To achieve this, spacings of 10 to 50 km would have been required. For such long spacings great difficulties arise, since special measures need to be taken to maintain the proper phase of the oscillations during their transmission through the line connecting the antennas.

To avoid such difficulties the Manchester group worked with a spe-

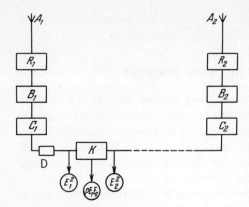

Fig. 71. Block diagram of the inter-ferometer designed by the Man-chester group (Hanbury Brown, Jennison, and Das Gupta 1952).

cial interferometer of original design, which is shown schematically in Fig. 71. Here A_1 and A_2 are the two antennas, each 500 m^2 in area. These antennas are connected to the inputs of two superheterodyne receivers R_1 and R_2, working independently. The receivers are tuned to a frequency of 120 Mc/s, with a bandwidth of 200 kc/s. The inter-mediate-frequency signal voltages at the receiver inputs are detected by the square-law detectors B_1 and B_2, and sent through the filters C_1 and C_2, of 1- to 2-kc/s passband width. The input voltages from both receivers are multiplied by the correlator K. If in addition the root-mean-square values of the input voltages, $\langle E_1{}^2 \rangle$ and $\langle E_2{}^2 \rangle$, are known, it is possible to compute the cross-correlation coefficient ρ (where $\rho \langle E_1 E_2 \rangle$ is the rms input voltage at the correlator). One low-frequency output, modulating a high-frequency carrier, is transmitted over a radio link. The two low-frequency outputs are made to arrive simultaneously at the correlator by inserting a delay D in the base-line. The phase of the two low-frequency signal voltages should be preserved through all the components; this problem is simpler than the corresponding one in typical interferometers using radio frequen-cies.

As the source drifts through the directional diagram of the inter-ferometer, the rms voltages $\langle E_1{}^2 \rangle$ and $\langle E_2{}^2 \rangle$ increase. These two quan-tities, and also $\rho \langle E_1 E_2 \rangle$, are recorded simultaneously.

It can be shown theoretically that the correlation coefficient ρ de-pends on the angular extent of the observed source, on the spacing l, and on the wavelength λ. If we assume that the source is a rectangle of side a, the expression giving ρ is

$$\rho = \frac{\sin^2(\pi a l/\lambda)}{(\pi a'/\lambda)^2}. \tag{7-1}$$

Choosing the proper value of l, it is possible to compute the smallest value of a. By performing a series of measurements with different lengths and orientations of the baseline it is possible to deduce not only the angular extent of the source, but also its shape and the brightness distribution.

Table 7 shows the results of the observations by the Manchester group. The angle φ, showing the orientation of the baseline, is measured from north to west. From Table 7 we see that the discrete source in Cygnus has a very asymmetric shape. If the orientation of the baseline is changed, its extent decreases from 2′10″ to 35″. The position angles of the main axes of this source are 90° and 120°. The results of the observations definitely contradict the simple idea that the source Cygnus A has an elliptical shape and uniform brightness (later on we shall consider the shape of this source in detail).

In contradistinction to the source in Cygnus, the discrete source Cassiopeia A is more or less symmetric. Its equivalent angular extent is about 4′.

Concurrently with the work at Manchester the same problem was taken up at Sydney by Mills (1952b).

If we assume that discrete sources are to be identified with nebulae (as Mills believed, apparently because of the identification of Taurus A with the Crab Nebula), then their angular extent would be of the order of 1′. In order to obtain a lobe 1′ wide at a frequency of 100 Mc/s, the spacing of a two-antenna interferometer should be 10 km. Exactly such an interferometer system was constructed by Mills, using two antennas—a larger, fixed one, and a smaller, portable one.

As at Manchester, Mills used a relay system. The signal from the small, portable antenna was transmitted back to the large, fixed one. The local-oscillator frequency was transmitted simultaneously with the signal from the portable antenna in order to preserve the phase of the reconstituted signal. To compensate for the transmission time of the signal, a time delay was introduced in the system. The equality of the signal levels was ensured by automatic gain controls.

The spacing was changed by moving the portable antenna. As the spacing was increased, the amplitude of the records shown on the interferometer's charts for the four strongest sources decreased con-

TABLE 7. Diameters of Cygnus A and Cassiopeia A.

Baseline		Cygnus A		Cassiopeia A	
Length (km)	Azimuth φ	Corr. coeff. ρ	Diam. a	Corr. coeff. ρ	Diam. a
0.30	349.°5	0.99 ± 0.10	<5′	0.96 ± 0.09	3′40″ (<5′50″)
2.16	113.0	0.30 ± 0.03	2′10″ ± 4″	0.08 ± 0.02	2′55″ ± 10″
2.16	235.5	0.79 ± 0.08	1′00″ ± 7″	<0.01	3′30″
3.99	177.0	0.79 ± 0.07	0′34″ ± 8″	0.07 ± 0.01	—

TABLE 8. Percentage modulation.

Source	Baseline (km)			
	0.29	1.02	5.35	10.01
Cygnus A	1	1	0.3	0.05
Taurus A	1	0.55	–	–
Virgo A	1	0.40	–	–
Centaurus A	1	0.30	–	–

tinuously. This shows that the angular extent of the sources was comparable to the width of the directional diagram of the interferometer. Table 8 shows the results of Mills's observations. The dashes indicate that the percentage modulation was less than one-tenth of the smallest fluctuation registered by the recording apparatus.

Mills's results clearly show that the angular extent of Cygnus A is considerably smaller than that of other sources. For example, with a spacing of 1.02 km, the percentage modulation for Taurus A, Virgo A, and Centaurus A is already considerably smaller. This shows that the angular extent of these sources is of the same order of magnitude as the width of the main lobe of the interferometer, that is, $\lambda/l \approx 10'$. On the other hand, the percentage modulation of Cygnus A when $l = 1.02$ km remains unchanged, which means that its angular extent is considerably less than 10'. The interferometric pattern of Cygnus A disappears only with the huge spacing of 10 km, long after the patterns of other sources. Without any further analysis we can conclude that the angular extent of Cygnus A is roughly one-fifth of that of the other sources investigated by Mills.

As already mentioned in Sec. 2, by means of interferometric studies using different spacings it is possible to deduce the true intensity distribution over the source. If the axis of the interferometer is oriented in an east–west direction, it can be proved theoretically that the intensity distribution over the source in right ascension is given by the relation

$$B_t = \int_0^\infty A_\omega \cos(\theta_\omega - \omega t)d\omega, \tag{7-2}$$

where A_ω is the amplitude of the Fourier component, and θ_ω its phase (see Sec. 2). If we assume that the brightness distribution within the source possesses circular symmetry, then $\theta_\omega = 0$, and from the observed values of A_ω we can compute B_t. Mills made this assumption when he set about to reduce his own observations. However, according to the English observations discussed above, Cygnus A has a complex, drawn-out shape, and the assumption of circular symmetry is obviously incorrect. Nevertheless, even with such an assumption we

TABLE 9. Angular diameters.

Source	Diameter
Cygnus A	$1'.1$
Taurus A	4
Virgo A	5
Centaurus A	6

can get some mean value for the angular extent of the source. As regards the other sources, it was shown by later investigations that the assumption of circular symmetry can serve very well as a first approximation.

Table 9 shows Mills's results. It lists the effective extent of the sources between points where their surface brightness decreases by a factor of two (on the assumption of circularly symmetric brightness distribution).

With respect to the source Centaurus A, as we pointed out before, Mills found that its angular extent was of the order of 20'. His interferometric observations allowed him to reach some conclusions about the complex structure of this interesting extended source. It has a bright central condensation about 6' in diameter, surrounded by a considerably less bright but rather extended corona. Further observations confirmed these first ideas.

In Cambridge, Smith (1952) also investigated the angular extent of the discrete sources in Cygnus and Cassiopeia. He used an interferometer with variable spacing (up to $400\,\lambda$), working at $\lambda = 1.4$ m, and the phase-switching method of detection. Near the antennas of the two-element interferometer he placed preamplifiers to avoid any difficulties due to attenuation in the transmission lines. Furthermore, to avoid the need of long-term stabilization of the gain of the portable antenna (the other antenna remained fixed), Smith measured the ratio of the amplitudes of two records obtained with different spacings. Each measurement gave a ratio of two Fourier terms, and afterward it was possible to derive the dependence of the Fourier components on the spacing. Such a method gives a rather coarse picture of the brightness distribution over the source (considerably less accurate than the methods discussed previously), since measurements with large spacings were not made.

Figure 72 shows the dependence of the ratio of the Fourier components on the spacing, expressed in terms of wavelengths, for the two sources Cygnus A and Cassiopeia A (represented by dots and crosses, respectively). Qualitatively these curves confirm Hanbury Brown, Jennison, and Das Gupta's and Mills's results regarding the relatively small angular extent of the Cygnus source. On the assumption of a

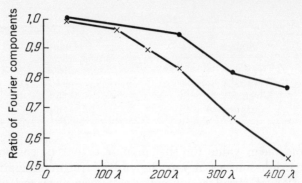

Fig. 72. Dependence of Fourier components on interferometer spacing for Cygnus A and Cassiopeia A (Smith 1952).

circularly symmetric brightness distribution, the effective extent of the Cassiopeia source is 5'.5, and that of the Cygnus source 3'.5. While the spacings used seemed quite adequate for the Cassiopeia source, for Cygnus A even the largest spacing (400 λ) was evidently too small. Therefore, while the determination of the angular extent of the Cassiopeia source is quite reliable, that for Cygnus A gives, essentially, only an upper limit to the true angular extent.

Mills (1953) continued his investigations on the intensity distribution over several strong sources. The observations were carried out with the variable-spacing interferometer described before. As already pointed out, the basic formula in Mills's method, Eq. (7-2), allows us to derive the intensity distribution over the source only in the case where this distribution is circularly symmetric. Essentially Eq. (7-2) gives only the integrated brightness, that is, the integral along a line perpendicular to the projection of the interferometer axis on the source. If the intensity distribution is not circularly symmetric, it is impossible to draw any simple conclusions above the brightness distribution from the observed integrated brightness. By making the more general assumption that the isophotes over the source are ellipses with a common center it is possible to derive an intensity distribution from three series of observations, using different orientations of the baseline. If the brightness distribution is still more complicated, a considerably larger number of observations is required (recalling that each series of observations at a given azimuth involves the use of several different spacings).

Figure 73 shows a sample from the records for Cygnus A and Virgo A with spacing 0.29 km (*a* and *c*) and 1.25 km (*b* and *d*). Evidently, increasing the spacing leaves the percentage modulation of Cygnus A practically unchanged, while that of Virgo A decreases markedly. This shows clearly that the angular extent of the first source is considerably smaller than that of the second.

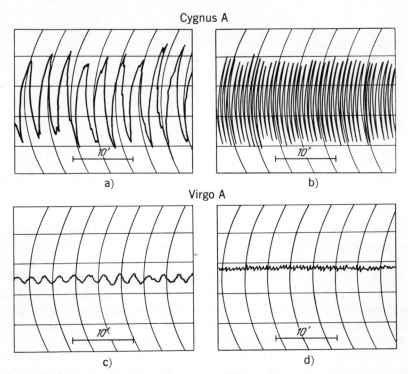

Fig. 73. Records produced by Cygnus A and Virgo A for two interferometer spacings (Mills 1953).

The observations were carried out at azimuth 90° (east–west) and, to a lesser extent, at azimuths 164° and 24°. The spacings at 90° were varied between 60 m and 10 km. For the other azimuths observations were carried out using only three spacings. The observations of Taurus A show that its brightness distribution does not possess circular symmetry (Fig. 74a). Figure 74b gives the integrated brightness

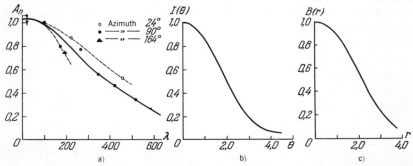

Fig. 74. Results of measurements of the angular dimensions of Taurus A (Mills 1953).

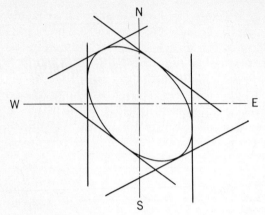

Fig. 75. Isophote for Taurus A, constructed on the basis of Mills's observations (1953).

distribution, and Fig. 74c gives the true brightness distribution along the east–west direction, obtained by integration of Abel's equation. From the observational data obtained with three different azimuths it is possible to derive rough isophotes. Figure 75 shows the isophote for Taurus A marking the region where the intensity is one-half of that observed at the center. The axes of this isophote are 5′.5 and 3′.5, and the position angle of the major axis is about 140°. The brightness temperature in the central region, at $\lambda = 3$ m, is about 4×10^6 °K.

Other sources were investigated by the same technique. Figure 76 shows the basic features of Virgo A with the interferometer oriented east–west. Isophotes derived from the observations show considerable ellipticity in the brightness distribution. The lengths of the two axes of the isophote joining the half-intensity points are 5′ and 2′.5, and the position angle of the major axis is about 50°.

Centaurus A gives a complicated picture. As already pointed out, the source is very elongated and has a bright condensation in the middle. New observations show that about 45 percent of the emission comes from the bright central condensation, whose effective extent in the east–west direction is about 5′. The weak outer region extends for 1°.5. Observations of the bright nucleus show its ellipticity clearly.

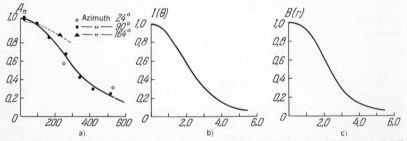

Fig. 76. Results of measurements of the angular dimensions of Virgo A (Mills 1953).

The size of the half-intensity isophote can be estimated very roughly as $6'.5 \times 3'$, the position angle of its major axis as $130°$. The brightness temperature at the center of the source is 4×10^6 °K, and in the brightest portions of the extended corona, $\approx 3 \times 10^4$ °K. Cygnus A was unfavorably located for observation, and so the results are uncertain. The brightness distribution over this source will be discussed later.

Since all the sources discussed so far have been identified with nebulae, it is of interest to compare the optical and radio brightness distribution. Such a comparison is shown in Fig. 77.

In the same year Jennison and Das Gupta (1953) investigated further the intensity distribution over Cygnus A. The observations were carried out at 125 Mc/s with the interferometer described above. In the case of an arbitrary intensity distribution over the source, the cross-correlation coefficient ρ is given by the expression

$$\rho^2 = \frac{F^2 \cos \psi + F^2 \sin \psi}{F^2 \cos (0)}, \qquad (7\text{-}3)$$

where $F \sin \psi$ and $F \cos \psi$ are the Fourier coefficients for the intensity distribution. In addition to the measurements already discussed, they used a series of measurements of the correlation coefficient at a position angle of $113°$, with different spacings. Figure 78 shows the results of their observations. This figure also includes Smith's (circles) and Mills's (squares) results.

The observed function $\rho^2(N)$, where N is the length of the spacing in terms of the wavelength, shows a secondary maximum. This suggests that Cygnus A must be composed of, at least, two separate emitting regions. Although it is not possible, from the curve in Fig. 78, to draw a simple conclusion about the intensity distribution, several additional considerations seem to indicate that Cygnus A is approximately symmetric, and comprises two emitting centers. The simplest distribution satisfying the observed curve, as well as previous observations, is shown in Fig. 79. The source is made up of two components of the same size ($51'' \times 30''$) and the same brightness, whose centers are $1' 28''$ apart. The position angle of the minor axis of the source is about $180°$. In Sec. 25 we shall attempt to interpret the peculiar intensity distribution of Cygnus A.

The intensity distribution in Taurus A, the Crab Nebula, is discussed in a paper by Baldwin (1954a), who carried out a series of measurements with two interferometers, working at $\lambda = 1.4$ m, with mutually perpendicular baselines. The spacings were varied from 10 λ to 300 λ east–west, and from 3 λ to 40 λ north–south. Baldwin's observational technique was largely similar to Smith's, and the intensity distribution was determined in the usual way.

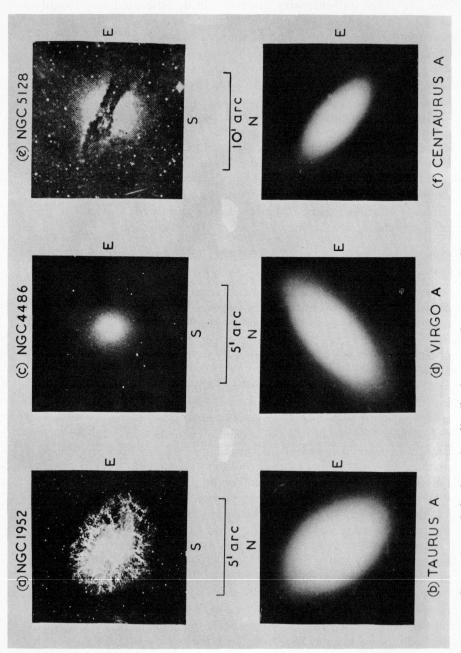

Fig. 77. Comparison of the intensity distribution over three intense discrete sources at optical (top) and radio wavelengths (Mills 1952)

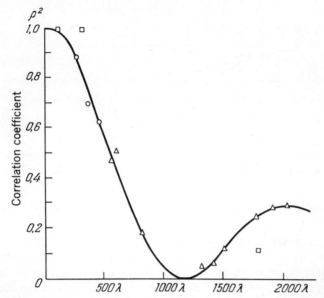

Fig. 78. Results of measurements of the intensity distribution over the source Cygnus A (Jennison and Das Gupta 1953).

The analysis of the observational data fails to indicate any departure from radial symmetry, in contradiction to Mills's results (see above). We should point out, however, that the 10-percent asymmetry in the intensity distribution found by Mills 2′.5 from the center of the emitting region is below the limit of the sensitivity in Baldwin's method.

Figure 80 shows the radial intensity distribution for Taurus A. The two curves mark the limits of the true brightness distribution. By the time of these observations it had been shown that, in a region extending 5′ to 3° in right ascension and 20′ to 10° in declination around the Crab Nebula, there is no source of radio waves whose flux could exceed 12 percent of the flux from Taurus A. Thus, according to the observations obtained at Cambridge, Taurus A is a radially symmetric, rather compact source, whose angular extent (between half-

Fig. 79. Intensity distribution (schematic) in Cygnus A (Jennison and Das Gupta 1953).

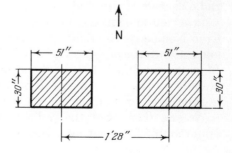

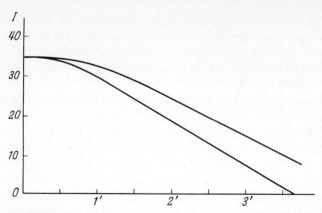

Fig. 80. Radial distribution of intensity in Taurus A (Baldwin 1954a).

intensity points) is about 5'. Also very important is the fact that the center of the radio emitting region is not more than 0'.5 away from the "center of gravity" of the Crab Nebula, as determined from optical observations, which means that the two points are coincident within observational error.

The intensity distribution over the Crab Nebula can be studied by another method. From time to time we have the opportunity to observe a very interesting phenomenon: a lunar eclipse of the Crab Nebula. The last ones took place in November 1955 and January 1956. From the observed radio-eclipse curves it is possible to deduce a rather accurate intensity distribution, and several interesting papers on this subject have recently been published. In particular, Costain, Elsmore, and Whitfield (1956) at Cambridge made interferometric measurements at $\lambda\lambda$ 3.7 m and 7.9 m. Figure 81 shows the intensity distribution at both wavelengths. From these observations it follows that the angular extent of the Crab Nebula at long wavelengths is appreciably greater than at short wavelengths. If this fact is confirmed, it will have considerable significance in any theoretical interpretation. We should keep in mind, however, that the observations at $\lambda = 7.9$ m were carried out with considerable difficulty. Another important result of this study is the agreement between the optical and radio centers of the Crab Nebula.

It is well known that annual observation of the occultation of the Crab Nebula by the solar corona is a powerful method for investigating the outermost layers of the solar atmosphere (Vitkevich 1951; Machin and Smith 1952). The study of the occultation of the Crab Nebula by the moon seems to be a reasonable method for determining an upper limit on the density of the lunar atmosphere (Link 1952).

We see, then, that the radio emission from Taurus A affords an opportunity for treating a variety of important astronomical problems, simply because of its fortuitous location near the ecliptic. This is quite apart from the exceptional intrinsic interest of the emission, a matter which we shall consider in detail in Chapter V.

Conway (1956) made an attempt to study the frequency dependence of the angular extent of discrete sources, in particular Cygnus A and Cassiopeia A. For this purpose he measured the angular extent at 500 Mc/s and compared the result with the observations at lower frequencies described above. His observations were obtained by means of an interferometer of variable spacing, up to 550 λ. Figure 82 shows the dependence in Cassiopeia A of the relative amplitude on the spacing, at different frequencies. From the curves it is apparent that the angular extent of this source at three different frequencies is practically the same. The same result was obtained for Cygnus A.

Several discrete sources were discovered in 1954 whose angular extent was considerably greater than that of Cygnus A, Taurus A, and other sources discussed previously. Hanbury Brown, Palmer, and Thompson (1954) discovered several discrete sources with the 218-ft

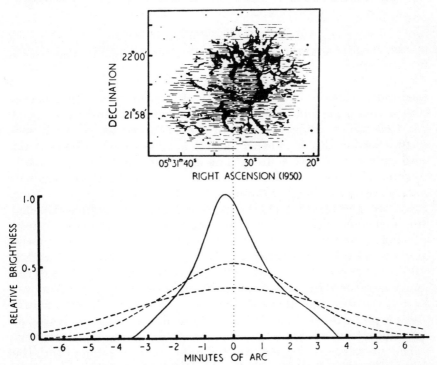

Fig. 81. Distribution of intensity across the Crab Nebula, projected onto the east–west axis, at λ = 3.7 m (full curve) and λ = 7.9 m (broken curves—two solutions). The dotted line corresponds to the center of gravity of the 3.7-m distribution (Costain, Elsmore, and Whitfield 1956).

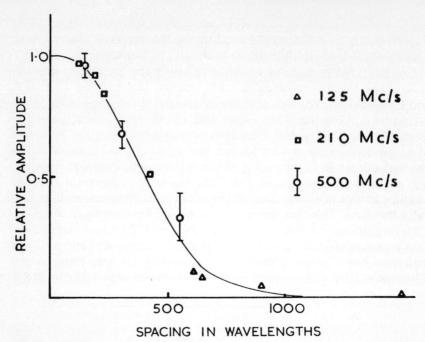

Fig. 82. Amplitude as a function of interferometer spacing for Cassiopeia A (Conway 1956).

fixed radio telescope of the University of Manchester, in combination with a small, portable antenna of 35-m² effective area, working at λ = 1.89 m. The angular extent of these sources, whose brightness was rather low, lay between 1° and 3°. Table 10 gives the relevant data concerning these sources. The last source listed in the table— Cygnus X—is observed in isophotes of cosmic radio waves as a secondary maximum (see, for example, Fig. 38).

Baldwin and Dewhirst (1954) discovered a very interesting extended source in Gemini. They used a radio interferometer of variable spacing, and made observations at 3.7 and 7.9 m. The coordinates of the center of gravity of this source are $\alpha = 6^h13^m37^s \pm 4^s$, $\delta = +22°38' \pm 5'$ (1950). According to their measurements the angular extent of the source is rather large, about 50'. The emitted flux at λ = 3.7 m was $F_\nu = 42 \times 10^{-25}$, and at λ = 7.9 m, 65×10^{-25} w m⁻² (c/s)⁻¹. Notice that the peculiar nebula IC 443 + S 40 is at the same place in the sky as this radio source (see Fig. 94), and has roughly the same angular extent.

The intensity distribution over the radio source associated with the nebula IC 443 + S 40 was studied by Rishbeth (1956) during its occultation by the moon on 1955 October 8. The observations were obtained with the Mills cross at λ = 3.5 m. Figure 83 shows the relative positions of the moon and the nebula at the moment of maximum phase of the occultation. (A photograph of the nebula is shown in Fig. 94.) Although at maximum eclipse only 0.1 of the surface of the source was occulted, the radio flux decreased by 20 percent. This shows the lack of symmetry in the surface brightness distribution over the source. From the observations we can deduce the following model of intensity distribution: a disk of uniform brightness 48′ in diameter whose center coincides with the optical center of the nebula (point Q in Fig. 83), on which there is superposed a source of relatively small angular extent, centered at point S. In the region of the small source the nebula IC 443 + S 40 is particularly bright in Hα. The ratio of the fluxes from the "small" and "large" sources is 3:4.

Elsmore and Whitfield observed this same eclipse from Cambridge. In their case the moon occulted the southern part of the nebula, where there is no Hα emission. The weakening of the flux was considerably smaller than that expected from a disk of uniform brightness.

About the same time, Bolton, Westfold, Stanley, and Slee (1954)

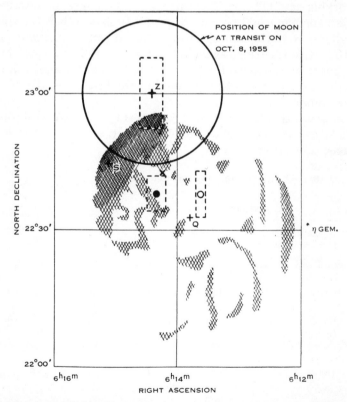

Fig. 83. Diagram of the nebula IC 443 and the associated radio source 06N2A (Rishbeth 1956).

TABLE 10. Angular diameters of discrete sources.

No.	Galactic coordinates		F_ν [10^{-25} w m^{-2} (c/s)$^{-1}$]	Angular diameter, a
	l	b		
1	100°5	+2°1	8	$1°.5 < a < 3°$
2	111.8	−0.3	6	$1°.5 < a < 3°$
3	121	−1.5	5	$1°.5 < a < 3°$
4	128.3	+3.8	8	$1°.4 \pm 0°.4$
5	56.5	+4.1	20	$1°.5 \pm 0°.4$
6	45.9	+0.8	30	$1° < a < 2°$

discovered several other discrete sources of large angular extent ($>1°$). For their observations they used three different arrangements. The first consisted of a 72-ft fixed paraboloid, whose directional diagram could be tilted with respect to the axis of the reflector by moving the mast supporting the feed. This radio telescope worked at $\nu = 150$ Mc/s. The second consisted of a sea interferometer with a lobe width of 1°, working at 110 Mc/s. The third was a two-antenna interferometer whose spacing could be varied from 4 to 22 λ. When the spacing was small, so that the source crossed the directional digram more slowly, they used the fringe-swinging method. In this way they showed that it is possible to measure the emitted flux as the source appears over the horizon, and investigate roughly the brightness distribution over each source.

This technique led to the discovery of twenty new extended sources. Table 11 gives the coordinates, angular extent, and flux for several of the sources thus observed.

TABLE 11. Fluxes and angular diameters at $\nu = 100$ Mc/s.

Source	Coordinates				F_ν [10^{-24} w m^{-2} (c/s)$^{-1}$]	Diameter
	Equatorial		Galactic			
	α	δ	l	b		
A	03^{h}17^m	−37°15′	206°	−56°	6	$\frac{1}{2}°$ to 1°
B	04 36	+20	145	−16	20	10° × 5°
C	05 10	−43 30	215	−35	15	1° to 2°
D	05 42	0	173	−14	15	10° × 5°
E	08 20	−42 15	227	− 2	15	1° to 2°
F	08 24	−44	230	− 2	70	5°
G	09 15	−10	210	+27	>10	−
H	12 40	+ 7	270	+69	25	4°
J	13 22	−43	274	+20	50	2°
K	16 51	−45	309	− 2	350	10° × 6°
L	17 41	−27 25	329	0	≫300	12° × 2°

The sources *B, D, K,* and *L* are significantly elongated along the galactic equator. Some of these sources had been observed before; for example, *A* is listed as Fornax A in Mills's catalogue, *E* as Puppis A in the same catalogue, and so on. Source *J,* or Centaurus A, has an extended corona. We shall refer to the other sources in Sec. 10.

An extensive investigation of discrete sources by interferometric methods was carried out by Ryle and his collaborators at Cambridge, during 1953 and 1954 (Shakeshaft, Ryle, Baldwin, Elsmore, and Thomson 1955). They determined the angular extent of a great number of sources. The interferometer (Ryle and Hewish 1955) consisted of four elements whose centers were at the vertices of a rectangle 580 m (157 λ) by 52 m (14 λ) (the wavelength chosen was 3.7 m). Each element consisted of a reflector shaped as a parabolic cylinder about 100 m in length and 12 m in width (Fig. 84). The dipole arrays were arranged along the focal line of the mirror. Each element in the interferometer had a directional diagram of width ±1° in right ascension and ±7°.5 in declination, and could be pointed to any given altitude. The observations were obtained by the phase-switching method.

The elements of the interferometer could be combined in two ways

Fig. 84. One of the four elements of the large Cambridge radio interferometer (Ryle and Hewish 1955).

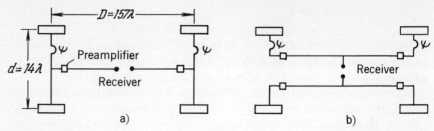

Fig. 85. Block diagram of the Cambridge radio interferometer (Ryle and Hewish 1955).

(Fig. 85). The diagram in Fig. 85a shows the antenna system connected so as to have a spacing of 157λ in an east–west direction. The resolving power of such an arrangement is about 20′; hence it permits a determination of the coordinates and of the flux from sources whose angular extent is less than 20′. Obviously with an interferometer of such high resolving power extended sources would not be detected. To study extended sources the elements of the interferometer were connected as illustrated in Fig. 85b. This system is equivalent to a two-antenna interferometer, whose baseline is oriented north–south. Figure 86 shows a typical record obtained with such an interferometer.

Figure 86a portrays the passage of the Crab Nebula, whose angular extent is considerably smaller than the width of the directional diagram. The record is typical for sources of small angular extent. In Fig. 86b a record of an extended source, the Perseus cluster of galaxies, is reproduced. Near the center of this cluster there is a source of smaller angular extent, namely the peculiar galaxy NGC 1275. The

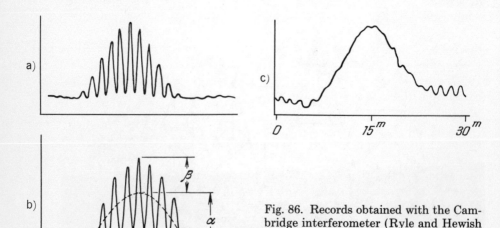

Fig. 86. Records obtained with the Cambridge interferometer (Ryle and Hewish 1955).

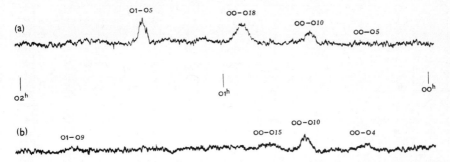

Fig. 87.   Records of discrete sources obtained with the Mills cross at Sydney (Mills and Slee 1957).

tracing therefore consists of a smooth maximum, corresponding to the extended source, and the oscillations due to the small source. Figure 86c corresponds to the extended source IC 443 in Gemini.

By means of the Cambridge interferometer 1936 sources were observed between declinations $+83°$ and $-38°$. The overwhelming majority of the sources had small angular diameters; only 30 sources had diameters between 20′ and 3°. For the 500 strongest sources the coordinates were established with a precision of $\pm 2^m$ in right ascension and $\pm 12'$ in declination. The coordinates of the weaker sources were less accurately known. The flux from the weakest sources was 7×10^{-26} w m^{-2} (c/s)$^{-1}$, although the exhaustive catalogue based on these observations lists only the sources for which $F_\nu \geq 2 \times 10^{-25}$ w m^{-2} (c/s)$^{-1}$.

The number of known discrete sources has been greatly increased as a result of this investigation by the Cambridge radio astronomers. Among these, a fairly large number of extended objects were detected.

Mills and Slee (1957) have published the first results of a survey of discrete sources carried out at Sydney, at $\lambda = 3.5$ m, using the Mills cross. The observations covered the region of sky $8^h > \alpha > 0^h$, $10° > \delta > -10°$. In this region 383 sources have been detected, as compared to 227 in the Cambridge survey. Figure 87 shows typical records obtained at Sydney at two different declinations.

Figure 88 shows the position of the sources observed at Sydney (black circles) and at Cambridge (open circles). The size of the circle is proportional to the flux. Extended sources are shown by a wavy line around them. There are striking discrepancies between the coordinates of the sources computed at Cambridge and at Sydney. Mills and Slee have shown convincingly that there are large errors in the Cambridge catalogue, due to instrumental effects.

The Sydney catalogue seems to be exhaustive for sources whose flux is greater than 2×10^{-25} w m^{-2} (c/s)$^{-1}$. The completion of this

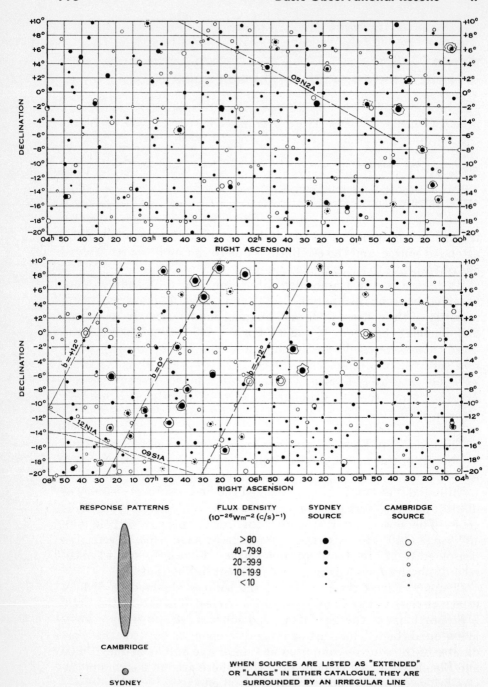

Fig. 88. A chart of the sources observed at Sydney and at Cambridge (Mills and Slee 1957). The half-power response contours are on the same scale.

catalogue for the entire zone of the sky accessible to the Mills cross (up to $\delta = +10°$) is anxiously awaited by all astronomers.

8. OPTICAL IDENTIFICATION OF DISCRETE SOURCES

Immediately after the discovery of discrete sources, the question of their nature was raised. This question is linked with the problem of identifying the discrete sources of cosmic radio waves with objects observed optically. However, in order to carry out this identification it is first necessary to determine accurately the coordinates of the source.

In 1946, when the first discrete source in Cygnus was discovered, its coordinates were only known to within $2°$. Under these circumstances, it is obvious that no serious attempt at identification could be made. Within 2 years, however, observations made in Australia were able to localize this source (and also a series of other relatively strong sources) to within $8'$. This was achieved by interferometric methods.

By 1951 the accuracy in the determination of the coordinates of discrete sources had increased considerably. Thus, for example, Smith (1951a) at Cambridge, by means of two independent interferometers —two cophased arrays working at 3.7 m, and two parabolic reflectors working at 3.7 m and 1.4 m—determined very accurately the coordinates of the four strongest sources. At that time it had not yet been established that discrete sources had finite angular extent, and hence the coordinates given by Smith can be regarded as those of the center of gravity of the extended source. By means of three different methods—(a) time of transit of the star over the antenna, (b) displacement of the collimation plane, and (c) a "double-transit" method—Smith made a very careful determination of the coordinates of the sources, taking atmospheric refraction into account. Table 12 shows the results of Smith's observations, giving the type of interferometer and the method used to determine the coordinates, as well as the mean values of the coordinates for the sources investigated.

During the following year Mills (1952c) made another careful determination of the coordinates of the six strongest sources. The observations were carried out at $\lambda = 3$ m, with the same interferometer that had been used previously to determine the coordinates and intensities of 77 sources (see Sec. 7). Table 13 gives the results of these observations.

It is very useful to compare Tables 12 and 13 and convince ourselves of the striking agreement between results obtained by different authors, using different methods, and at different wavelengths. For example, the coordinates of Cygnus A given by Smith differ by only 1^s in right ascension and $0'.5$ in declination from those given by Mills. If

TABLE 12. Coordinates of some discrete sources.

Source	Wavelength (m)	Observational method[a]		$\alpha(1950)$		$\delta(1950)$	
Taurus A	3.7	S		$05^h31^m35^s$	$\pm5^s$	$22°05'$	$\pm40'$
	3.7	P		05 31 35	±5	22 11	±20
	1.4	P		05 31 34.5	±2	22 09	±5
	1.4	DCP		—		22 01	±3
			Mean	*05 31 34.5*	*±3*	*22 04*	*±5*
Virgo A	3.7	S		12 28 17	±5	12 50	±1
	3.7	P		12 28 18	±5	12 46	±20
	1.4	P		12 28 18.8	±2	12 33	±15
	1.4	DCP		—		12 31	±15
			Mean	*12 28 18*	*±3*	*12 37*	*±10*
Cygnus A	3.7	S		19 57 45	±5	40 37	±5
	3.7	P		19 57 45	±5	40 32	±5
	1.4	P		19 57 45	±1	40 34.5	±1
	1.4	DCP		—		40 35.3	±0.7
			Mean	*19 57 45*	*±1*	*40 35*	*±1*
Cassiopeia A	3.7	S		23 21 13	±5	58 30	±3
	3.7	P		23 21 12	±3	58 32	±2
	1.4	P		23 21 12.0	±1	58 32.1	±0.7
	1.4	DCP		—		58 32.9	±0.5
	1.4	DT		—		58 30	±3
			Mean	*23 21 12*	*±1*	*58 32.1*	*±0.7*

[a] S, synphase antenna; P, paraboloid; DCP, displaced collimation plane; DT, double transit.

we remember that 5 years before these investigations the coordinates of the Cygnus source (the only one known at the time) had been measured with a probable error of 2°, the enormous progress achieved by the radio-astronomical methods of investigation becomes apparent.

Although the sources in Taurus, Virgo, and Centaurus could be identified with the nebulae NGC 1952, 4486, and 5128 immediately after their detection, a long time elapsed before it was possible to find any optically observed objects corresponding to the two strongest radio sources, Cygnus A and Cassiopeia A. Immediately after the discovery of these strong sources a large number of hypotheses as to their nature were formulated. The impossibility of identifying these strong sources with any optical object opened the way to many different fantastic theories. Several authors, for example, assumed that these sources were very close to us, within the limits of the solar system. Ryle (1949) attempted to measure the annual parallax of several of the strongest sources. These observations gave negative re-

TABLE 13. Mills's observations of discrete sources.

Source	$\alpha(1950)$	$\delta(1950)$	Angular Diameter
Fornax A	$03^h19^m30^s \pm 6^s$	$-37°18' \pm 3'$	$20' \pm 5'$
Taurus A	05 31 29 ± 2.5	$+22$ 00 ± 3	<10
Hydra A	09 15 46 ± 4	-11 55 ± 8	<15
Virgo A	12 28 15.5 ± 2.5	$+12$ 44 ± 6	<10
Centaurus A	13 22 30 ± 4	-42 46 ± 2	25 ± 3
Cygnus A	19 57 44 ± 2.5	$+40$ 35 ± 1.5	<10

sults; considering their accuracy it was only possible to conclude that the distance to the discrete sources exceeded 0.02 pc. Smith (1951*b*) sought to determine the parallax and proper motion of Cygnus A, Cassiopeia A, Virgo A, and Taurus A. After a long series of very careful interferometric observations at 3.7 m and 1.4 m, he was not able to obtain any positive results. Hence he concluded that Cygnus A and Cassiopeia A were at least 0.5 pc away, which meant that they could not be within our solar system.

The analysis of photographs of the region of the sky around Cygnus A led Dewhirst (1953) and other English investigators to the conclusion that there was a very faint object (16^m or 17^m) coincident with the source. On blue photographs of the Cassiopeia A region obtained in England, it was possible to detect a very faint nebula about $2'.5$ from the "center of gravity" of the source, whose angular extent was not yet known.

In September 1951, after Smith had made an accurate determination of the coordinates of Cygnus A and Cassiopeia A, Baade and Minkowski (1954*a*) photographed the regions of the sky where these objects are located with the 200-in. Hale reflector at Palomar Mountain.

When the region around Cassiopeia A was photographed in blue light, the plate showed a small, arch-shaped nebula, $2'.8$ in extent. The position of the center of gravity of Cassiopeia A did not coincide with the nebula, but was $2'$ south of it. We should point out that when Baade and Minkowski made these observations it had not yet been established that Cassiopeia A and Cygnus A had an extended angular dimension.

Red plates ($\lambda\lambda$ 6400 to 6700 A) gave a completely different picture. South of the "arch" referred to above, it was possible to detect several "fragments" of nebulosity of a remarkable shape, never observed before. Some of these fragments are elongated streaks about $25''$ long, while others have a stellar appearance; some are very bright, others barely visible. None of these fragments of nebulosity was observed on the blue plates.

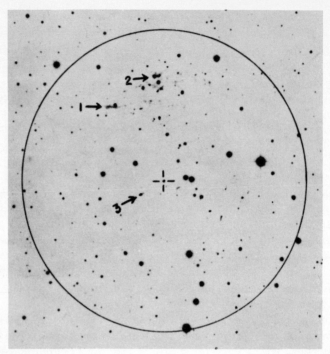

Fig. 89.   Photograph of the radio nebula Cassiopeia A (Baade and Minkowski 1954a).

The total area covered by the nebulous wisps is contained within a circle 6'.3 in diameter (Fig. 89). The coordinates of the center of the disk are $\alpha = 23^h21^m12^s \pm 1^s$, $\delta = +58°32'.1 \pm 0'.7$, which coincide exactly with the coordinates of the center of gravity of the radio source as measured by Smith (Table 12). Smith's value for its angular diameter was 5'.5. Such striking agreement in the values of the coordinates and the angular extent proves beyond doubt that the remarkable nebulosity discovered by Baade and Minkowski is responsible for the discrete radio source Cassiopeia A.

Using a nebular spectrograph at the prime focus of the 200-in. reflector, Baade and Minkowski obtained spectra of three filaments in the nebulosity. The dispersion was 220 A/mm in the red, 132 A/mm in the violet (third to fifth order). In Fig. 89 the arrows point toward the filaments for which spectra were obtained.

Figure 90 shows a spectrum of the arch. This spectrum is unique. It shows only the well-known nebular lines due to doubly ionized oxygen ($\lambda\lambda$ 4959 and 5007 A), and the red neutral oxygen doublet ($\lambda\lambda$ 6300 and 6364 A), a familiar feature in the spectra of the night sky and aurorae. The Hα line of hydrogen is absent; this means that the nebula is excited mainly by collisions, rather than by radiation.

The most striking characteristics of the spectrum shown in Fig. 90 are the enormous widths of the emission lines and the complexity of their structure. Each line has one strong component and three weaker "satellites," displaced to the red; actually, each line constitutes a very broad band with intensity maxima. Such great line widths can only be explained as a Doppler effect, where the velocities of the emitting atoms are not thermal, but "macroscopic" in character. The velocity dispersion, as derived from the line profiles, attains fully 5000 km/s (from -1500 to $+3500$ km/s). The detailed profiles of [O I] and [O III] do not agree. This could mean that the excitation in different regions of the filament is different.

Filament *1* shows an entirely different spectrum. It corresponds to one of the wisps of nebulosity observed on the red-sensitive plates (Fig. 91). This spectrum shows Hα and two lines of [N II] at λλ 6548 and 6584 A, the latter ones being 1.5 to 4.5 times as intense as the hydrogen line. The λ 6300 A [O I] line is also present. The line widths correspond to a velocity dispersion of about 400 km/s, much smaller than the value obtained for filament *2*. The spectrum of filament *3* resembles that of *1*.

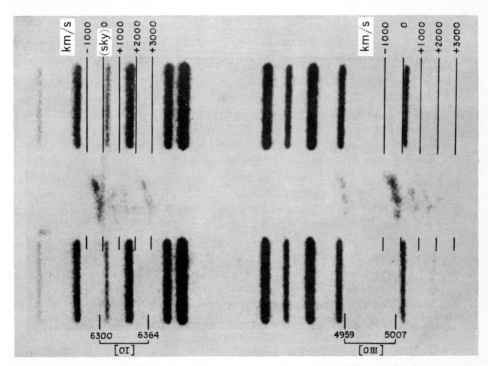

Fig. 90. Spectrum of the arch (filament *2*, Fig. 89) in the radio nebula Cassiopeia A (Baade and Minkowski 1954*a*).

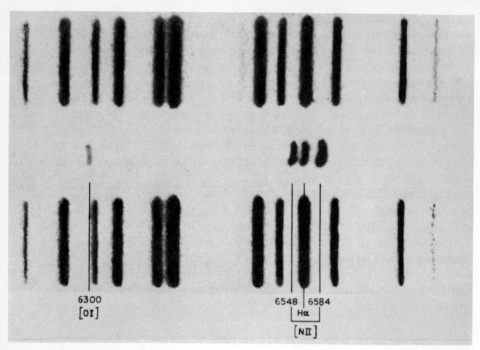

Fig. 91. Spectrum of a wisp (filament *1*, Fig. 89) in the radio nebula Cassiopeia A (Baade and Minkowski 1954*a*).

At the end of 1953 Baade and Minkowski photographed the nebula once more. In spite of the short time that had elapsed between observations, a comparison of these new plates with those obtained in 1951 shows strong changes in intensity and marked motions in the filaments that form the northern arch of nebulosity. On the other hand, the red fragments of nebulosity do not show any perceptible changes. More spectra were also obtained confirming the results of the earlier analysis; while the arch showed radial velocities of the order of 2000 km/s, the red fragments of nebulosity gave radial velocities not exceeding 50 km/s.

At the 1958 Paris symposium on radio astronomy Minkowski presented his new results concerning the interpretation of the optical observations of Cassiopeia A. These results differ strongly from those published before. From an analysis of fast-moving filaments he determined that the expansion velocity of the nebula is 7400 km/s. Furthermore, he places it at a distance of 3400 pc, in excellent agreement with radio observations at 21 cm. The physical conditions in the nebula are very peculiar: the electron temperature of the filaments, computed, as usual, from the intensity ratio of $\lambda\,4363$ [O III] to $N_1 + N_2$, is about $2 \times 10^4 \,^\circ$K; the electron concentration, computed from the intensity ratio of the [S II] lines, varies between 10^3 cm^{-3} and 10^6 cm^{-3} for different filaments; the mean mass of each filament is 10^{30} gm; and their total number is about 200. Minkowski assumes that the mass of the "invisible" part of Cassiopeia A is approximately ten times the "visible" mass, so that the total mass of this remarkable nebula should at least equal that of the sun.

The front side (relative to the observer) of the nebula expands considerably more slowly than the back side. This may be explained as the consequence of a braking effect of the interstellar gas.

In Cassiopeia A, together with the rapidly expanding filaments, there are fragments of nebulosity moving relatively slowly, with velocities which do not exceed 100 km/s. From their spectra it is found that in these fragments $T_\varepsilon < 10\,000°$K and $n_\varepsilon \approx 10^3$ cm^{-3}. Minkowski assumes that they are parts of the nebula which have been almost completely stopped by the resistance of the interstellar medium. If this is true, however, it would imply that the globules present in the medium have a density of 10^5 cm^{-3}. These fragments would also have to contain a hundred times as much neutral hydrogen as ionized hydrogen, according to Minkowski's estimate. It seems to us that the slow-moving fragments are remnants of the interstellar gas surrounding the exploding star, strongly compressed by the hot gas ejected during the outburst.

According to Minkowski the time of expansion, since the explosion took place, is only 256 ±14 yr. This would place the supernova event at the beginning of the 18th century, at the time of Newton! Why then did it pass unobserved? The reason is the very high interstellar absorption, which reaches 6^m at the center of the nebula according to Minkowski's estimate. This means that if its absolute magnitude at maximum was -15^M or -16^M, the apparent magnitude was only $+3^m$ or $+2^m$, and it remained unnoticed (concerning the identification of Cassiopeia A with a supernova see also Sec. 19).

According to Minkowski, Cassiopeia A is apparently a type II supernova, while the supernova of A.D. 1054 belongs to type I (population II). This means that exploding stars are old, first-generation objects whose masses are relatively small. This type of supernova is characterized by the descending branch of the light curve, which is very nearly exponential. For several supernovae of this type the light curves are almost identical. Burbidge et al. (1956) assume that the energy released by these supernovae is due to the radioactive decay of californium; the half-life of Cf^{254} agrees exactly with the exponential light curve.

Type II supernovae belong to the relatively young galactic population forming the "flat" system. We would expect these stars to be considerably more massive, so that in principle it is quite possible that the mass of Cassiopeia A considerably exceeds that of the Crab Nebula. If the velocity of the gas is $\simeq 7 \times 10^8$ cm/s, the total kinetic energy of the ejected envelope may be $\simeq 5 \times 10^{50}$ erg, or a hundred times as great as in the Crab!

At the position of the radio source Puppis A ($\alpha \simeq 8^h$, $\delta \simeq -42°$, angular extent $\simeq 1°$) Baade and Minkowski found a system of filaments covering an elliptical area of $50' \times 80'$, which agrees roughly with the angular extent of the radio source. They obtained spectra of several filaments, showing Hα, [N II], and possibly other lines. The lines were rather wide, with random velocities of about 150 to 200 km/s. As in the case of the nebula in Cassiopeia, if there is an expansion its velocity is smaller than the velocities of the internal motions.

The most striking results were obtained by direct photography and spectrographic measurements of the region of the sky around Cygnus A. Let us first point out that this region, in spite of its proximity to the galactic equator, does not suffer from much absorption.

The radio source Cygnus A seems to be located in the middle of a large cluster of galaxies. The brightest galaxies in this cluster belong, in general, to types E and S0, and are of photographic magnitude 17. The position of the remarkable nebula identified with the radio source is $\alpha = 19^h 57^m 44^s.49$; $\delta = +40°35'46''.3$ (1950), practically coinciding with the coordinates of Cygnus A, as given by Smith and Mills (Tables 12 and 13).

The object identified with Cygnus A is actually a pair of colliding galaxies. Figure 92 shows a photograph of this object. The bright central region ($5'' \times 3''$) is surrounded by a fainter elliptical region of dimensions $18'' \times 35''$, whose major axis has a position angle of about $150°$. In the photograph one can distinctly see the two nuclei of the colliding galaxies, perturbed by tidal forces. Both colliding galaxies

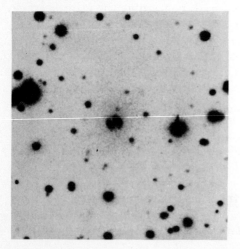

Fig. 92. The radio nebula Cygnus A, a pair of colliding galaxies (Baade and Minkowski 1954a).

are late spirals, similar to our own. The spatial orientation of the galaxies is face-to-face.

Spectra of the central regions of the colliding galaxies (their nuclei) were obtained by means of the nebular spectrograph at the prime focus of the 200-in. reflector, and at the Newtonian focus of the 100-in. In these spectra strong forbidden lines are clearly seen, while the continuum is very weak, barely above the limit of detection. Photoelectric measurements show that more than 50 percent of the total light is in the emission lines. The lines are displaced to the red, showing a velocity of recession of 16 830 km/s. From the red shift and using 180 km/s per 10^6 pc as the value of Hubble's constant, the distance to the nebulae is 10^8 pc. From an analysis of spectra taken in 1955 we can deduce that the relative radial velocity of the colliding galaxies is about 3000 km/s.

Another peculiarity of the spectra is that hydrogen is relatively faint compared to nitrogen and oxygen. From the width of the emission lines the velocity spread caused by random motions is found to be about 1000 km/s. The line λ 3727 A [O II] shows an anomalous extent, covering the full length of the slit. The diameter of the source deduced from this emission line is somewhat larger than the visible extent of the nebulae.

Taking interstellar absorption into account, the photographic magnitude of the galaxies is $m_{pg} = 15^m.05$. Their distance is 10^8 pc, their absolute photographic magnitude is $M_{pg} = -20^M$. The angular extent of $35''$ indicates that the linear extent of the source is of the order of 10^4 pc.

Thus, the strong radio waves from Cygnus A seem to be the result of the face-on collision of two galaxies, each of which resembles our own stellar system. The radio power generated by this process is fantastically great: the flux of radio waves from Cygnus A observed at the Earth rivals that from the sun. But the distance of the sun is only a few light minutes, while the colliding galaxies in Cygnus are 300 000 000 light years away, and the flux is proportional to the square of the distance!

The radio flux from Cygnus A is large absolutely, not just relatively. The integrated flux density between $\nu = 22$ Mc/s and $\nu = 3000$ Mc/s is

$$\int F_\nu d\nu \approx 10^{-10} \text{ erg cm}^{-2} \text{ sec}^{-1},$$

while the flux emitted in the optical region by an object of 15^m is 10^{17} times as small as that of the sun, that is, 1.3×10^{-11} erg cm^{-2} sec^{-1}. In other words, the absolute value of the radio flux from Cygnus A, integrated over the spectrum, is six times as great as the optical flux.

After the discovery of the colliding galaxies in Cygnus, Baade and Minkowski detected a similar phenomenon in Perseus. This cluster of galaxies contains the rather bright galaxy NGC 1275, which Baade and Minkowski identified with the radio source located in that region. The spectrum of NGC 1275 consists of emission lines, superimposed on a strong continuum. The great width of the emission lines indicates a large dispersion in the random velocities, approaching 4500 km/s. It is interesting to note that NGC 1275 belongs to a group of galaxies, investigated by Seyfert (1943), which show bright emission lines in their nuclear regions. The line widths indicate a dispersion in the random velocities of up to 10 000 km/s. Among these galaxies only NGC 1275 appears to be a radio source. This fact shows that the existence of large random velocities is not sufficient to produce strong radio emission, although all the optical objects identified with radio sources do show these large velocities. The galaxy NGC 1275 is at a distance of $\approx 2.5 \times 10^6$ pc. Recently Mills found that NGC 1068, with wide emission lines at the nucleus, also appears to be a radio source, although considerably weaker than NGC 1275.

Near NGC 1275 there is a rather extended source of radio waves, identified with a large cluster of galaxies in Perseus. In order to resolve the question whether the cluster as a whole or the peculiar nebula NGC 1275 is responsible for the radio emission, observations were carried out at Cambridge with an interferometer of variable spacing. When progressively larger spacings were used, the flux from the source decreased, but not significantly. From this it was concluded that three-fourths of the radio flux observed in this region of the sky comes from a source of small angular extent, namely the peculiar galaxy NGC 1275, and one-fourth from the remainder of the cluster.

Thus several of the discrete sources have been identified with peculiar galaxies. Among them we may list the strong sources Cygnus A, Virgo A (identified with NGC 4486), Centaurus A (NGC 5128), the source identified with NGC 1275, and others. In particular, Shklovsky and Kholopov (1952) identified the discrete source Fornax A with the rather bright galaxy NGC 1316. The coordinates of this galaxy are $\alpha = 3^h18^m24^s$, $\delta = -37°20'$; this position agrees very well with that of the radio source (see Table 13). Additional support for this identification comes from the fact that the Fornax source has a relatively large angular extent, about 20'. The angular extent of NGC 1316 is about 10'. It is natural to expect that the "radio diameter" should be larger than the optical, since this is also the case with other sources. The morphological character of NGC 1316 (the presence of a dark equatorial band) is similar to that of the radio galaxy NGC 5128, identified with Centaurus A. Identification of Fornax A with the galaxy NGC 1316 was made independently by de Vaucouleurs (1953a).

Figure 93 shows a photograph of the galaxy NGC 5128. The ellipse surrounds the central and brightest region of the radio source Centaurus A, according to Mills's interferometric observations. There is a striking coincidence between the bright radio region and the wide dark band, which we may regard as equatorial, and which runs across the galaxy. We should also point out that this galaxy is a relatively nearby object. It is one of the brightest, with $m_{pg} = 7^m.2$. Evidently it can only be a few hundred kiloparsecs from us.

The galaxy NGC 4486 ($m_{pg} \approx 10^m$) is one of the brightest in the Virgo cluster. Figure 183 (Sec. 25) shows a photograph of this galaxy taken with the 200-in. Palomar reflector. In the outer part of the galaxy it is possible to resolve starlike objects, apparently globular clusters. This elliptical galaxy, as was already known 40 years ago, shows a remarkable peculiarity which we shall describe in Sec. 25. It is natural to identify the strong radio waves from NGC 4486 with this peculiar feature.

Shain (1958a) has recently reported on radio observations of NGC 5128, 1316, and 4486 performed in Australia with the well-known cross. For NGC 5128 isophotes were made at $\nu = 85.5$ and 19.7 Mc/s; the two sets give similar pictures: an elongated, extended source with brightness largely concentrated toward the center. Its huge extent is quite remarkable ($7°.5 \times 2°.5$, according to the outer isophote). Assuming the distance to the source to be 750 kpc, the linear extent of the emitting region becomes 100 kpc by 30 kpc, and it is even larger at the lower frequency. The total flux at $\nu = 19.7, 85.5,$ and 1390 Mc/s is respectively 280, 87, and 6×10^{-24} w m^{-2} (c/s)$^{-1}$. Since the contri-

Fig. 93. The radio galaxy NGC 5128 (Baade and Minkowski 1954b). The ellipse bounds the region of greatest radio intensity.

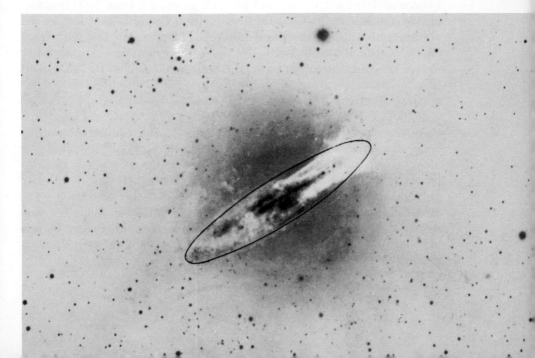

bution of the central bright region to the total flux is noticeably smaller at lower than at higher frequencies, we can assume that the spectra of the central and extended regions are different. It may be significant that the major axis of the source is approximately perpendicular to the dark wide band running across it.

The source Fornax A identified with NGC 1316 is clearly elliptical; its angular extent at $\nu = 85.5$ Mc/s is $0°.5 \times 0°.8$. If we assume it to be at a distance of 2 Mpc, its maximum extent becomes $\simeq 100$ kpc. Comparing several properties of NGC 5128 and 1316, Shain arrives at the conclusion that the two radio galaxies are alike.

Together with these peculiar galaxies, a whole series of discrete radio sources were identified with common galaxies of different types, and also with clusters of galaxies. We should note, however, that the radio flux from normal galaxies is considerably smaller than that from peculiar galaxies. Thus, for example, the flux from one of our nearest neighbors, M 31, is about one-fifth that from the peculiar galaxy NGC 4486, although the optical flux from M 31 is 100 times that from NGC 4486. In Chapter VI we shall make a detailed analysis of the radio waves arising in objects located beyond our stellar system.

Aside from galaxies, both normal and peculiar, discrete sources of cosmic radio waves have in many cases been identified with special gaseous nebulae, located within the Galaxy, and also with normal ionized diffuse nebulae. Almost all of the peculiar galactic nebulosities identified with discrete sources follow the pattern of the Crab Nebula, the remnants of the supernova of A.D. 1054. The strongest discrete source, Cassiopeia A, has been shown by Baade and Minkowski (1954a) to be associated with a peculiar galactic gaseous nebula. In this group of discrete sources we should also include Puppis A, which, as we pointed out before, Baade and Minkowski have identified with fragments of peculiar nebulosity. In the same group are the extended source in Gemini, identified with the remarkable nebula IC 443 + S 40 (Fig. 94) and the extended source in Auriga, also identified with a peculiar nebula, whose position is near α Aurigae, at $\alpha = 4^h57^m$, $\delta = +46°30'$.

As the sensitivity and resolving power of the equipment are increased, many other similar objects will undoubtedly be detected. We must keep in mind that extended sources of radio waves at low galactic latitudes are relatively weak compared to the general galactic background, and hence they become difficult to detect. Recently, for example, an extended source of radio waves, about $3°$ in diameter, was detected in the region of the Veil Nebula in Cygnus. Figure 95 is a photograph of this nebula, showing the position of the new discrete source.

Fig. 94. The radio nebula IC 443 + S 40; on the right is the star η Geminorum. The photograph is from the National Geographic Society–Palomar Observatory *Sky Survey Atlas*.

The peculiar nebulae identified with this type of galactic discrete source show a large dispersion in the velocities of their internal motions, as well as a complicated structure. Thus, for example, the velocity dispersion in IC 443 + S 40 is around 100 km/s, as can be deduced from the spectral lines of this nebula. In the filaments of the Cygnus nebula the velocity dispersion reaches 75 km/s. It is tempting to suppose that these large velocity dispersions are related in some way to the anomalously strong radio emission from these objects.

What is the nature of the peculiar galactic nebulae identified with discrete sources? As for the Crab Nebula, it is now known that it constitutes the remnants of a supernova outburst. Investigations of the filaments of the Veil Nebula indicate that it may also be the remnant of a supernova explosion of some 10 000 years ago. Also the latest observations of Minkowski have definitively shown that the nebula associated with Cassiopeia A originated from a supernova. We have every reason to assume, then, that peculiar nebulae which are supernova remnants are also strong radio sources (see Sec. 19). This conclusion is of great significance for our understanding of the supernova process,

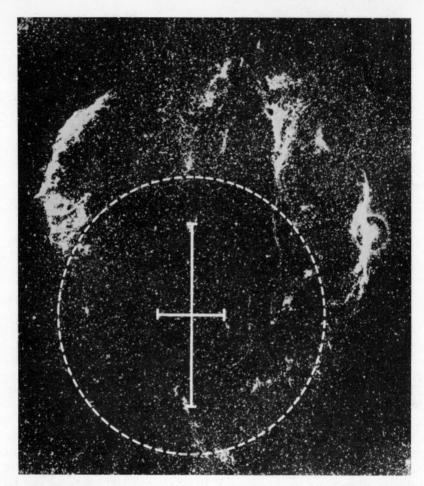

Fig. 95. The Veil Nebula in Cygnus. The broken circle bounds the region of radio emission.

and also bears on a number of other problems, such as the origin of cosmic rays.

Another class of galactic objects with which discrete sources have been identified is the ionized diffuse nebulae known as H II regions. The radio emission from these objects seems to be thermal; we shall discuss this point in detail in Sec. 9. Here we shall only point out that in the meter range these sources are very weak, but in the decimeter and centimeter range their strength is comparable to, and sometimes exceeds, that of the sources associated with ejecta of supernovae or peculiar galaxies.

Thus, all discrete radio-wave sources identified up to the present time are associated with nebulae, either galactic or extragalactic. Up to now (1958) not a single source has been identified with a star. Of course we do not exclude the possibility that when weaker sources are detected a number of them may be identified with some kind of peculiar star; theoretically this is quite possible. However, it is beyond doubt that the overwhelming majority of discrete sources are not stars. Thus the term "radio star," often applied to discrete sources in general, is definitely a misnomer. The term arose before it became apparent that radio sources had measurable angular diameters. The assumption was that discrete sources were peculiar stars with anomalous emission in the radio range, whose combined emission we observed as a general galactic background, just as we observe the Milky Way as the sum of the optical emission from billions of stars. It seems, however, that this is entirely false. Very likely, a much more appropriate name for the discrete sources would be "radio nebulae."

Dewhirst (1958) has recently reported on several new identifications of discrete sources with optically observed objects, mainly galaxies. They are based on comparisons between the original plates of the National Geographic Society–Palomar Observatory *Sky Survey Atlas* and the new catalogues of discrete sources compiled at Sydney and Cambridge; we cannot yet be fully confident that they are correct. With fair certainty Hydra A was identified with two very close elliptical galaxies of $m_{pg} = + 15^m.9$, $M_{pg} = - 20^M.1$. If this identification is confirmed, Hydra A would be similar in nature to Cygnus A. One source was reliably identified with the cluster of elliptical galaxies NGC 6161. Seven sources were more or less confidently matched either with double or with groups of very close elliptical galaxies. Optically these objects were all very faint (as a rule $m_{pg} > 16$), and therefore their radio magnitudes (see Sec. 24) have large negative values, between $-6^m.2$ and -10^m.

We should keep in mind that the number of radio sources so far identified with optical objects is relatively small. Thus, for example, of the 1936 sources in the Cambridge catalogue, about 100 are more or less reliably identified with optical objects, and in another 250 cases there is approximate agreement between the observed coordinates of the radio sources and optical objects. Therefore, together with individual identifications, it is significant to make statistical analyses of the distribution of radio sources in both position and brightness. Such analyses would give us a general idea of the spatial distribution of discrete sources.

The first serious attempt to make a statistical analysis was that of Mills, who used for this purpose his catalogue of 77 discrete sources (Mills 1952a). By means of lines of constant galactic latitude he divided the region of the sky investigated into ten equal strips. He then grouped the strips into two identical zones, differing only in the

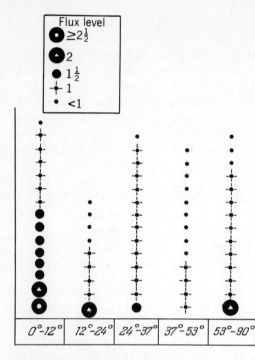

Fig. 96. Distribution of sources in galactic latitude (Mills 1952a).

sign of the galactic latitude. In Fig. 96 the discrete sources in the five resulting strips are indicated by means of circles of different sizes. The diameter of each circle corresponds to the magnitude of the flux emitted by the source, expressed on a logarithmic scale [so-called "flux levels," obtained from the relation $L = \log(F_\nu \times 10^{25})$, where F_ν is the spectral flux density]. Mills did not include the sources located in longitudes $50° < l < 130°$, in order to avoid the spurious results they would have introduced into the statistical analysis owing to the difficulties involved in their observation. The diagram in Fig. 96 shows that the strongest sources are concentrated to the galactic equator, while the weaker sources are distributed more or less isotropically.

Seeliger's theorem in stellar astronomy is well known. We can express it thus: if any emitting objects (for example, stars) are evenly distributed in space, then the ratio of the number N_m of such objects whose apparent magnitude is less than m to the number N_{m-1} whose apparent magnitude is less than $m - 1$ is 3.98:

$$N_m/N_{m-1} = 3.98.$$

It follows that the number N of objects whose emitted flux is greater than F_ν is given by the expression

$$N(>F_\nu) \propto F_\nu^{-1.5}.$$

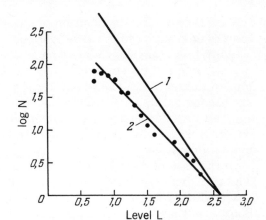

Fig. 97. Dependence of log N on L for the sources observed by Mills (1952a).

Figure 97 shows this relation on a logarithmic scale for all the sources investigated by Mills. The abscissa shows the emission level L, and the ordinate the common logarithm of the number of sources whose emission level is greater than L. From the figure we see that the points fall on the straight line 2 ($N \propto F_\nu^{-1}$), rather than on 1 ($N \propto F_\nu^{-1.5}$). This indicates that the distribution of sources is nonuniform.

Figure 98 shows the value of log $N(L)$ for two kinds of sources: the first group includes all sources at a galactic latitude $|b| < 12°$, the second group those at $|b| > 12°$. It is quite obvious that the resulting straight lines are entirely different. The sources close to the galactic equator (line 1) show strong departures from uniform distribution; on the other hand, within the observational error the sources at high galactic latitudes (line 2) seem to be distributed uniformly. Both curves rapidly approach a horizontal asymptote for small values of F_ν. This is due to the limited sensitivity of the receiving apparatus: the weak sources are obliterated by stronger ones and cannot be observed.

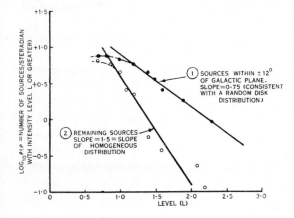

Fig. 98. Dependence of log N on L for (1) low-latitude, (2) high-latitude sources (Mills 1952a).

Thus, the spatial distribution of the stronger and weaker discrete sources of cosmic radio waves is opposite to that observed for stars. The brighter stars show almost no concentration toward the galactic plane, while fainter stars do. This concentration increases as we go toward fainter stars. The reason for this relation is clear: on the average the faint stars are farther away from us than the bright ones, and, since all stars considered belong to our highly flattened stellar system, on the average the galactic concentration of faint stars should be more pronounced.

In the case of discrete sources the situation is quite the opposite. The statistical regularities found by Mills led him to believe that the observed discrete sources could be classified into two classes or types. Sources of Class I are not far from the galactic plane. These are strong sources located in our own stellar system, distributed far apart from one another and at large distances from the sun (otherwise their galactic concentration would not be noticeable). The distance to the nearest of these sources can be roughly estimated on the basis of the following simple considerations. The slope of the line $N(F_\nu)$ for Class I sources is approximately the same as that for stars of fifteenth magnitude (these are mostly main-sequence stars of types A to G). On the average we can assume that they are at a distance of 1 to 2.5 kpc. Therefore, we can make the very approximate assumption that the nearest discrete sources are at a distance of 1000 pc. The distance to the Crab Nebula—a typical example of a Class I source—is 1100 pc, in good agreement with our results.

As for Class II sources, their spatial distribution can be considered more or less uniform, and in principle this could be explained in two ways. (*a*) Sources of this type are very close objects, and their "radio luminosity" is relatively small compared to that of Class I sources. They are so close to us that no concentration to the galactic plane can be detected (as is the case for bright stars). (*b*) These sources are outside the Galaxy and their radio luminosity is many times as great as that of Class I sources. On the basis of Mills's data we cannot distinguish between the two possibilities, although he was inclined to favor the second.

Today, as more and more discrete sources become identified with extragalactic nebulae, the proposal that Class II sources are of extragalactic origin seems to be the correct one.

The division of the discrete sources into two classes, first proposed by Mills, is of deep significance. Class I sources, which as we have seen are peculiar nebulae, are as a rule remnants of supernova explosions. (If we include under Class I *all* discrete sources in our Galaxy forming a highly flattened system, then to the supernova ejecta we should add the H II regions, which are strong emitters at decimeter and cen-

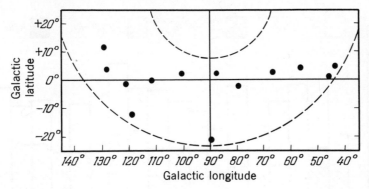

Fig. 99. Distribution of discrete sources (Hanbury Brown and Hazard 1953*b*).

timeter wavelengths. Here, however, we shall accept as Class I sources only those galactic objects identified with peculiar nebulae, that is, with supernova ejecta. We shall refer to sources identified with H II regions in Sec. 10.) Class II sources are peculiar galaxies ("radio galaxies") such as Cygnus A, NGC 1275, 4486, and 5128, and the like.

Thirteen sources observed by Hanbury Brown and Hazard (1953*b*) with the 218-ft fixed reflector at Jodrell Bank (see Sec. 6) show very clearly the division of sources into two classes. In Fig. 99 we see that the majority of these relatively strong sources are concentrated around the galactic equator. Of the thirteen sources, whose flux averages 5×10^{-25} w m^{-2} (c/s)$^{-1}$, ten lie in the zone $|b| < 5°$. The area of this zone is around 1000 deg^2, while the entire area observed was about 3000 deg^2. Obviously, these strong sources concentrated to the galactic equator must be regarded as Class I.

Important statistical results were obtained at Cambridge from an analysis of the abundant material collected in their catalogue (Shakeshaft *et al.* 1955). First it was observed that the spatial distributions of extended sources and sources with small angular diameters are different. Figure 100 is a histogram showing the distribution of sources of small angular extent as a function of latitude. The ordinate axis gives the number of sources per unit solid angle for different values of the flux. These sources have a roughly isotropic distribution, in agreement with the distribution of Class II objects.

A statistical analysis of extended sources is considerably more difficult, since their number is smaller. For one thing, as we mentioned earlier, the very concept of "extended source" is quite indefinite. Figure 101 is a histogram showing the distribution of "strong" $[F_\nu > 10^{-24}$ w m^{-2} (c/s)$^{-1}]$ and "weak" extended sources. "Strong" sources, which can be identified with Class I objects, show a marked concentration to the galactic equator. In the case of "weak" extended sources there is strong observational selection, since they are difficult to observe at

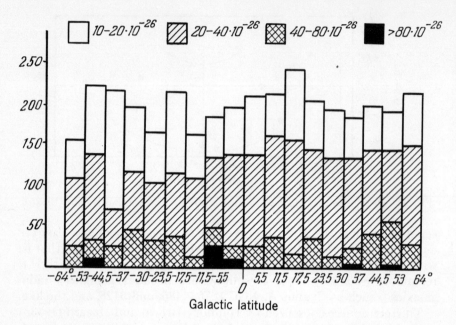

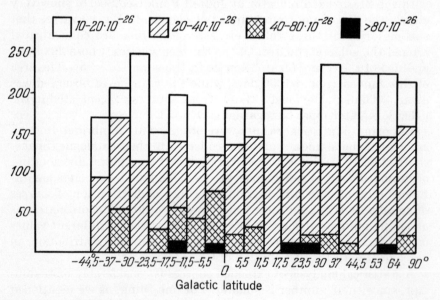

Fig. 100. Histograms of sources with small angular diameters (Shakeshaft *et al.* 1955). The ordinate is the number of sources per steradian. The upper histogram is constructed for galactic longitudes from $l = 340°$ to $l = 30°$, the lower from $l = 90°$ to $l = 195°$.

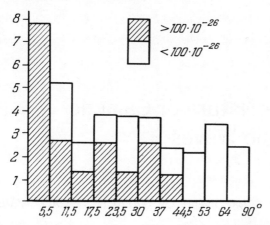

Fig. 101. Histogram of extended sources (Shakeshaft *et al.* 1955). The ordinate is the number of sources per steradian.

low galactic latitudes, where there is a bright background of radio emission showing a complicated structure. The number of such objects at low galactic latitudes must therefore be greater than that shown by the Cambridge catalogue. The general results of the statistical analysis confirm Mills's separation of discrete sources into two classes.

Mills presented new statistical data on the distribution of Class I sources at the Paris symposium in 1958, based on recent observations obtained with a cross at 3.5 m. Thirty-nine sources whose fluxes exceed 10^{-25} w m^{-2} (c/s)$^{-1}$ were observed in the interval $170° < l < 358°$, the overwhelming majority being nonthermal emitters. The number and intensity of these sources were considerably lower toward the anticenter than toward the center, and they were identified by Mills with the remnants of supernovae of type II. In the direction of the anticenter we observe, as a rule, less bright and more extended radio nebulae.

Thus, we see that statistical methods of investigation corroborate the conclusions as to the nature and spatial distribution of discrete sources—conclusions based on many observations of different objects made by radio astronomers and astrophysicists.

III

The Nature of Galactic Radio Waves

9. THE FLAT COMPONENT OF GALACTIC RADIO WAVES

In this chapter we shall deal with the problem of the physical processes responsible for the observed radio emission from the Galaxy.

We have shown in Sec. 5 that cosmic radio waves must be divided into several components. First, cosmic radio waves comprise the galactic and the extragalactic. The galactic radio waves may in turn be separated into at least three components: (*a*) radiation from the "disk" of our Galaxy (the Oort–Westerhout distribution), and from its spherical corona; (*b*) radiation from sources forming the "flat" component of the *general* field of galactic radio waves—for example, from the secondary maxima in Cygnus and at the anticenter (see Sec. 4); (*c*) the combined radiation from discrete sources of Class I, which also form a strongly flattened system within the Galaxy (Sec. 8). In the present section we shall be concerned only with the investigation of the flat component of galactic radio waves (see Sec. 5).

One might initially assume that the flat component of the general field could be regarded as a compound effect of the sources of Class I. But such a point of view can scarcely be accepted, for the spectra of Class I sources are sharply distinguished from the spectrum of the flat component of the general field. As is evident from the data cited in Sec. 6, the spectra of these sources are characterized by a quite strong increase in spectral flux density with decreasing frequency: the effective temperatures of the sources depend on the frequency by a law of the form

$$T_b \propto \nu^\gamma,$$

where γ lies within the interval $[-2.5, -3]$. Only for the Crab Nebula in the meter wavelength range is $\gamma = -2.2$; in the decimeter range, however, its spectrum becomes considerably steeper. On the other hand, we have pointed out in Sec. 5 that the brightness temperature of the flat component of galactic radio waves exhibits a frequency dependence

$$T_b \propto \nu^{-2},$$

if the frequency is not too low. Thus the spectra of sources comprising the flat component are distinct from typical Class I spectra. The latter are very like the spectrum of the spherical component of galactic radio waves, for which, as we have seen in Sec. 5,

$$T_b \propto \nu^{-2.82},$$

resembling the spectrum of discrete Class I sources.

Since the spectra of Class I sources are so strikingly different from the spectrum of the flat component of galactic radio waves, the latter clearly cannot be considered to be determined by the combined radiation of Class I sources.

Generally speaking, may the radio waves originating in the *disk* be regarded as resulting from an aggregate of Class I sources? We here show that this circumstance is also impossible. According to Fig. 97, the radiation flux from the nth source (in order of decreasing observed intensity) may be expressed as

$$F = F_0 n^{-0.75}, \tag{9-1}$$

where F_0 is the flux from a mean source corresponding to the "level of radiation" for which $\log N$ (Fig. 98) is equal to zero. From Fig. 97 we find that the corresponding level L_0 is ≈ 2.5; since by definition $L_0 = \log (10^{25} F_0)$, we have $F_0 = 5.6 \times 10^{-23}$ w m^{-2} (c/s)$^{-1}$. Integrating over all Class I sources we obtain

$$F_{tot} = F_0 \int_0^N n^{-0.75} dn = 4 F_0 N^{1/4}. \tag{9-2}$$

Since this flux refers to one steradian (for in Fig. 98 the quantities refer to one steradian) it is possible, knowing N, to compute the brightness temperature. The order of magnitude of N may be estimated from the distance to the nearest sources, if we assume a homogeneous distribution of sources. In this event

$$N \propto (R/a)^2,$$

where R is the extent of the Galaxy in the given direction, and a is the distance to the nearest source. The brightness temperature determined by the combined radiation of all Class I sources within the galactic plane in a given direction will be

$$T_b' = 4 F_0 (\lambda^2 / 2k)(R/a)^{1/2}. \tag{9-3}$$

If we take $R = 20$ kpc in the direction of the galactic center, and if $a \approx 1$ kpc, $\lambda = 3$ m, and $F_0 = 5.6 \times 10^{-23}$ w m^{-2} (c/s)$^{-1}$, we find

$$T_b \approx 300°\text{K},$$

whereas the brightness temperature of the flat component of galactic radio waves at this wavelength, and in the region of the galactic center, is in the neighborhood of 4000°K.

To reduce the error arising from our assumption of a homogeneous distribution of Class I sources, analogous calculations may be carried out for various galactic longitudes. In every case the computed T_b is found to be significantly less than the observed. It should be noted that a decrease in a in Eq. (9-3) would not be critical for the results of the calculations, as T_b depends only weakly upon a.

In this manner the radiation from the disk cannot be reduced to the combined effect of Class I sources.

What then are we to regard as the source of the flat component? Even in the earliest stages of the development of radio astronomy, it was conjectured that ionized interstellar gas might account for radio waves of measurable power. Such a gas, as will be shown below, possesses a nonvanishing optical thickness for radiation in the radio range; hence, by Kirchhoff's law, it must radiate within that range. This radiation is inherently thermal, so that it is not especially difficult to develop a complete quantitative theory for it.

It was Jansky who first suggested that interstellar gas might be the source of cosmic radio waves. In due course Reber (1940) gave support to this idea. At the end of the 1940's and in the early 1950's, many authors developed the theory of an interstellar origin for cosmic radio waves (Reber and Greenstein 1947; Piddington 1951; and others). But by 1947 it had already become clear that this mechanism was far from adequate for a full explanation of cosmic radio waves. Even apart from such discrete sources as Cygnus A and Cassiopeia A, whose radiation is singularly impossible to interpret as the thermal radiation of an ionized gas, a large portion of the general radiation from the Galaxy cannot be regarded as thermal. This will be particularly clear from the following considerations: at wavelengths $\lambda \approx 10$ m, T_b exceeds 100 000°K over extensive portions of the sky. If the radiation were inherently thermal, then it would follow from the laws of thermodynamics that the brightness temperature of the sky could not exceed the kinetic temperature of ionized interstellar gas, that is, 10 000° to 12 000°K. When this became evident, the "interstellar" theory of the origin of galactic radio waves was abandoned by most investigators (Unsöld 1949; Westerhout and Oort 1951; and others), and hypotheses were intensively worked out which attributed all observed galactic radio waves to the combined radiation from postulated "radio stars."

In 1952, on the basis of an analysis of observational data, the present author divided the galactic radio waves into two components, the *spherical* and the *flat,* differing markedly in their spectra; he substantiated the assumption that only the flat component could be interpreted as thermal radiation from ionized clouds of gaseous interstellar matter. Brightness temperatures $T_b \geq 100\,000°$K were typical for the spherical component of galactic radio waves, while, according to the observations at hand, T_b was less than $10\,000°$K for the flat component. It is important to recall that the clouds of interstellar matter in our Galaxy themselves form a flat subsystem.

We pass to a detailed exposition of the theory of emission and absorption of radio waves by an ionized gas. An analogous theory is of fundamental importance for problems of thermal radio waves from the sun, and also for the effects of the terrestrial atmosphere on the propagation of radio waves. However, in the latter two cases it is usually necessary to consider departures of the refractive index of the medium from unity, as well as the influence of magnetic fields. For cosmic radio waves, these last two circumstances may be neglected in the overwhelming majority of cases.

If a source of thermal radio waves is an ideal black body at temperature T, then its brightness temperature is of course equal to T. But cosmic sources of thermal radio waves are not in fact ideal black bodies; furthermore, their temperatures are not constant over all space. As a result, the measured brightness temperature will, generally speaking, be lower than the temperature of the source. It will also depend on other physical properties of the body, such as the density distribution of the matter.

Let r denote the coefficient of reflection of a medium at temperature T, so that $p = 1 - r$ is the coefficient of absorption. Then, by radiation theory,

$$B_{\text{em}} = pB_\nu(T), \qquad (9\text{-}4)$$

where B_{em} is the intensity of emission, and $B_\nu(T)$ the Planck function, which, for radio frequencies, may be approximated by the Rayleigh–Jeans function:

$$B_\nu(T) = 2kT_\nu^2/c^2. \qquad (9\text{-}5)$$

Consider a mass of ionized gas whose temperature varies along the line of sight. Then the equation of transfer for radiation across an area of 1 cm^2 may be written down in the customary form:

$$dI_\nu = -k_\nu I_\nu ds + J_\nu dV = -k_\nu I_\nu ds + J_\nu \cdot ds \cdot 1 \qquad (9\text{-}6)$$

or

$$dI_\nu/d\tau_\nu = -I_\nu + J_\nu/k_\nu = -I_\nu + B_\nu(T), \qquad (9\text{-}7)$$

where

$$d\tau_\nu = k_\nu\, ds.$$

Integrating the equation of transfer, we obtain

$$I_\nu(\tau_\nu) = \int_0^{\tau_\nu} B_\nu[T(t_\nu)]e^{-(t_\nu - \tau_\nu)}\,dt_\nu + I_0 e^{-\tau_\nu}, \qquad (9\text{-}8)$$

where I_0 is the intensity of any radiation which may be incident on the side of the source opposite the observer. The optical depth τ_ν increases along the trajectory of the ray. At the point with optical depth $\tau_\nu^{(1)}$ the intensity of the radiation will be equal to

$$B_\nu[(\tau_\nu^{(1)})] = \int_0^{\tau_\nu^{(1)}} B_\nu[T(t_\nu)]e^{-\tau_\nu}\,d\tau_\nu + I_0 e^{-\tau_\nu^{(1)}}. \tag{9-9}$$

If the optical depth $\tau_\nu^{(1)}$ of the source is large, or if $I_0 = 0$, then the last term in Eq. (9-9) vanishes. In this event we have, by the definition of brightness temperature,

$$T_b = \int_0^{\tau_\nu^{(1)}} T(\tau_\nu)e^{-\tau_\nu}\,d\tau_\nu. \tag{9-10}$$

Equation (9-10) gives the brightness temperature of a gaseous body which is not ideally black. If $T = $ const and $\tau_\nu^{(1)} \to \infty$, then $T_b = T$ ($=$ const), so that in this case the body may be regarded as ideally black. If the gaseous source is isothermal (that is, if $T_b = $ const), then

$$T_b = T(1 - e^{-\tau_\nu^{(1)}}). \tag{9-11}$$

Thus at small optical depths,

$$T_b = T \cdot \tau_\nu^{(1)}; \tag{9-12}$$

this is just the case which obtains in ionized diffuse nebulae at centimeter and decimeter wavelengths since, as we shall demonstrate, $\tau_\nu^{(1)} \propto \nu^{-2}$. If the optical depth is large but the temperature does not remain constant, then Eq. (9-10) will determine T_b as a weighted mean of T. Provided that grad T is small, a calculation indicates that 80 percent of the radiation emanates from the region where $0.1 < \tau < 2.3$. If the linear dimensions of that region are not large and if grad T is also small, then the brightness temperature will approximate the local temperature prevailing in the same region. The local temperatures obtaining at various frequencies will in general be different; that is, different strata of the source will be observed at different frequencies. This is the situation for thermal radio waves from the solar atmosphere. A similar situation may be expected to obtain for the interstellar medium, if observations are made at certain (sufficiently low) frequencies; but such observations have not thus far been carried out.

We therefore conclude that a major problem in the theory of thermal radio waves from cosmic sources is the determination of their optical depths as a function of frequency, and the derivation of the physical properties of ionized interstellar gas. The determination of optical depths in turn rests on the calculation of absorption coefficients for an ionized gas (a plasma).

The basic physical theory of the absorption coefficient for radio waves in an ionized gas has been developed in connection with the theory of the propagation of radio waves in the ionosphere [see, for example, the monograph of V. L. Ginzburg (1949)]. We shall now summarize the fundamental relations.

To begin with, we remark that the problem is deliberately treated as a classical one, since $h\nu/kT \ll 1$ for radio wavelengths. However, it should be noted that in astrophysics the same problem is solved quantum-mechanically ("free–free" or "hyperbolic" transitions). In both cases the results of the calculations must of course be essentially coincident. In the sequel, following Ginzburg, we shall confine ourselves to the classical interpretation.

Consider the propagation of an electromagnetic wave of circular frequncy ω through a conducting medium (an ionized gas) of electrical conductivity σ and dielectric constant ϵ. From the standpoint of the theory of electrons, σ and ϵ are determined by the motion of free electrons in the electric field of the wave. One may readily show that the bound electrons may be neglected. We further assume that the plasma is *homogeneous;* this condition will evidently be fulfilled if $\lambda \gg N_\epsilon^{-1/3}$, where N_ϵ is the concentration of free electrons. In interstellar gas clouds $N_\epsilon \approx 10$ cm^{-3}, so that the condition will be fulfilled to sufficient accuracy for $\lambda > 3$ to 5 cm. But even for shorter wavelengths the relations we are about to introduce may be justified through statistical smoothing.

The total current density arising from the motion of electrons will be

$$J = \varepsilon \sum_{k=1}^{N} \frac{dr_k}{dt}. \tag{9-13}$$

Since the velocity dr_i/dt of the ions is much less than that of the electrons, the ion current need not be considered. According to the theory of electrons the total current J is the sum of the "true" current J_{true} and the polarization current dP/dt:

$$J = J_{\text{true}} + dP/dt = J_{\text{true}} + i\omega P. \tag{9-14}$$

Furthermore, by definition we have

$$D = \epsilon E = E + 4\pi P,$$

whence

$$P = \frac{\epsilon - 1}{4\pi}\, E.$$

Substituting this expression into Eq. (9-14) and recalling that $J_{\text{true}} = \sigma E$, we have

$$J = \left(\sigma + i\omega \frac{\epsilon - 1}{4\pi} \right) E.$$

Equating this to Eq. (9-13) we obtain:

$$J = \left(\sigma + i\omega \, \frac{\epsilon - 1}{4\pi} \right) E = \epsilon \sum_{k=1}^{N} \frac{dr_k}{dt}. \tag{9-15}$$

To find dr_k/dt we must solve the equation of motion of an electron in the field of a radio wave of frequency ω:

$$m \, d^2 r_k/dt^2 = \epsilon E_0 e^{i\omega t} - g \, dr_k/dt. \tag{9-16}$$

The last term in Eq. (9-16) describes the damping of the oscillations of the electron as a result of collisions with neighboring particles. Because of these collisions the electron returns to the neighboring particles the energy it has borrowed from the electromagnetic field; hence the energy of the particles is increased. In this way the final effect of the collisions is merely to transform the electromagnetic energy resident in the wave into thermal energy of the medium. But this process constitutes the true absorption of radiation.

The quantity $g \, dr_k/dt$ represents the average change of momentum per unit time as a result of collisions. Since in the mean the electron transfers to a particle a momentum of the order of $m \, dr/dt$ in each collision, the change of momentum over 1 sec will be

$$g \, dr/dt = \nu_{\text{coll}} \, m \, dr/dt, \tag{9-17}$$

where ν_{coll} is the collision frequency. The equation (9-16) of motion of the electron may now be written

$$m \, d^2 r/dt^2 + m\nu_{\text{coll}} dr/dt = \epsilon E_0 e^{i\omega t}. \tag{9-18}$$

A particular integral of this equation will be

$$\frac{dr}{dt} = \frac{(\epsilon/m) \, E_0 e^{i\omega t}}{i\omega + \nu_{\text{coll}}},$$

whence

$$\sum_{k=1}^{N} \frac{dr_k}{dt} = \frac{N(\epsilon^2/m) \, E_0 e^{i\omega t}}{i\omega + \nu_{\text{coll}}}.$$

Introducing this into Eq. (9-15) we find:

$$J = \left(\sigma + i\omega \, \frac{\epsilon - 1}{4\pi} \right) E = \frac{N(\epsilon^2/m)\nu_{\text{coll}} E}{\omega^2 + \nu^2{}_{\text{coll}}} - i \, \frac{N\omega(\epsilon^2/m)E}{\omega^2 + \nu^2{}_{\text{coll}}}. \tag{9-19}$$

Comparing the real and imaginary parts of Eq. (9-19) we have:

$$\sigma = \frac{\epsilon^2 N \nu_{\text{coll}}}{m(\omega^2 + \nu^2{}_{\text{coll}})}, \qquad \epsilon = 1 - \frac{4\pi\epsilon^2 N}{m(\omega^2 + \nu^2{}_{\text{coll}})}. \tag{9-20}$$

In all cases of interest for radio astronomy, $\omega^2 \gg \nu^2{}_{\text{coll}}$, greatly simplifying Eq. (9-20); we may take

$$\epsilon = 1 - 3.18 \times 10^9 \, N\omega^{-2}, \qquad \sigma = 2.53 \times 10^8 \, N\nu_{\text{coll}}\omega^{-2}. \qquad (9\text{-}21)$$

The quantities ϵ and σ have a well-defined physical meaning only for static or almost static fields. For the rapidly varying fields of radio waves, ϵ and σ must be replaced by the index of refraction n and the index of absorption k of the medium. In that event, it will be recalled that upon solving Maxwell's equations for the propagation of a field in an absorbing medium the complex dielectric constant is introduced:

$$\epsilon' = \epsilon - i(4\pi\sigma/\omega) = (n - ik)^2, \qquad (9\text{-}22)$$

and that Maxwell's equations then assume the same form as for dielectrics. If the wave is propagated through an infinite homogeneous medium in the z direction, then

$$E = \text{const} \cdot e^{\pm i(\omega/c)z\sqrt{\epsilon}} \, e^{i\omega t} = \text{const} \cdot e^{\pm i(\omega/c)kz} \, e^{i\omega t}. \qquad (9\text{-}23)$$

This expression defines the absorption coefficient $\kappa = 2\omega k/c$. According to Eq. (9-22),

$$4\pi\sigma/\omega = 2nk, \qquad \epsilon = n^2 - k^2,$$

whence

$$n = \{\tfrac{1}{2}\epsilon + [(\tfrac{1}{2}\epsilon)^2 + (2\pi\sigma/\omega)^2]^{1/2}\}^{1/2},$$
$$k = \{-\tfrac{1}{2}\epsilon + [(\tfrac{1}{2}\epsilon)^2 + (2\pi\sigma/\omega)^2]^{1/2}\}^{1/2}; \qquad (9\text{-}24)$$

here the square roots have the positive sign, since n and k are real and positive quantities.

In practice the following condition is always fulfilled:

$$|\epsilon| \gg \frac{4\pi\sigma}{\omega} = \frac{4\pi e^2 N_\epsilon \nu_{\text{coll}}}{m\omega^3}, \qquad (9\text{-}25)$$

so that

$$n = \epsilon^{1/2} = [1 - 4\pi e^2 N/(m\omega^2)]^{1/2},$$

$$\kappa = \frac{4\pi\sigma}{nc} = \frac{4\pi e^2 N \nu_{\text{coll}}}{mc\omega^2[1 - 4\pi e^2 N/(m\omega^2)]^{1/2}} \qquad (9\text{-}26)$$

$$= \frac{0.106 \, N\nu_{\text{coll}}}{\omega^2[1 - 3.18 \times 10^9 \, N/\omega^2]^{1/2}}.$$

The problem is now reduced to the determination of ν_{coll}.

Under the conditions prevailing in the interstellar medium, electron–proton and electron–electron collisions may take place. In the latter case the particles will be equally charged and will have the same accelerations, oppositely directed. Dipole radiation from one charge will be "canceled" by dipole radiation from the other; only a small quadratic effect remains, and this may be neglected. Another possibility is an ion–electron collision, in which the dipole radiation from one charge reinforces the corresponding radiation from the other. But the radiation from an electron will be much greater, since its mass is far less than that of a proton. The very term "collision" presupposes an interaction that diminishes rapidly with distance, whereas in our case Coulomb forces, falling off relatively slowly with distance, are operative. Hence in calculating ν_{coll} we shall solve the problem of electron scattering by the Coulomb field of a heavy nucleus.

We proceed from the laws of conservation of energy and angular momentum. In a system of polar coordinates, these laws may be written in the form

$$\tfrac{1}{2}m\left[\left(\frac{dr}{dt}\right)^2 + r^2\,\frac{d\varphi}{dt}\right] - \frac{\epsilon^2}{r} = W = \tfrac{1}{2}mv_\infty^2, \qquad (9\text{-}27)$$

$$r^2\,\frac{d\varphi}{dt} = bv,$$

where $v_\infty \equiv v$ is the velocity of the electron at infinity and b is the impact parameter (Fig. 102).

By substituting $dr/dt = (dr/d\varphi)\,(d\varphi/dt)$, the time may be eliminated from Eqs. (9-27), yielding

$$r = a(\epsilon\cos\varphi - 1)^{-1}, \quad a = mv^2b^2\epsilon^{-2}, \quad \epsilon^2 = 1 + m^2v^4b^2\epsilon^{-4}. \quad (9\text{-}28)$$

Thus the trajectory of the electron is a hyperbola with the ion at one of its foci. It is known that the angle φ_0 between the asymptotes of the hyperbola can be determined from the equation

$$\cos\varphi_0 = 1/\epsilon.$$

The total deflection $\theta = \pi - 2\varphi_0$ of the electron may now be obtained from the expression

$$\tan\tfrac{1}{2}\theta = \cot\varphi_0 = (\epsilon^2 - 1)^{-1/2} = \epsilon^2/(mv^2b); \qquad (9\text{-}29)$$

this formula relates the impact parameter b to the angle of deflection θ. In every such "collision" the momentum of the electron changes by the fraction $(1 - \cos\theta) = 2\sin^2\tfrac{1}{2}\theta \approx 2\tan^2\tfrac{1}{2}\theta$ of its original value (restricting ourselves to small deflections). The aggregate of all the changes of momentum experienced by the electron in all collisions with ions in unit time will be

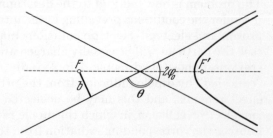

Fig. 102. Scattering by a
Coulomb center.

$$mv_{\text{coll}}\,dr/dt = m(dr/dt)\cdot 2\pi N_i v \int\limits_0^\infty b(1-\cos\theta)\,db,$$

whence

$$\nu_{\text{coll}} = 2\pi N_i v\cdot\varepsilon^4 m^{-2} v^{-4}\int\limits_0^\infty b^{-1}\,db,$$

where N_i is the concentration of ions.

Our approximation has taken *all* collisions into account. The integral diverges at both limits: this is a consequence of the assumptions we have made. Formally, we replace the limits of the integral by b_1 and b_2, say, imposing the restriction that $b_1 < b < b_2$. The integral is now replaced by the logarithmic factor $\ln (b_2/b_1)$. There remains an integration over a Maxwellian distribution of electron velocities. Introducing the mean value

$$\frac{1}{\langle v^3\rangle} = \frac{(32\pi)^{1/2}}{3}\left(\frac{m}{kT}\right)^{3/2}$$

and returning to Eq. (9-26), we finally have

$$\kappa(\nu) = \frac{1}{n}NN_i\,\frac{4\cdot 2^{1/2}}{3\,\pi^{1/2}}\,\frac{1}{(mkT)^{3/2}}\,\frac{\varepsilon^6}{cv^2}\,\ln\frac{b_2}{b_1}. \qquad (9\text{-}31)$$

To specify the limiting impact parameters b_1 and b_2, we now consider all the approximations which we have made in the course of the calculation.

(1) We have assumed that the deflection θ of the electron is small; that is, its orbit is almost a straight line. This is equivalent to assuming that the potential energy at distance b is considerably less than the kinetic energy:

$$\frac{Z\varepsilon^2}{bmv^2}\ll 1.$$

But for small b the reverse is true; thus if $b_1{}^{(1)} = Z\varepsilon^2/(mv^2)$, then for $b < b_1{}^{(1)}$ our approximation is certainly not justified.

(2) We have implicitly regarded ω as smaller than the reciprocal of the duration of a single collision; that is, $\omega b/v \ll 1$. This provides an upper bound $b_2{}^{(1)} = v/\omega$ on b. The meaning of this bound is that for $b > b_2{}^{(1)}$ the collisions are so slow that there will be practically no radiation at frequencies greater than, or of the order of, ω.

(3) A point model of an electron has been used. If b is smaller than the de Broglie wavelength, this model is invalid. We shall assume that the effective "cut-off" of the impact parameter occurs at one-half of the de Broglie wavelength; thus we take $b_1{}^{(2)} = h/4\pi mv$.

(4) There is one more circumstance which provides a bound on b: the screening of the field of the ion by the fields of other ions and of electrons. Let an ion of charge ε be located at the origin of coordi-

nates. The potential of the ion and the charges screening it satisfies the Laplace equation with spherical symmetry:

$$\Delta\varphi = \frac{d^2\varphi}{d\rho^2} + \frac{2}{\rho}\frac{d\varphi}{d\rho} = -4\pi f(\rho),$$

where ρ is the distance from the ion, and $f(\rho)$ is the charge density. By the Boltzmann formula, the concentration of positive ions at a point with potential φ will be

$$N^+(\varphi) = N_i e^{-\varepsilon\varphi/kT},$$

where N^+ ($\varphi = 0$) $= N_i$; that is, in the absence of a field, the concentration $N^+ = N_i = N$. The electron concentration at the same point will be

$$N^-(\varphi) = N e^{+\varepsilon\varphi/kT}.$$

Hence the charge density of the screening particles will be equal to

$$f(\rho) = \varepsilon N(e^{-\varepsilon\varphi/kT} - e^{\varepsilon\varphi/kT}) = -2\varepsilon^2 N\varphi/kT,$$

since $\varepsilon\varphi \ll kT$. We now have the following equation for the desired potential:

$$\frac{d^2\varphi}{d\rho^2} + \frac{2}{\rho}\frac{d\varphi}{d\rho} - \frac{8\pi\varepsilon^2 N\varphi}{kT} = 0,$$

whence

$$\varphi = \varepsilon\,\frac{e^{-\rho/D}}{\rho}, \qquad D = \left(\frac{kT}{8\pi^2\varepsilon^2 N}\right)^{1/2}.$$

For $\rho > D$ the potential becomes vanishingly small, while for $\rho < D$ it deviates only slightly from the Coulomb potential. It is reasonable to suppose that $b_2^{(2)} < D$.

We consequently have two upper limits and two lower limits on b. Under the conditions in the interstellar medium, $b_2^{(2)} > b_2^{(1)}$ because of the small particle concentration. On the other hand, in the solar corona $b_2^{(2)} < b_2^{(1)}$; furthermore, $b_1^{(1)} \approx 10\,b_2^{(2)}$.

The absorption coefficient for diffuse ionized interstellar matter may now be written in its final form:

$$\kappa_\nu = \frac{8}{3\pi\,2^{1/2}}\frac{\varepsilon^6}{(mkT)^{3/2}}\frac{1}{c}\frac{N_i N_\varepsilon}{\nu^2}\ln\left[\frac{(2kT)^{3/2}}{13.5\,\nu\varepsilon^2\,m^{1/2}}\right]$$

$$= 9.8 \times 10^{-3}\frac{N_i N_\varepsilon}{T^{3/2}\nu^2}\left[19.8 + \ln\left(\frac{T^{3/2}}{\nu}\right)\right]. \tag{9-32}$$

Here we have taken $b_2^{(1)}/b_1^{(1)} = (2kT)^{3/2}/(13.5\nu\varepsilon^2\,m^{1/2})$ in accordance with the more detailed calculations of Ginzburg (1949).

The optical depth of the medium will be equal to

$$\tau_\nu = \int \kappa_\nu ds = 9.8 \times 10^{-3}T^{-3/2}\nu^{-2}[19.8 + \ln(T^{3/2}/\nu)]\int N_i N_\varepsilon ds. \tag{9-33}$$

In the expression for κ a Maxwellian velocity distribution was used; it follows that the temperature in Eq. (9-33) is a *kinetic* temperature. The temperatures entering into the basic equations (9-10) and (9-12) are therefore also kinetic temperatures. Since the ion (proton) concentration is approximately equal to the electron concentration,

$$\int N_i N_\varepsilon ds \approx \int N_\varepsilon^2 ds.$$

In astrophysics the latter quantity is known as the *emission measure,* and is denoted by $\mathscr{E}$; thus $\mathscr{E} = \langle N_\varepsilon^2 \rangle\, l$. The electron concentration is usually expressed as the number per cubic centimeter, and the length l of the radiating column of gas is expressed in parsecs. The significance of the "emission measure" is that it determines the intensity of the optical radiation of a nebula in the Balmer lines. The emission mechanism for these lines (under the conditions of ionized interstellar gas) resides, as is well known, in the recombination of electrons and protons, leading to the formation of excited hydrogen atoms, and in the subsequent cascade transitions. According to Menzel (1937), the equilibrium concentration which obtains in processes of this nature for excited hydrogen atoms in the nth quantum state will be

$$N_n = b_n(T) \cdot N_i N_e h^3 n^2 (2\pi m k T)^{-3/2} \exp(\chi/n^2 kT);$$

the dimensionless parameters $b_n(T)$ are tabulated by Baker and Menzel (1938). For $n = 3$, the upper level for Hα emission, $b(10\,000°\text{K}) = 0.089$.

The intensity of the hydrogen lines is defined in the customary way as

$$I_\text{H} = \frac{1}{4\pi} N_n A_{n2} \cdot h\nu_{n2} ds = \text{const} \cdot \int N_i N_e ds, \qquad (9\text{-}34)$$

where A_{n2} is the Einstein coefficient for spontaneous transitions and $h\nu_{n2}$ is the energy of a quantum of radiation. It is clear from Eq. (9-34) that the intensity of the optical line radiation is proportional to the emission measure $\mathscr{E}$. At sufficiently high frequencies, the intensity of thermal radio waves from clouds of ionized interstellar gas must therefore be proportional to the intensity of their optical radiation in the Balmer lines. For $T = 12\,000°\,\text{K}$, the kinetic temperature of an H II zone, the following equation holds (for an optically thin layer):

$$\begin{aligned} I_{\nu,\,\text{radio}} &= B_\nu(T)[1 - e^{-\tau_\nu(T)}] \approx B_\nu(T)\tau_\nu(T) \\ &\approx 6.1 \times 10^{-14} I_{\text{H}\alpha} = 1.32 \times 10^{-6}\,\mathscr{E}. \end{aligned} \qquad (9\text{-}35)$$

Since the intensity of the radio waves determines the brightness temperature, the latter will also be proportional to $\mathscr{E}$ (for an optically thin layer); we then have

$$T_\text{b} = \frac{c^2}{2k\nu^2} I_{\nu,\,\text{radio}} = \frac{\lambda^2}{2k} I_\nu = \frac{\lambda^2}{2k} B_\nu(T)\tau_\nu(T) \propto \mathscr{E}. \qquad (9\text{-}36)$$

Assigning numerical values to the quantities entering into Eq. (9-36) we obtain, for $T = 12\,000°\,K$:

$$T_b = 4.8 \times 10^{-6}\lambda^2\, \mathcal{E}. \tag{9-37}$$

Equation (9-35) shows that the intensity of the thermal radiation from an optically thin layer of ionized gas is independent of the frequency [to within a logarithmic factor that depends very weakly on frequency—compare Eq. (9-32)], while the brightness temperature in Eq. (9-37) is proportional to the square of the wavelength. But this is exactly the spectrum which is observed for the flat component of galactic radio waves; we have discussed this at length in Sec. 5.

It may therefore be concluded that the character of the spectrum of the thermal radio waves from an ionized gas is identical with that of the flat component of galactic radio waves; that is, the observations are in full qualitative agreement with a mechanism of thermal radio emission from ionized interstellar gas.

We shall now examine the matter quantitatively: we shall inquire whether interstellar space contains enough ionized gas to provide the observed intensity of cosmic radio waves in the regions of the secondary maxima. For one of the brightest emission objects in the sky, the central portion of the Orion Nebula, $\mathcal{E} \approx 8 \times 10^6$. For very bright diffuse nebulae such as NGC 6523 (M 8) and NGC 6618 (M 17), $\mathcal{E} \approx 10\,000$ to $30\,000$. Ionized clouds of interstellar gas have emission measures between 1000 and 5000; in low galactic latitudes, such clouds sometimes extend for many degrees. Finally, the faintest of all emission regions in the sky, barely distinguishable from the background luminosity of the night sky and detected only at the limit of sensitivity of modern photoelectric equipment, have emission measures near 400.

Table 14 gives, as a function of emission measure, and at various wavelengths, values of the brightness temperature of thermal radio emission from ionized interstellar gas. The table has been computed

TABLE 14. Brightness temperature (°K) of thermal radio emission from ionized gas clouds.

Emission measure	Wavelength λ (cm)				
	33	75	150	300	370
100	0.5	2.6	10.4	40.2	62
400	2	10.4	40.2	160.8	250
1 000	5	26	104	416	640
5 000	25	130	520	2 040	3 200
10 000	50	260	1 040	4 000	5 800
30 000	150	780	3 050	10 000	11 000

from Eq. (9-37). For high $\mathcal{E}$, $\tau > 1$; that is, the layer is no longer optically thin.

In the region of the secondary maxima of radio emission in Cygnus (the extended source Cygnus X) is found a large group of bright diffuse nebulae, a major portion of which is screened from us by strongly absorbing dust clouds. In Fig. 103 we present a mosaic photograph of this region, obtained by G. A. Shajn and V. F. Gaze. The emission measure of the large group of nebulae shown in this photograph exceeds 5000. According to Table 14, the brightness temperature for such an emission measure is $25°$K at 33 cm; this is just the brightness temperature which is observed in the region (see Sec. 5). However, it is necessary to keep in mind that the observations give a value of the brightness temperature smoothed over the lobe of the antenna ($\approx 2°$). On the other hand, the distribution of brightness over the Cygnus emission nebulae is strongly distorted by the absorption of light due to cosmic dust. It is therefore difficult to establish a detailed correlation between the brightness distribution of radio and of optical emission, although a broad relation does unquestionably prevail. When large precision reflectors of high resolving power are put into service, one of the most urgent tasks will be the investigation of an interdependence between the surface brightness of the thermal radio emission from nebulae and their optical radiation at Hα.

The presence of a large number of bright nebulae in the direction of Cygnus is manifestly not fortuitous. It is notable that this direction forms almost a right angle with the direction to the galactic center. Evidently a spiral arm of our Galaxy extends in the direction of Cygnus, the sun being found in the same spiral arm, near the inner edge. The Cygnus X source will be discussed further in Sec. 10.

In the region of the southern constellations Vela and Carina another extended maximum of radio emission appears along the galactic equator; its radiation is also thermal in character [see, for example, the isophotes of Allen and Gum (1950) at $\lambda = 1.5$ m, which are reproduced in Fig. 27]. The Vela–Carina region is located in a part of the celestial sphere almost diametrically opposite to the Cygnus region. In this direction we are evidently observing the same spiral arm as in Cygnus, but in the opposite sense.

In the anticenter region there is found a rather weak secondary maximum of radio emission whose spectrum patently demonstrates its thermal character (compare Table 4, Sec. 4). Strömgren and Hiltner have detected an extended field of Hα emission in this region. The emission measure of the region is just sufficient to ensure the production of thermal radio waves of the observed intensity. It is not precluded that this secondary maximum of thermal radio emission and the associated optical hydrogen emission region represent a fragment of a spiral arm of our Galaxy located beyond the sun.

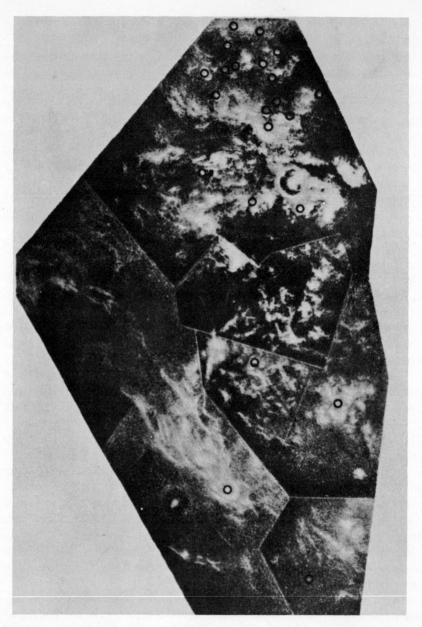

Fig. 103. A mosaic of photographs of the Cygnus nebulosity, obtained by G. A. Shajn and V. F. Gaze.

In 1952 the author constructed from these data a rough map of the spiral structure of the Galaxy (Shklovsky 1952*a*). In the same year Morgan and his collaborators derived the spiral structure of the Galaxy from an analysis of the distribution of the H II emission regions observed optically. The two methods of determining the spiral structure are equivalent since, by Eq. (9-35), the intensity of the thermal radio emission of an ionized gas is proportional to its optical emission in the Hα line. The results obtained were in fact coincident in their general features; in particular, the spiral arm passing through the Cygnus region was outlined distinctly both in our data and in Morgan's.

The optical data have the advantage over the radio data that they were obtained with instruments of very high resolving power. But the radio data are not burdened by the influence of interstellar absorption, and could be carried to very remote portions of the Galaxy. At the present time the most effective means of investigating the spiral structure of the Galaxy is by an analysis of the hydrogen radio line at $\lambda = 21$ cm, which we shall treat in detail in Chapter IV.

As pointed out in Sec. 5, there had long been an animated discussion whether the radio emission of the "disk" is intrinsically thermal at relatively high frequencies (> 1000 Mc/s). In that section we emphasized that recently obtained data on the spectral composition of this radiation have compelled us to reject this idea. It must be added that the half-width of an intense band of radio emission extending along the equator may, in the region of the galactic center, be as much as 6° or 7°, and as a rule will be no less than 3° or 4°. On the other hand a cloud of ionized interstellar gas forms a very flat system, ≈ 150 pc in thickness. If the sources in the disk could be identified with thermal emission from ionized interstellar gas, the half-width of the intense equatorial band would be considerably smaller.

A decisive argument in favor of a hypothesis that the radio emission from the disk arises in ionized interstellar gas would be the detection, at low frequencies, of a *dark band* extending along the equator in the vicinity of the galactic center. The presence of such a band would necessarily be the result of absorption of the spherical component of galactic radio waves by H II zones. This is a consequence of the following considerations: Suppose there were no H II regions in the Galaxy which could be responsible for thermal emission and for the absorption of radio waves. In this event, the brightness temperature T^* of the sky in the direction toward the galactic center would be determined by the temperatures T_1 of the spherical component of the galactic radio emission, and T_0 of the extragalactic radiation; that is,

$$T^* = T_1 + T_0 = \int t_1 dR + T_0, \qquad (9\text{-}38)$$

where $t_1(R)$ is that portion of the observed brightness temperature arising from sources found in the interval $(R, R + dR)$ which is due to the spherical component. The presence of an H II zone would result in additional emission of a thermal character and, secondly, in the absorption of radio waves from the spherical component and from extragalactic sources. The brightness temperature would then be altered, assuming the form

$$T = T_e(1 - e^{-\tau}) + \int_0^{\infty} t_1 e^{-\tau} \, dR + T_0 e^{-\tau}, \tag{9-39}$$

where T_e is the kinetic temperature of the H II zone.

The change in the brightness temperature introduced by the presence of the H II region would be equal to

$$T - T^* = T_e(1 - e^{-\tau}) - \int_0^{\infty} t_1(R)(1 - e^{-\tau}) \, dR - T_0(1 - e^{-\tau}). \tag{9-40}$$

The second term on the right in Eq. (9-40) depends on the distribution $t_1(R)$ of the sources of the spherical component, and on the distribution of the H II region along the line of sight; the latter requires a computation of the quantity

$$\tau = \int_0^R \kappa_\nu \, dR.$$

In Sec. 5 we established that $t_1(R)$ decreased extremely slowly with distance from the galactic center. On the other hand, the strong concentration of the disk radiation to the center compels the conclusion that the H II regions responsible for that radiation are also strongly concentrated to the center of our stellar system. Equation (9-40) may therefore be rewritten in the following form, with sufficient accuracy:

$$T - T^* = T_e(1 - e^{-\tau}) - (T_1 + T_0)(1 - e^{-\tau})$$
$$= (1 - e^{-\tau})(T_e - T_1 - T_0). \tag{9-41}$$

It follows from Eq. (9-41) that as long as $T_1 + T_0 < T_e$, $T - T^*$ will be positive. But as soon as $T_1 + T_0$ exceeds $\approx 12\,000°$K $(= T_e)$, $T - T^*$ will become negative; that is, in the corresponding region of the sky a dark band will be observed. Now $T_1 + T_0 \approx 12\,000°$K at $\lambda \approx 4.5$ m; hence the dark band predicted by the theory must be sought at frequencies below 60 or 65 Mc/s.

Observations carried out by Blythe (1957b) at the sufficiently low frequency of 38 Mc/s and with good directivity (see Sec. 5) have indicated that a dark band along the galactic equator is absent. The band of radio emission extending along the equator remains bright, as is evident from the isophotes reproduced in Fig. 54.

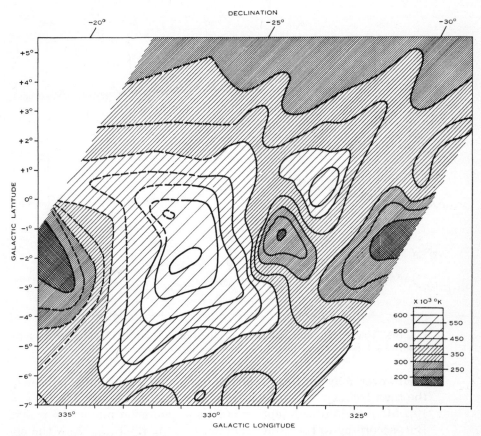

Fig. 104. Isophotes of brightness temperature near the galactic center, at $\lambda = 15.2$ m, $\nu = 19.7$ Mc/s (Shain 1957).

As a result of investigations conducted by Shain (1957), who used a cross antenna system at a very low frequency (19.7 Mc/s) and with high directivity (width of main lobe $1°.4$), it has finally become necessary to abandon the attempt to identify the high-frequency radiation from the disk with the thermal emission from H II zones. In Fig. 104 are presented isophotes of radio emission obtained by Shain for the direction of the galactic center.

Shain's observations clearly show the dark band along the galactic equator which the theory of thermal radio emission by an ionized gas predicts. Figure 105 gives tracings representing the radiation received at three different declinations. Very deep minima are clearly shown; these appear as the galactic equator traverses the antenna beam. Thus there can be no doubt that within the galactic plane there exists ionized interstellar gas which strongly absorbs low-frequency cosmic radio waves.

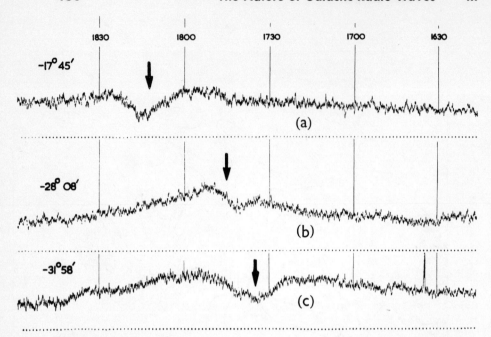

Fig. 105. Records obtained at three declinations (Shain 1957). The arrows indicate the sidereal time of passage of galactic latitude $b = -1°.3$.

However, may not this gas be responsible for a substantial portion of the high-frequency radiation from the disk? By Eq. (9-33), for $\nu = 19.7$ Mc/s the optical depth $\tau = 1$ at an emission measure $\mathcal{E} = 800$. But, according to the same formula, $\tau_{\nu=19.7} = 1200\tau_{\nu=600}$. Now the observations of Piddington and Trent at $\nu = 600$ Mc/s indicate that $T_b = 50°$K in the neighborhood of the galactic center (at $b = 8°$). If we regard this radiation as thermal emission from H II zones whose kinetic temperature is $10\,000°$K, then, by Eq. (9-37), $\mathcal{E} \approx 5000$. For so high a value of $\mathcal{E}$, $\tau_{\nu=19.7} = 6$, so that total absorption must be observed at $b = 8°$ and $\nu = 19.7$ Mc/s. It is evident from Fig. 104 that absorption has scarcely set in at $b = 4°$, while at $b = 8°$ it is entirely inconsequential. Taking $\tau_{\nu=19.7} = 0.5$ at $b = 4°$, it may be shown that even at this low a galactic latitude the proportion of thermal radiation is in any case < 8 percent.

In isolated regions along the equatorial belt, thermal emission from H II zones may be of considerable importance. In this event it will be induced by clusters of diffuse nebulae. Thus, for example, there are visible in the isophotes of Fig. 104 three regions of strong absorption in the area around the galactic center: each of these regions has been reliably identified with certain groups of nebulae. Figure 106 gives a schematic distribution of H II zones observed by Gum (1955) near the

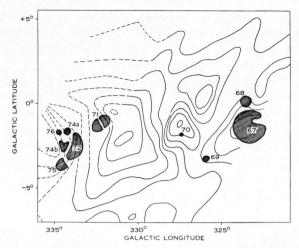

Fig. 106. Distribution of H II regions observed by Gum (Shain 1957).

galactic center. It is clear that the absorption regions in the isophotes of Fig. 104 correspond to H II zones.

Another effect of H II zones is their absorption of the long-wave portion of the spectra of discrete sources. It is well established that "breaks" are present in this portion of the spectra (see Sec. 6). A number of authors have shown that this phenomenon may be explained both qualitatively and quantitatively by absorption in H II zones, consisting either of interstellar matter or of ionized gas, and located within the discrete sources themselves. Emission measures of the order of several hundred would be adequate.

10. DISCRETE SOURCES OF THERMAL RADIO EMISSION

The flat component of galactic radio waves is on the whole a consequence of emission by ionized interstellar gas; we have given a qualitative and quantitative demonstration in the foregoing section. The interstellar gas is, however, distributed inhomogeneously over the Galaxy, exhibiting a strong tendency to concentrate into discrete clouds and intricate nebulosity. Because of this "clumpiness" in the distribution of interstellar H II regions, the flat component must similarly evince an inhomogeneity in brightness distribution. Although with radio telescopes of low resolving power this fine structure in the flat component is lost (the antenna lobes exert a strong smoothing influence), isophotes secured with high resolution do reveal some detail. We may cite the isophotes reproduced in Fig. 34 (Sec. 3), obtained at $\lambda = 75$ cm with a beamwidth of $2°$; several maxima appear near the galactic center, and all have been identified with individual emission nebulae. Minor maxima are similarly apparent in isophotes obtained at $\lambda = 50$ cm (Fig. 35, Sec. 3) and at $\lambda = 22$ cm (Fig. 45, Sec. 5).

Several years prior to the detection of these maxima, it had been predicted theoretically that ionized diffuse nebulae must individually emit thermal radio waves of measurable power (Shklovsky 1952a). The existence of this thermal emission is a necessary consequence of the relatively large dimensions of the nebulae, and of their high emission measures. Consider a typical bright diffuse nebula of apparent size $20' \times 20'$ and of mean emission measure $\approx 10\,000$. The Lagoon Nebula M 8 (NGC 6523) is an object of this sort; photographs of this and several other nebulae are presented as Figs. 107, 108, and 109. The radiation flux F_ν from such a nebula is given by

$$F_\nu = 2kT_b\lambda^{-2}\Omega = 2kT_e\lambda^{-2}(1 - e^{-\tau_\nu})\Omega, \qquad (10\text{-}1)$$

where the solid angle Ω subtended by the nebula is 3.4×10^{-5} sterad, and T_e is taken as $12\,000°$K. From Eqs. (9-37) and (9-35) we find the

Fig. 107. The Orion Nebula, a source of thermal radio emission.

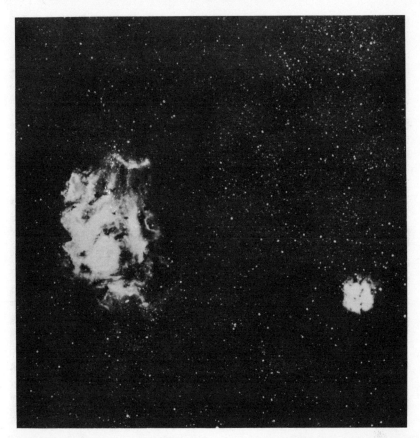

Fig. 108. The Lagoon and Trifid Nebulae, sources of thermal radio emission.

following relation between the optical depth τ_ν and the emission measure $\mathcal{E}$:

$$\tau_\nu = 4 \times 10^{-10}\lambda^2\mathcal{E}. \tag{10-2}$$

In Table 15 we have given the optical depth, brightness temperature, and radiation flux of our standard nebula as computed for various frequencies. Some of the characteristics of thermal emission by ionized gaseous nebulae are evident from the table: $F_\nu = $ const as long as $\tau_\nu \ll 1$; but as the optical depth approaches unity, the radiation flux falls off as frequency decreases. At sufficiently low frequencies $F_\nu \propto \lambda^{-2}$, in accordance with the Rayleigh–Jeans law. The dependence of flux on frequency is plotted in Fig. 110; the spectra of discrete sources must conform to this curve if the sources are to be interpreted as H II regions.

The spectrum of the discrete source Centaurus A is also plotted in Fig. 110. At low frequencies, sources of thermal emission will radiate

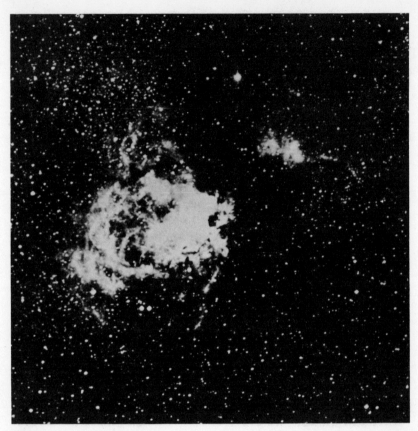

Fig. 109. The diffuse gaseous nebula NGC 6618, a source of thermal radio emission.

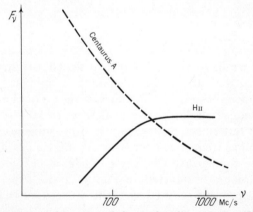

Fig. 110. The spectrum of sources of thermal radio emission (H II regions). For comparison, the spectrum of the discrete source Centaurus A is given as a broken curve.

TABLE 15. A standard gaseous nebula.

Frequency ν (Mc/s)	Optical depth τ_ν	Brightness temperature (°K)	Flux F_ν $[10^{-25}$ w m^{-2} (c/s)$^{-1}]$
3000	0.0004	4.8	4.5
1000	0.0036	43	4.5
400	0.022	270	4.5
200	0.088	1 080	4.5
100	0.35	3 600	3.7
50	1.40	9 200	2.4
25	5.6	12 000	0.8

far more weakly than such discrete sources of nonequilibrium radiation as the ejecta of supernovae, colliding galaxies, and other peculiar objects. But in the decimeter and centimeter range the extended diffuse nebulae will be among the more powerful discrete sources.

After the theory had been developed, Haddock, Mayer, and Sloanaker (1954a, 1954b) detected a number of discrete sources at $\lambda = 9.4$ cm, using the 50-ft paraboloid of the Naval Research Laboratory. Identification was made with bright diffuse nebulae, including the Orion Nebula M 42 (NGC 1976), the Omega Nebula M 17 (NGC 6618), the Lagoon Nebula M 8 (NGC 6523), the Trifid Nebula M 20 (NGC 6514). The thermal radio-wave flux from these sources was soon measured at $\lambda \approx 21$ cm with the same reflector (Hagen, McClain, and Hepburn 1954); the results are given in Table 16. Mills (1952a) had earlier observed the source 17-2 B at $\lambda = 3$ m, and the writer identified this with the Lagoon Nebula. Using the optically determined

TABLE 16. Thermal emission from gaseous nebulae.

Nebula NGC	Nebula Name	$\alpha(1950)$	$\delta(1950)$	Flux F_ν $[10^{-25}$ w m^{-2} (c/s)$^{-1}]$ $\lambda = 9.4$ cm	$\lambda = 21$ cm	$\lambda = 75$ cm
IC 1805		2^{h}30^m	+61°14′	11	—	—
1976	Orion	5 33	− 5 27	45	42	—
6188		16 36	−48 55	7	—	23
6334		17 17	−36 0	25	—	33
6357		17 22	−34 9	40	—	51
6514	Trifid	17 59	−23 2	11	42	28
6523	Lagoon	18 1	−24 23	20	—	—
6604		18 15	−12 16	10	9	—
6611	M 16	18 16	−13 48	16	21	—
6618	Omega	18 18	−16 14	68	82	—
7000	N. America	20 55	+44 45	$T_b = 2°$K	$T_b = 4°$K	—

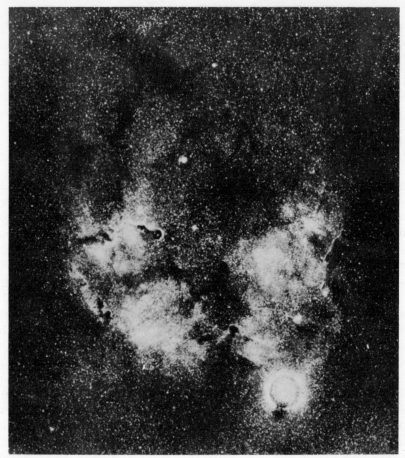

Fig. 111. The large diffuse nebula IC 1396, a source of thermal radio emission.

angular dimensions and emission measure, he showed that the radiation flux from this source should be equal to $F_\nu \approx 1.5 \times 10^{-24}$ w m^{-2} (c/s)$^{-1}$ at $\lambda = 3$ m. This agrees with the flux observed at this wavelength.

We thus note that individual ionized diffuse nebulae can actually be observed in the meter range even though conditions are more favorable for decimeter observations. For example, Hanbury Brown and Hazard had observed a discrete source located at $\alpha = 21^h37^m \pm 2^m$, $\delta = +56°30' \pm 1°30'$ with the Jodrell Bank 218-ft fixed reflector, at $\lambda = 1.89$ m; in 1954 the writer obtained an identification with the large but not very bright diffuse nebula IC 1396. The nebula is round, about 2° in diameter, and quite compact. The photograph shown in Fig. 111 was made by Shajn and Miss Gaze. A radio flux of $F_\nu = 5 \times 10^{-25}$ w m^{-2} (c/s)$^{-1}$ was observed. From the observed 1° radius of the nebula, we find the rather low brightness temperature of 65°K at $\lambda = 1.89$ m. The formulas given above readily yield an emission measure ≈ 500, averaged over the disk of the nebula; this is in accord with optical observations.

In 1954 McGee, Slee, and Stanley (1955), observing at $\nu = 400$ Mc/s ($\lambda = 75$ cm) with an 80-ft fixed reflector, detected a number of sources of thermal emission which had previously been observed in the United States at higher frequencies. The radiation flux from these sources is tabulated in the last column of Table 16, and the spectra, clearly thermal in character, are shown in Fig. 112.

Early in 1955 radio emission from the brightest diffuse nebulae—the Orion and the Omega Nebulae—was detected in the Soviet Union at $\lambda = 3.2$ cm (Kaidanovsky, Kardashev, and Shklovsky 1955). The data fully confirm the theoretical predictions as to the nature of the high-frequency spectra of these sources. Radiation from these sources

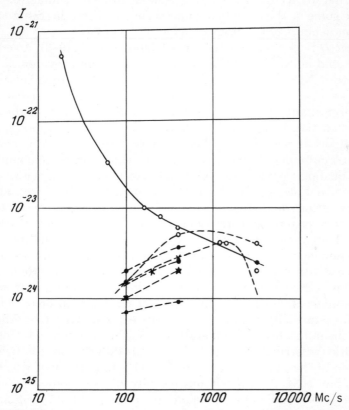

Fig. 112. The spectra of several sources of thermal radio emission, given as broken curves (McGee, Slee, and Stanley 1955). The solid curve represents the spectrum of Centaurus A.

was also detected in the United States in 1955 at $\lambda = 3.15$ cm, but the results have not yet been published.

In 1955 G. A. Shajn and the author identified as H II regions some of the extended sources of cosmic radio waves which had been discovered by Bolton *et al.* (see Sec. 7). We have pointed out in Sec. 7 that the term "discrete source of radio waves" is a relative one. Those of the discrete sources referred to which lie in low galactic latitudes actually represent coarse structural detail in the flat component of galactic radio waves. It is no coincidence that they are extended along the galactic equator.

We may compute the mean brightness temperatures of the sources from their angular size and their radiation flux. Thus the mean brightness temperatures of the sources *B, D, F, K,* and *L* (Table 11, Sec. 7) are respectively 420°, 310°, 290°, 6200°, and $\approx 12\,000°$K at $\lambda = 3$ m.

In order to secure identifications with diffuse nebulae, we shall analyze some of the extended galactic sources listed in Table 11. No emission nebulosity has been found at the point *B*.

Source *D* is located near the Orion Nebula (NGC 1976). A complex of dark and emission nebulae found in the region is evidently associated with NGC 1976. The chart in Fig. 113 shows the distribution of the nebulae and also of the hot O and B0 stars (designated by filled circles) and the B1 to B5 stars (crosses). Because all these objects are at the same distance from the sun (roughly within observational error), in a sufficiently high galactic latitude ($-17°$), and confined to a quite narrow band in longitude, there can be no doubt that we are dealing with a united physical system comprising substantial masses of hot stars, dark matter, and ionized gas. The rectangle in Fig. 113 outlines an extended ($10° \times 5°$) source of radio waves found within this system; it corresponds to a mass of ionized hydrogen, rendered unobservable optically by intervening dark matter. This H II region, source *D,* cannot be a distant object (it must in fact be included within the Orion Nebula complex), for its galactic latitude ($-14°$) is relatively high; similarly, background radiation from distant nebulae cannot be a significant factor. It should be noted that the mean emission measure of the source (as derived from the observed mean brightness temperature) has the very low value of 700. The optical distribution of the H II clouds differs from the true distribution in the manner shown by Fig. 113. In particular, it is interesting that NGC 1976 itself lies entirely outside source *D,* and is a much weaker radio object. Equations (9-36) and (9-37) will help to explain this circumstance. In a nebula whose emission measure is as great as that of NGC 1976, the brightness temperature T_b is comparable to the kinetic temperature T_e, although $T_b < T_e$ however

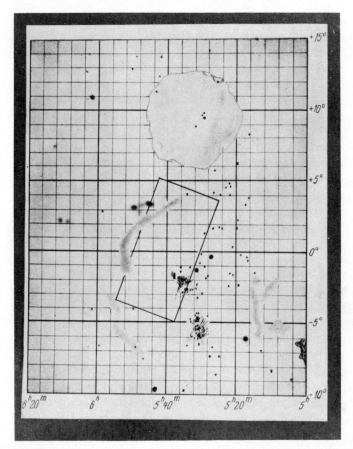

Fig. 113. Distribution of nebulae and hot stars in the vicinity of the Orion Nebula. The rectangle bounds the region of radio emission.

great the emission measure (and hence the optical brightness at Hα) may be. On the other hand, the angular size of this (optically) bright nebula is comparatively small, resulting in a small radio flux, whereas an extended but optically faint nebula would yield a considerable radio flux.

Source F lies in the northern portion of the extensive nebulosity centered near $\alpha = 8^h23^m$, $\delta = -46°$, whose brighter regions were first noted by Shapley (1940). In point of fact, this is the largest of all known nebulae: it is no less than $13° \times 10°$ in extent, and has a marked fibrous texture, especially in the vicinity of the radio source F. The nebulosity is ≈ 200 pc distant (Gum 1952). The "center of gravity" of source F is located two or three degrees from the brightest part of the nebulosity, near the galactic equator. Errors in the radio observations may well be responsible for this discrepancy, but it is not evident whether the large discrepancy in angular size should also be ascribed to errors, or whether the radio source simply occupies the northern part of the nebulosity. The brightness temperature observed for source F leads to a mean emission measure of ≈ 6700. In Allen and Gum's isophotes of galactic radio waves at $\lambda = 1.5$ m (Fig. 27, Sec. 3), a well-defined secondary maximum emerges near $l = 230°$, $b = 0°$. The radio flux may be evaluated directly from the isophotes; it is approximately 8×10^{-23} w m^{-2} (c/s)$^{-1}$, about the same as at $\lambda = 3$ m (Table 11, Sec. 7). This implies that the emission arises from an optically thin layer of ionized gas. We may regard source F as reliably identified with the nebulosity.

Source K occupies a region at $l = 304°$, at the edge of which is found a group of nebulae comprising NGC 6164, 6165, and 6188; near $l = 312°$ is located the bright nebula NGC 4628, which adjoins some remarkable filamentary nebulosity of the peripheral type (size $\approx 5°$). The mean emission measure of 20 000 for source K is quite large. The emission nebulae we have listed may give rise to a significant part of the radiation from source K, especially in view of the presence of interstellar absorbing matter. Yet the majority of the radio waves emitted by the source appear to be excited by masses of ionized interstellar gas located far within the Galaxy. This interpretation is substantiated by the presence of a noteworthy dark nebula extending along the galactic equator; most of source K lies within the dark nebula. We conclude, then, that the correspondence between source K and the ionized gas observable optically is not very satisfactory, but may be explained on the basis of the absorption of light.

In general, however, a relation must unquestionably obtain between radio and optical (Hα) radiation from emission nebulae, as is best evidenced by the current detection of radio waves from a number of bright emission objects. We have seen that complications arise when extended radio sources are considered, although it remains highly probable that their radio emission originates in gaseous matter. If we accept this, we must then expect their radio emission to follow a characteristic spectrum: the intensity per unit frequency interval must be independent of frequency. Thus at short wavelengths these sources will exhibit increasing contrast to the over-all background of galactic radio waves.

One might well have expected Table 11 to include a strong radio source corresponding to the bright and very large ($7°.5 \times 6°.5$) η Carinae nebulosity. (The source might be even more extensive if, as seems likely, IC 2944 and other nebulae are associated with the η Carinae complex.) Repeated inspection of the area may confirm the presence of an extended radio source. Mills, Little, and Sheridan (1956c) have in fact recently detected such a source near η Carinae; Fig. 114 shows

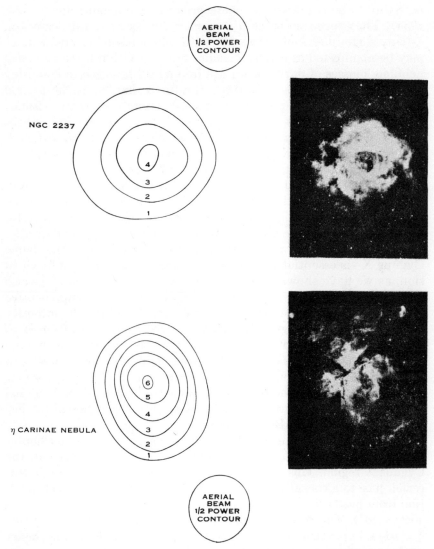

Fig. 114. Isophotes and photographs of NGC 2237 and 3372 (Mills, Little, and Sheridan 1956c).

their isophotes in this region, and also for the Rosette Nebula NGC 2237. Photographs of these objects are also reproduced in Fig. 114.

In summary, the nebulosity responsible for extended sources of radio waves may assume the form of either (a) a very large emission nebula, such as that near η Carinae ($\approx 8° \times 7°$), or the Vela–Puppis nebula centered at $\alpha = 8^h24^m$, $\delta = -44°$ (at least $13° \times 10°$); or (b) a group of nebulae. Groups of emission nebulae like that in Orion are probably to be regarded as a widespread phenomenon (Shajn 1953, 1954). They often embrace a region of the sky several degrees in extent, sometimes more than 10; with small resolving power, they may be manifested as extended radio sources. Within these groups of nebulae, masses of dark matter are frequently interposed, obscuring the optically observable ionized gas. It is likely that "invisible" ionized hydrogen is not infrequently of greater significance than "visible" hydrogen; at any rate, masses of optically unobservable ionized gas are always present in groups of nebulae which are associated with dark obscuring matter. Source D in Orion is a good illustration; it is a gigantic unified conglomeration of gas, dust, and stars.

Observations carried out at relatively high frequencies with antennas of adequate directivity will undoubtedly permit a more detailed study of extended radio sources. They will be of particular value for the investigation of the great complexes of hot stars, gas, and dust.

The secondary maximum in the galactic radio emission—the source Cygnus X—is essentially an extended source of thermal radio emission, as we have shown (Sec. 4) by analyzing its spectrum (see also Secs. 5 and 9). Davies (1957) has recently concluded a comprehensive investigation which confirms this result. He examined the intensity distribution in the Cygnus X region at seven frequencies from 92 to 1425 Mc/s. The isophotes obtained are given in Fig. 115; an antenna of high directivity was employed, whose main lobes are represented on the same scale. A consideration of the spectrum of this source now decisively establishes the thermal character of its radiation. As a final demonstration, the source has been observed in absorption at the low frequency of 22 Mc/s (Burke 1956).

Research by Davies at the 21-cm hydrogen radio line (see Chapter IV) has revealed a marked deficiency of neutral hydrogen in the Cygnus X region. This indicates that the interstellar hydrogen in this region has, to a considerable degree, been ionized by ultraviolet radiation from hot O stars concentrated in this area (in galactic longitudes 40° to 48°). The intensity maximum for Cygnus X (at $\alpha = 20^h30^m$, $\delta = 41°$), in particular, appears to be associated with a large cluster of O stars whose angular diameter is 0°.5.

Cygnus X has an emission measure of 6000. Despite the quite irregular intensity distribution over the source, no regions of high brightness but of small angular size ($\approx 1°$) have been detected; an independent interferometric survey at 158 Mc/s and with spacing of 56 λ fails to reveal such regions. The ionized interstellar gas comprising the source amounts to several million solar masses. To the great complexes of interstellar gas, dust, and hot stars we may, then, assign Cygnus X as well (compare Fig. 103).

At low frequencies we may expect to observe *absorption* of cosmic

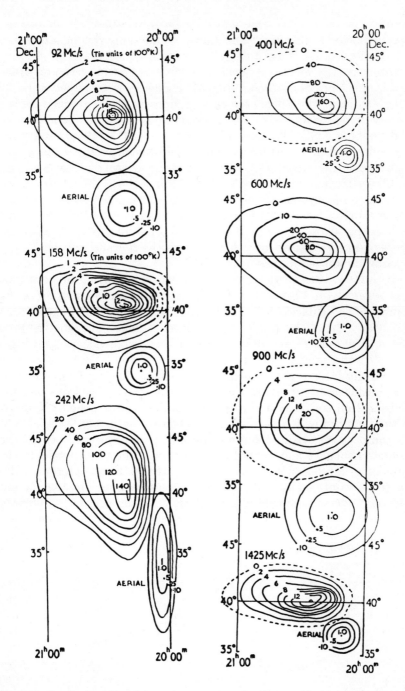

Fig. 115. Isophotes of brightness temperature for Cygnus X at seven frequencies (Davies 1957).

radio waves as they traverse ionized gaseous nebulae. This phenomenon is entirely analogous to the "dark" band along the galactic equator which has already been detected at low frequencies.

Absorption by ionized gaseous nebulae was first observed by Mills,
Little, and Sheridan (1956a, 1956c). Using a cross-shaped antenna
array at 3.5 m, they observed in emission six of the ionized diffuse
nebulae which had earlier been observed at high frequencies. The
bright nebula NGC 6357 [$\alpha = 17^h18^m.4$, $\delta = -34°3'$ (1950)] was, however, observed in absorption. Figure 116 illustrates the variation of
brightness temperature with sidereal time at several declinations. The
brightness temperature drops as this region of the sky (curve e) passes
through the main lobe of the antenna. We conclude that the electron
temperature of the nebula is in any case less than the 7800°K brightness temperature; it is apparently in the neighborhood of 6000°K.
These observations evidently demonstrate the *thermal* nature of radio
emission by ionized diffuse nebulae.

Shain (1957) has carried out, at 19.7 Mc/s, important observations
of H II regions in absorption. His isophotes of radio emission in the
vicinity of the galactic center (Fig. 104, Sec. 9) distinctly show three
dark areas, the effects of absorption by H II zones. The smallest area
(the central one) represents a group of very bright gaseous nebulae,
projected on the galactic center from a distance of ≈ 3 kpc from the
sun. Strong interstellar absorption conceals the group from our view
to a considerable extent. In the future, the study of gaseous nebulae
in absorption will yield a new method of determining their distances.
At sufficiently low frequencies such nebulae completely "screen" cosmic radio waves arising from nonthermal sources located behind the
nebulae; hence we may regard that portion of the brightness temperature of a gaseous nebula in excess of 10 000°K as determined by
sources of nonthermal galactic radio waves between the nebula and
the sun. Given the nonthermal galactic emission per unit volume in

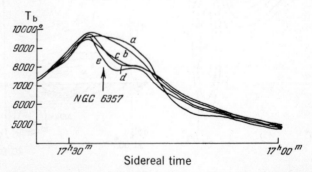

Fig. 116. Brightness temperature as a function of sidereal time for five sections,
at different declinations, through NGC 6357 (Mills, Little, and Sheridan 1956a).

the solar neighborhood, we would clearly be able to deduce the true distance.

At the 1958 Paris conference, Shain reported on his most recent investigations, attacking a problem which is the reverse of the foregoing one. He found the dependence of T_b in H II regions (observed in absorption at 19.7 Mc/s) upon the distance to these regions. By evaluating the distance to the gaseous nebulae from the optical data, he could obtain the characteristic nonthermal galactic radio emission in the solar neighborhood. It was found that near the sun T_b rises by 100 K deg/pc. This value leads to the nonthermal emission per unit volume in the solar neighborhood, according to the formula

$$2kT_b/\lambda^2 = l\epsilon_\nu/4\pi;$$

at $\nu = 19.7$ Mc/s, we have $\epsilon_\nu = 5 \times 10^{-38}$. Taking the spectrum of nonthermal radio emission into account, we find that $\epsilon = \int \epsilon_\nu d\nu \approx 7.5 \times 10^{-30}$ erg cm^{-3} sec^{-1}. This value is 10 to 20 times as small as the mean for the "halo" (see Sec. 23).

The examination of discrete sources of thermal radio emission constitutes a powerful new method for studying diffuse interstellar matter. The method is especially advantageous, as we have already noted, because of its independence of interstellar extinction of light—so great an impediment to optical observations of ionized interstellar gas. Fully as important is the concomitant ability of radio astronomy to reach objects at very great distances from us, including those located on the opposite side of the galactic center.

An effective way to investigate interstellar extinction arising from cosmic dust is to observe the same diffuse nebulae concurrently in Hα light and at radio frequencies using antennas of high resolving power. Taking the simplest case as an illustration, suppose that we have a uniform distribution of ionized gas intermixed in constant proportion with absorbing dust. Further, let $\epsilon_{H\alpha}$ and ϵ_ν denote respectively the emission per unit volume in the Hα line and at radio frequencies; and let k_α denote the absorption coefficient of the dust. Then the ratio of the radio intensity to the Hα intensity will be

$$\frac{I_\nu}{I_{H\alpha}} = \frac{\int_0^{l_1} \epsilon_\nu ds}{\int_0^{l_1} \epsilon_{H\alpha} e^{-k_\alpha s}\, ds} = \frac{\epsilon_\nu l_1}{(\epsilon_{H\alpha}/k_\alpha)(1 - e^{-k_\alpha l_1})} . \tag{10-3}$$

But since ϵ_ν and $\epsilon_{H\alpha}$ are proportional to $N_i N_e$, we have

$$\frac{I_\nu}{I_{H\alpha}} = \frac{f(T_\varepsilon)\tau_1(\alpha)}{1 - e^{-\tau_1(\alpha)}}, \tag{10-4}$$

where $f(T_\varepsilon)$ is a known universal function and $\tau_1(\alpha) = k_\alpha l_1$. If there is no absorption of light, then $I_\nu/I_{H\alpha} = f(T_\varepsilon)$ merely. Equation (10-4) shows explicitly that the optical depth $\tau_1(\alpha)$, and hence the interstellar extinction of light, may be determined from a measurement of the ratio $I_\nu/I_{H\alpha}$.

We see, then, that the construction of large radio telescopes of high resolving power may be expected to facilitate a thorough study of ionized diffuse nebulae and of interstellar extinction. In conjunction with 21-cm investigations of H I zones (see Chapter IV), radio astronomy will enable us to obtain remarkably detailed information on the nature of diffuse interstellar matter.

11. THE RADIO-STAR HYPOTHESIS

The spectra of the spherical and disk components of the galactic radio emission differ fundamentally, as we have emphasized in Sec. 4, from the spectrum of the flat component. In the preceding section we have discussed the manner in which the flat component is formed from thermal emission by interstellar H II regions. But the spherical and disk components have another peculiarity; their brightness temperature is exceptionally high at low frequencies. At 9.15 Mc/s, for example, $T_b \approx 10^6$ °K. At meter wavelengths the radiation flux from the spherical component, integrated over the whole sky, greatly exceeds the flux from the flat component; in fact, the latter cannot be detected at all with radio telescopes of low resolving power.

Even in 1948 it had been recognized, on the basis of spectrum analysis, that at least a significant part of the spherical component must be ascribed to nonthermal (nonequilibrium) emission. This becomes especially clear if we take the enormous brightness temperatures at low frequencies into account. Since the kinetic temperature T_e of the interstellar medium does not exceed 10 000° to 15 000°K in any of the available data, an assumption that the component in question represents thermal emission by interstellar gas would force T_b to be less than T_e, in sharp contradiction to observation.

The suggestion had been made as early as 1946 that cosmic radio waves result from the aggregate radiation of separate stars in the Galaxy. There is an analogue in optical astronomy, in that the Milky Way represents the integrated effect of stellar radiation. Similarly, it was very natural for the notion of a "radio Milky Way," an aggregate of starlike objects or "radio stars," to come forward at a definite stage in our understanding of the nature of cosmic radio waves.

An impulse toward the further consideration of a radio-star hypothesis had been the parallel development of solar radio astronomy. Solar radiation, as recognized in 1946, comprises two components:

(a) from the quiet sun, and (b) from the disturbed sun. The first component constitutes thermal emission by the outer layers of the solar atmosphere (by the corona, at meter wavelengths). The brightness temperature of this component is determined by the kinetic temperature of the corresponding layers; at meter wavelengths it is of the order of 10^6 °K. The second component is of great intensity, is rather variable with time, and depends critically on solar activity.

Now it has been evident that the combined thermal emission of stars in our Galaxy could not yield the cosmic radio waves in question. Pikelner (1950) has shown that, in the coronas of very different types of stars, the kinetic temperature must be of the same order as in the solar corona. If we regard the radio intensity of a source as proportional to its brightness temperature (Rayleigh–Jeans law), we may write

$$T_b \propto W\langle T_* \rangle ; \tag{11-1}$$

here T_b is the brightness temperature of the sky resulting from the combined radiation of all the stars in the Galaxy, T_* is the mean brightness temperature of a stellar disk, and W is the total dilution coefficient of the stars ($\approx 10^{-14}$). If $T_* = 10^6$ °K, then $T_b \approx 10^{-8}$ °K, which is some ten or fifteen orders of magnitude smaller than the observed brightness temperature of cosmic radio waves. The radio-star concept can therefore be retained only by restricting ourselves to non-equilibrium radiation of discrete stars.

During major solar outbursts, the radio brightness temperature may attain 10^{13} °K, or even more. However, such a high level of solar radio emission will prevail for only some 10 min of an entire 11-yr cycle. The mean brightness temperature of the sun over a cycle may be estimated at 10^8 to 10^9 °K. Thus in a single great outburst the sun may emit a hundred times as much energy as the quiet sun emits in 11 yr!

Yet our sun can hardly be the most powerful radio source of all the stars. It is quite possible, even probable, that certain giant stars with nonstable atmospheres, and peculiar stars in general, are exceptionally active emitters of radio waves—far more active than the sun. This circumstance had been viewed as an eminent supporting argument for a radio-star origin of cosmic radio waves.

A brilliant confirmation of the radio-star hypothesis seemed to be at hand when, in 1946, there were discovered discrete sources of rapidly fluctuating intensity. Some investigators had at once expressed the opinion that these discrete sources were the nearest radio stars. The idea survived stubbornly for quite a long time. As more and more optical identifications—peculiar nebulae and galaxies—were secured for powerful discrete sources, it began to become clear that by no means all the sources could be regarded as "radio stars" responsible for the general field of galactic radio waves. The opportunities for identification of discrete sources with "radio stars" were still further narrowed by the resolution of these sources into Classes I and II. It was seen that the attempts at identification of "radio stars" would have to be confined to Class II sources, for Class I sources represent such peculiar galactic objects as the Crab Nebula. Objects of this type are so rare that their combined radiation would be entirely inadequate to explain the observed intensity of galactic radio emission.

Furthermore, it had no sooner been demonstrated that the intensity fluctuations of discrete sources were of ionospheric origin (see Sec. 5), and that the radio flux from all sources investigated (including the Class II sources) was actually constant to high accuracy, when the radio-star hypothesis met new difficulties. Nonequilibrium radio emission by stars, especially by peculiar stars, must under all conditions be variable to a considerable degree; this is demonstrated by the observed character of solar radio emission.

The radio-star hypothesis was confronted with yet another difficulty when it was understood that the sources of the spherical component of the galactic radio emission are negligibly concentrated to the galactic center. None of the subsystems of stars in the Galaxy is known to exhibit such a distribution.

But all these difficulties emerged gradually. The radio-star hypothesis occupied a commanding position between 1946 and 1953. It was vigorously developed in papers by Unsöld (1949), by Westerhout and Oort (1951), and by various other investigators.

Unsöld had worked out the consequences of an assumption that the "radio stars" were stars of high activity—prolific in spot formation and flare production, and having strongly variable magnetic fields accompanied by corpuscular streaming. Unsöld proposed that not only do galactic radio waves originate in stars of this type, but in fact these stars act as powerful generators of cosmic rays. Proceeding from the available data on the generation of primary cosmic rays by the sun, he secured relations with the radio emission by the disturbed sun. Despite the refutation of many of Unsöld's ideas in the light of more recent developments in the field, his suggestion as to the bearing of galactic radio waves on primary–cosmic-ray sources has proved a highly fruitful one. Conspicuous among the achievements of radio astronomy has indeed been the establishment of such an interrelation— one which, however, differs completely from that proposed by Unsöld.

One consequence of a radio-star hypothesis would be the following: If we regard the "radio stars" and ordinary stars as being equally numerous, then Eq. (11-1) would at once imply that the brightness temperature of the sources is of the order 10^{18} °K, to provide for the observed intensity of cosmic radio emission. The radio brightness of the hypothetical objects, averaged over time, would thus exceed that of the sun by some ten orders of magnitude, and would even exceed the emission during the greatest outbursts by five orders. We cannot but regard this immense value for the radio luminosity as a further objection to a radio-star hypothesis.

Several authors have repeatedly analyzed the observed intensity of galactic radio waves in an effort to deduce statistically the population of "radio stars" in the Galaxy. The methods of Westerhout and Oort (1951) are of interest in this connection. These authors express the brightness temperature T_b of a given region of the sky as

$$T_b = n_0 \int (n/n_0) A T_s ds = n_0 A T_s \int (n/n_0) ds, \qquad (11\text{-}2)$$

where A denotes the area of a "radio-star" disk in square parsecs, T_s the brightness temperature of the disk, and n_s the concentration of "radio stars" in the solar neighborhood. We may evaluate the integral $\int (n/n_0) ds$ as soon as we know the spatial distribution of the sources of the spherical component of the galactic radio emission (see Fig. 57, Sec. 5). In particular, at the galactic poles Fig. 57 yields a value of 7.3 kpc for the integral. At a wavelength of 3 m, $T_b = 450°K$ at the poles, so that by Eq. (11-2) we have

$$n_0 A T_s = 0.062 °K/pc.$$

The quantity $n_0 A T_s$ is a most important characteristic number for "radio stars" near the sun, since it permits the derivation of the radio luminosity, the concentration in space, and the distance of the "radio stars." Oort and Westerhout have shown that the flux F_N from the Nth "radio star" (in order of decreasing flux), and its distance r_N, are given by:

$$F_N = (4\pi/3N)^{2/3} e^{\sigma^2/4} n_0^{-1/3} \, 2k\nu^2 c^{-2}(n_0 A T_s), \qquad (11\text{-}3a)$$

$$r_N = (4\pi/3N)^{-1/3} e^{\sigma^2/4} n_0^{-1/3}; \qquad (11\text{-}3b)$$

here σ denotes the dispersion in the absolute radio luminosities of the "radio stars."

These are very general formulas. They provide a description of the state of affairs which would obtain if a number of discrete sources, distributed in a given way and possessing a given luminosity dispersion, were responsible for the brightness of the sky. Even in optical astronomy Eqs. (11-3) would be perfectly applicable, for the Milky Way represents the compound radiation by the stars of the Galaxy. Interstellar extinction of light would, however, invalidate the formulas, since it has not been allowed for, although interstellar absorption does not affect radio waves of sufficiently high frequency.

Since intense Class II sources (such as Virgo A and Centaurus A) have been explicitly barred from consideration as "radio stars," the flux from the strongest "true radio star," so to speak, may not in any event exceed 3×10^{-24} w m^{-2} (c/s)$^{-1}$ at $\lambda = 3$ m (the more intense sources have all been identified with nebulae and galaxies). If we suppose that $F_s < 10^{-24}$ w m^{-2} (c/s)$^{-1}$ and that the dispersion $\sigma = 0$, Eq. (11-3a) yields $n_0 = 0.015$ pc^{-3}, whence by Eq. (11-3b), $r_s = 3.1$ pc. The concentration of "radio stars" in the solar neighborhood would therefore be only one-sixth that of ordinary stars. If we take into account the spherical distribution in the Galaxy which the "radio stars" presumably follow, we find that their total population would be 3×10^{11}, greater than the number of ordinary stars in the Galaxy. Even a very small luminosity dispersion would sharply increase the population of "radio stars": for example, if the dispersion in log F is ± 1, then $n_0 = 0.8$ pc^{-3} and $r_s = 12$ pc, whence the total population of "radio stars" becomes $\approx 10^{13}$!

If the nearest "radio stars" are not Class II sources at all, then their intensity must not exceed that of the weakest Class II sources. But by diminishing F_N we would increase the population of "radio stars" still further, in accordance with Eq. (11-3a). If we remove the restriction that the discrete sources shall be stellar in character, this difficulty is avoided.

In summary, we find that an explanation of the general field of galactic radio waves in terms of emission from an aggregate of "radio stars" would imply an excessive population of these objects in our stellar system. It has become increasingly clear that the majority of discrete sources cannot be deliberately labeled "radio stars."

This is not at all to suggest that there might not be true starlike objects—genuine radio stars—among the weaker sources, even among those now observable. For suppose there is a star that emits, at radio wavelengths, nonequilibrium radiation of a power similar to that of the sun during its most violent outbursts. The solar flux observed at the Earth may at such moments approach 10^{-15} w m^{-2} (c/s)$^{-1}$ at $\lambda \approx 5$ m. At a distance of 1 pc the flux would be 4×10^{10} times less, or 2.5×10^{-26} w m^{-2} (c/s)$^{-1}$. Large radio telescopes now in operation might be able to measure such a flux at the limit of sensitivity, so that outbursts of radiation on the nearest stars might just be detected. But outbursts on peculiar stars might be considerably stronger than for the sun; though the stars may be tens or hundreds of parsecs distant, their nonequilibrium radiation might well be measurable, in principle. Emission of this sort would evidently be highly variable (true variability, as distinguished from ionospheric effects), since variability of flux is a condition that a discrete source be a true radio star.

Kraus, Ko, and Stoutenburg (1955) have reported the interesting discovery of a variable source of radio waves. Observations were conducted at 242 Mc/s with the helical antenna of The Ohio State University. Large fluctuations were detected in the intensity of a weak source in Hydra, at $\alpha = 8^h19^m \pm 1^m$, $\delta = +8° \pm 3°$. These fluctuations do resemble the scintillation of discrete sources which is due to the ionosphere (Sec. 5). However, the fact that the source was observed far above the horizon ($h \approx 60°$), and that the observations were made at a rather high frequency, renders an ionospheric origin improbable. The prevailing intensity of the source, in the absence of fluctuations, was $\approx 1 \times 10^{-25}$ w m^{-2} (c/s)$^{-1}$; during the 2- or 3-min fluctuations the intensity rose by a factor of five or six. At times no fluctuations were observed at all, but sometimes they were very prominent; they were detected both in the evening and during the night. No correlation was found with solar or ionospheric activity. The authors were of the opinion that these considerations all indicated that a true radio star had been under observation. More extensive study will undoubtedly shed further light on this enigmatic phenomenon.

The radiation flux from powerful discrete sources remains constant, in general, to within 5 to 10 percent (Sec. 5). But there is another source in Hydra, the quite intense Hydra A [$\alpha = 9^h15^m46^s$, $\delta = -11°55'$ (1950)], which constitutes a noteworthy exception to this rule. Mills had found as early as 1952 that the intensity of Hydra A differed by 30 percent on two records of its interference pattern; nothing similar had been detected on any of the other sources observed with the same interferometer.

Since 1954 the Australian observers have systematically examined the variation of the flux from Hydra A (Slee 1955). To eliminate errors which might arise from changes in the sensitivity of the apparatus and from other instrumental effects, powerful sources such as Taurus A were concurrently examined for flux variation. It was found from a series of 200 observations that the variability of Hydra A is far greater than that of the comparison sources. As expected, the variability of the latter was purely an instrumental effect.

In Fig. 117 we reproduce sample tracings for Hydra A and Taurus A, made on two different dates. The tracings reveal a variation of 20 percent in the flux from Hydra A, while that from Taurus A remains constant. Figure 117 also shows that the intensity variations of Hydra A proceed quite slowly; at any rate, the flux remains almost constant over about an hour. The extensive observations permitted the compilation of histograms for the magnitude of the flux from Hydra A and from the comparison source Virgo A (Fig. 118). It will at once be noted that, whereas the flux from Virgo A remains constant to within 10 percent, the flux from Hydra A varies widely. An interesting feature is that no periodic flux variations were observed; the variations seem entirely random. The flux remains approximately

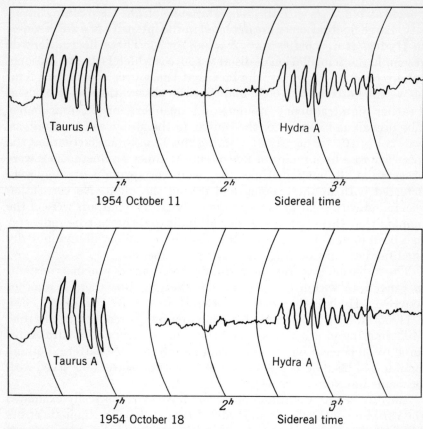

Fig. 117. Records of radio emission from the discrete sources Taurus A and Hydra A (Slee 1955).

constant for 6 to 24 hr. One may assert with considerable confidence that Hydra A is actually variable. This investigation was conducted with far more abundant observational material than had been used by Kraus and his colleagues. The intensity fluctuations themselves differed in character in the two cases: the Americans observed short-period bursts, while the Australians found quite slow random variations.

With the 200-in. Hale reflector, Minkowski discovered two galaxies, very close together, at the position of Hydra A. It is still not certain whether the galaxies are in collision (like Cygnus A), or even whether they may be identified with the radio source. According to interferometric measurements by Carter (1955), the angular dimension of Hydra A is one or two minutes of arc, so that the object is surely nonstellar. V. L. Ginzburg's suggestion that the intensity fluctuations might arise in the interstellar medium is still a possibility.

Seeger, Westerhout, and Conway (1957) have recently made a new

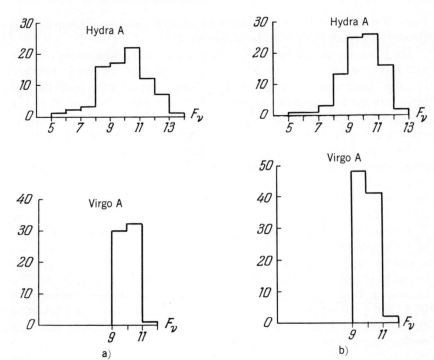

Fig. 118. Histograms showing the number of observations of fluxes F_ν for the discrete sources Hydra A and Virgo A (*a*) when rising, (*b*) when transiting (Slee 1955).

investigation of the suspected variable source Hydra A, using the 25-m Dwingeloo paraboloid at a wavelength of 75 cm. Although the observations were extended over a fairly long period (32 d), the flux variations did not exceed the error of measurement (10 percent). At the International Scientific Radio Union conference in the summer of 1957, Denisse also reported a failure to confirm the variability. As a result the highly important question as to the true variability of Hydra A still remains open. What is required now is several new series of careful observations, preferably at different frequencies.

Attempts to identify discrete sources with stars have continued even in the past few years. Thus, for example, Naqvi and Tandon (1955; Tandon 1956) have sought identifications with M dwarfs. Upon comparison of M-dwarf catalogues with discrete-source catalogues (such as that of Kraus) they found a large number of coincidences. But taking into account the low precision of existing discrete-source catalogues and the great abundance of red dwarfs, these identifications can hardly be regarded as convincing. At one time (1950) the author of this book had also attempted to correlate discrete sources with the red dwarfs closest to the sun. Although he was able to establish a few coincidences of coordinates, he soon became convinced that the attempt would have to be abandoned.

Yet generally speaking red dwarfs are especially interesting objects, and might in fact be powerful generators of radio waves. In particular, the "flare stars" (prototype: UV Ceti), of spectral type dM3e and later, are very numerous. Professor P. P. Parenago believes that the red emission dwarfs constitute a considerable fraction of the entire population of red dwarfs—the most abundant stars in the Galaxy. It is possible that essentially all red emission dwarfs are flare stars to some extent. During the flares of the UV Ceti stars, which last for several minutes, the luminosity rises by a factor of about ten. The flares occur from once to several times a week. Thus the flare phenomenon on such stars is of the same general character as on the sun. The latter is, however, associated with large and very intense bursts of radio emission.

Let us suppose that the radio power during such flares is proportional to the power of the optical radiation emitted in the process. For the surplus optical emission (both line and continuous emission) in large solar flares, this power is $\simeq 3 \times 10^{27}$ erg/s. We shall take the absolute magnitude of a red dwarf as 14 between flares, and 12.5 during flares. Then the power of the optical radiation in a red-dwarf flare will be $\simeq 10^{30}$ erg/s, some 300 times as great as for such large solar flares as the one of 1956 February 23. If now such a dwarf is $\simeq 10$ pc distant from the sun, its radio flux would be detectable with large radio telescopes (see above).

Another class of nonstable stars from which we might expect bursts of radio emission is the T Tauri family (stars associated with dust nebulae). Although T Tauri stars are not common in the Galaxy, perhaps being numbered in the tens of thousands, they may be even more powerful sources of radio waves than the UV Ceti stars; the power of their flares is $\simeq 3 \times 10^{32}$ erg/s.

It is possible that there are discrete sources which would have been identified with stars were it not for the fact that the short duration of their outbursts has caused them to escape detection. If large radio telescopes could be pressed into service in a radio patrol of the sky—making systematic observations of a number of stars, such as nonstable and magnetic stars—some true radio stars might be revealed, provided that their radiation during outbursts is sufficiently intense. Radio astronomy will surely attack this interesting problem in the near future. But in any event the proportion of stellar radio waves in the general field of galactic radio waves must be insignificant, especially if the expected radio intensity in bursts on true radio stars is smoothed over time.

It should furthermore be kept in mind that the radio spectrum of the disturbed sun, which must be analogous to the radio spectrum of flare stars, differs sharply from the spectrum of the nonthermal galactic radio emission. The latter is of the same character in the galactic halo as it is in the disk and the nucleus. The agreement in the spectra of nucleus, disk, and halo argue in favor of an identical nature of the radio sources in all parts of the Galaxy (see Sec. 5). The red dwarfs, on the other hand, are by no means distributed like the sources of radio waves forming the halo. In consequence, even apart from the halo, not even the radio emission of disk and nucleus can be reduced to radiation by any class of star.

We have indicated above that all the objections to a radio-star hypothesis for the origin of galactic radio waves are common to any "discrete" hypothesis—namely, any hypothesis which seeks to reduce the general field of galactic radio waves to emission by some type of discrete source.

An alternative to a discrete hypothesis would be a "continuous" hypothesis, an assumption that sources of nonequilibrium galactic radio emission are *continuously* distributed in interstellar space. Because the nonequilibrium and the spherical components coincide, it would be necessary to postulate, at great distances from the galactic plane, a continuous distribution of radio sources within the interstellar medium.

Whereas the flat component of galactic radio waves is localized in the interstellar medium (Sec. 9), the spherical, on the other hand, is very "patchy," corresponding to the patchy distribution of the interstellar H II regions. What is the mechanism of the emission by the spherical component? It is not related (or very little related) to stars or to clouds of interstellar gas, as is evident from the distribution of the sources of this component. But it is also clear that a continuous space distribution of the sources must be intimately associated with some material agent.

Alfvén and Herlofson (1950) and Kiepenheuer (1950) independently, and almost simultaneously, made a first proposal of a mechanism of nonequilibrium radio emission whose source resided in interstellar space. We refer to emission by relativistic electrons in interstellar magnetic fields. It is no exaggeration to say that the introduction and development of the theory of this mechanism have constituted a decisive advance in our understanding of the nature of cosmic radio waves. In Sec. 13 we shall present the theory of radio emission by the spherical component, as based on this mechanism. But we must first make a preliminary inquiry into the subject of interstellar magnetic fields, so important for the theory.

12. INTERSTELLAR MAGNETIC FIELDS

The problem of interstellar magnetic fields has in recent years developed into one of the most vital of modern astrophysics. Electromagnetic processes assume a distinctive character when operative in cosmic objects. This is the result of two circumstances: high conductivity of matter in stars and in the interstellar medium, and the mobility of the latter.

It can be shown that the relaxation time t for magnetic flux in a conductor of dimension R and conductivity σ is given by

$$t \approx \sigma R^2/c^2. \qquad (12\text{-}1)$$

The conductivity of interstellar gas is $\approx 10^{12}$ sec^{-1}; hence if, for an interstellar gas cloud, $R \approx 10$ pc $\doteq 3 \times 10^{19}$ cm, $t \approx 10^{23}$ yr! It is almost as though interstellar gas were a superconducting fluid: in a superconductor a large value of the relaxation time t would arise from a very high value of σ or, as is the case for cosmic objects, from very high values of R.

On the other hand, like all compressible gases the interstellar medium is mobile. Interstellar gas is in a state of continual motion— in part a regular motion, such as that due to galactic rotation, but mainly a turbulent motion. Since superconductivity is a property of this moving diffuse medium, a singular circumstance arises: the magnetic lines of force become "frozen" into, or "glued" to, the moving gas.

It is as though the motion of the interstellar gas "tangles" or "twists" the magnetic lines of force. If the kinetic energy density $\frac{1}{2}\rho v^2$ of the interstellar medium is much greater than its magnetic energy density $H^2/8\pi$, the magnetic field will have very little influence on the motion of the clouds. In twisting the lines of force, the random motions of the clouds tend to intensify the magnetic field, since the continual twisting of the lines of force merely serves to increase the number of lines through unit cross-sectional area, and this is nothing but the measure of the magnetic field intensity.

As the magnetic-energy density approaches the kinetic-energy density of the medium, the magnetic field begins to gain mastery over the character of the motion. If the motion takes place in a homogeneous magnetic field in which $H^2/8\pi > \frac{1}{2}\rho v^2$, even small deformations in the lines of force of the moving gas will generate magnetic fields in such a manner as to oppose further deformations. The motion of the gas will then be almost completely under the control of the magnetic field; it will proceed primarily along the lines of force of the external field.

If the twisting of magnetic lines of force continues for a sufficiently long time in a turbulent conducting gas, the kinetic and magnetic energy densities will become equal. It is not meaningful to speak of an initial value of the magnetic field, at a time when the twisting process sets in. One may always assume that at least a very weak magnetic field exists; for example, a magnetic field of insignificant intensity may arise from local kinetic-temperature or pressure gradients in the interstellar medium. A weak field will continually strengthen through interaction with a conducting medium in random motion, and will eventually attain the equilibrium value

$$H = (4\pi\rho)^{1/2}v, \tag{12-2}$$

where v is the mean speed of the random motions of the gas.

We see, then, that general theoretical considerations lead us to expect magnetic fields in interstellar space. The intensity of these fields may be evaluated by Eq. (12-2) if we accept observationally derived values of $\rho \approx 10^{-23}$ gm/cm^3 for the density of matter in interstellar gas clouds, and $v \approx 10^6$ cm/s for the random velocity of individual clouds. These data yield a value of $\approx 10^{-5}$ gauss for H. This field must be oriented chaotically: to a first approximation the magnetic field must be homogeneous within each cloud.

So weak a magnetic field cannot be measured directly (as by Zeeman splitting of spectral lines). Indirect methods are essential in an observational determination of the intensity and the character of interstellar magnetic fields.

One indirect method is an analysis of the problem of the retention of primary cosmic rays within the Galaxy. We may use the observed isotropy of primary cosmic rays as a basis for three possible assumptions as to their localization in space: (a) they are of solar origin, and are constrained by magnetic fields to lie within a relatively limited region (10^3 to 10^4 AU, say) about the sun; (b) they are localized in the Metagalaxy; or (c) they are localized in the Galaxy, and are retained within it by magnetic fields which ensure the observed isotropy of the cosmic rays.

Hypothesis a can hardly be a justifiable one, despite the presence of active areas on the sun which, as is now firmly established, are sources of primary cosmic rays. In particular, a solar hypothesis would be unable to explain the occurrence in the primary rays of particles carrying energies up to 10^{18} ev. Using the expression

$$r = E/(300\,H) \tag{12-3}$$

for the radius of curvature of the path of such a particle in a magnetic field (here the energy E is in electron volts), we would have $r \doteq 3 \times 10^{20}$ cm $\doteq 100$ pc if $H = 10^{-5}$ gauss. Furthermore, the field must bear a highly special character to ensure the retention and the isotropy of the cosmic rays.

Hypothesis *b,* the suggestion of an extragalactic origin for primary cosmic rays, seriously conflicts with certain energy considerations. If we fill a volume large compared with our Galaxy with cosmic rays of the energy density observed at the Earth, the cosmic-ray energy enclosed within the volume will exceed by four orders of magnitude the radiation energy within the same volume. This means that primary cosmic-ray sources are required whose energy is four orders greater than the energy of radiation of all the stars; it is difficult to conceive of such sources.

The difficulty is removed by assuming a galactic localization of cosmic rays. Primary cosmic rays generated in some type of source will, on account of the presence of a magnetic field, be retained within the Galaxy over a period of time which is determined by collisions of cosmic-ray particles with interstellar atomic nuclei. The collisions lead to disintegration into showers of particles, and to a substantial decrease in the energy of the rays. Let us take 0.1 cm^{-3} as the mean concentration of interstellar atoms. Since the effective cross section for a cosmic-ray collision is $\approx 2.5 \times 10^{-26}$ cm^2, the "lifetime" of cosmic-ray protons will be very great—some 4×10^9 yr. As a result the sources need only be of relatively low power to produce cosmic rays of the observed density.

To retain particles of 10^{18}-ev energy within the Galaxy, the intensity of the magnetic field must be at least 10^{-7} gauss according to Eq. (12-3). Pikelner (1953) has considered the problem of retention of cosmic rays, obtaining two important results. First, the intensity of the magnetic field must be $\geq 3 \times 10^{-6}$ gauss in order to retain cosmic rays within the Galaxy for $\approx 10^8$ yr. Second, it seemed necessary to assume that randomly oriented magnetic fields were found not only in clouds of interstellar gas (and thus in the vicinity of the galactic plane), but also in gaps between the clouds. Otherwise, to confine cosmic rays to the Galaxy would be like an attempt to hold water in a sieve; cosmic-ray particles would pass almost unimpeded into extragalactic space if their velocity vector were directed at a sufficiently great angle to the galactic plane.

But chaotic magnetic fields cannot exist apart from interstellar matter; the approximation $H^2/8\pi \approx \frac{1}{2}\rho v^2$ must be fulfilled. Since $\rho < 10^{-25}$ gm/cm^3 in the interstellar medium, the speed of random motions in the highly rarefied medium between the clouds must be of the order of 30 to 100 km/s, in order that the field may attain $\approx 10^{-5}$ gauss (and such fields are necessary if the cosmic rays are to be retained for a long time). Such a high-velocity dispersion would, however, induce the rarefied gas to rise to vast distances above the galactic plane. Consequently the highly rarefied intercloud gas, and the randomly oriented magnetic fields associated with it, must assume the form of a gigantic spherical corona surrounding the nucleus of our Galaxy.

The agreement between the characteristics of this magnetized gaseous corona and the spherical component of the galactic radio emission is highly significant. We shall return to this matter in Sec. 13.

The presence of magnetic fields in interstellar space was demonstrated in 1949, when Hiltner (1949, 1951), Hall (1949), and Hall and Mikesell (1949, 1950) made the important discovery that the light of some stars is *polarized*. The degree of polarization is usually very small (2 or 3 percent), although in some cases it may rise to 8 or 10 percent. Quite distant giant stars have been observed to be polarized. The plane of vibration of the electric vector forms a small angle with the direction of the galactic equator, in most cases. This effect is most readily observed in those parts of the sky where the line of sight intersects the spiral arms of the Galaxy at a sufficiently great angle (Fig. 119). In those directions where the line of sight extends along a spiral arm, the direction of vibration is very poorly correlated with the direction of the galactic equator (Fig. 120).

Since the directions of polarization are found to be almost the same for many stars in the same part of the sky, it is natural to seek a cause for the polarization not in the stars themselves, but in the interstellar medium traversed by their light. Now absorbing interstellar dust exerts an influence as the light proceeds through the medium. If this absorption differs for light rays with differently oriented electric vectors, the stellar radiation should be partially polarized after its passage through the interstellar dust. In this event the degree of polarization should be correlated with the absorption, and hence with the reddening of the stars (for the absorption coefficient is inversely proportional to the wavelength). In Fig. 121 is plotted the dependence of degree of polarization on absorption for a group of stars in the direction of the galactic center. A correlation undoubtedly exists despite the large scatter. The observations suggest only that the presence of absorption is necessary for starlight to be polarized, not that it is sufficient.

We may therefore regard it as reliably established that polarization of starlight is caused by cosmic dust. The polarization may be thought of as arising in the following way. Let us suppose that the dust grains are elongated rather than spherical; then light will be absorbed more strongly by the grain if the direction of vibration of its electric vector is parallel to the axis of the grain. If the axes of the grains are randomly oriented, polarization will evidently not be observed. The observational results may be interpreted if we assume that the axes of

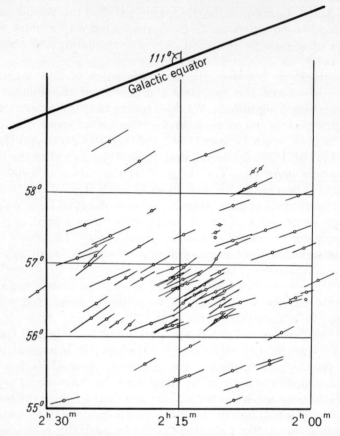

Fig. 119. Polarization of stellar radiation in a region where the line of sight intersects a spiral arm. The plane of vibration of the electric vector is approximately the same as the galactic plane.

the grains are oriented normal to the spiral arms. Thus within a given cloud, the orientation of the axes will be almost the same, whereas the orientation will differ from cloud to cloud. If a ray of light traverses several clouds, a "smoothing" may therefore take place, so that the final degree of polarization will appear to be insignificant even though the light has suffered considerable absorption.

The quantitative theory of the polarization of light upon passage through a cloud of identically oriented dust grains is very intricate; it can be treated only by means of a series of approximating assumptions, and the problem still awaits a complete solution. The most thorough investigation yet made in this area is that of Davis and Greenstein (1951); according to their calculations, the degree of polarization observed may be accounted for by ice needles, if they are

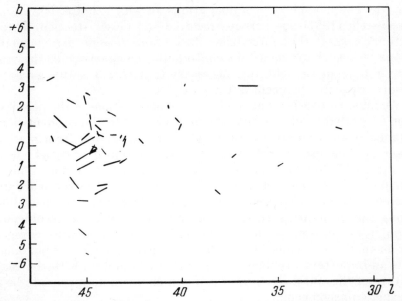

Fig. 120. Polarization of stellar radiation in a region where the line of sight extends along a spiral arm (Hiltner 1951).

sufficiently well aligned. To a first approximation, the degree of polarization is then independent of wavelength, in agreement with observations.

Why should solid interstellar particles be aligned at all? Of all the mechanisms which have been proposed, only interstellar magnetic fields will guarantee this orientation. There are several hypotheses

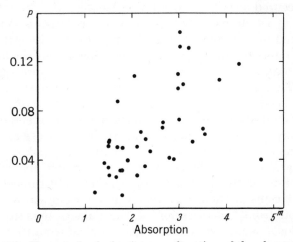

Fig. 121. Degree of polarization as a function of the absorption.

for a magnetic alignment of dust grains, of which that of Davis and Greenstein (1951) has perhaps received the most attention. These authors suggest that interstellar dust grains possess paramagnetic properties; such grains will be continually remagnetized as they rotate in a magnetic field. It is necessary that $H \simeq 3 \times 10^{-5}$ gauss to ensure adequate alignment of the particles.

We must emphasize that the foregoing observations of the polarization of starlight indicate that the dust grains are almost identically aligned over the entire trajectory of the light ray, from star to observer. The distances often exceed 1 kpc, so that if a magnetic field is responsible for the alignment of the grains it must remain almost homogeneous over a vast span of the Galaxy. The observational data therefore favor the presence of a regular magnetic field extending along the spiral arms. In turn, the spiral arms may be graphically thought of as magnetic tubes of force. But the exact nature of regular magnetic fields embedded in the spiral structure remains puzzling; it is an important problem, one intimately bound up with the very nature of the spiral structure itself, which, at root, appears to be a magnetic phenomenon.

G. A. Shajn (1955) has developed a noteworthy method for studying the interstellar magnetic field. He has directed attention to the existence of highly elongated bright nebulae (such as those observed in Hα light) as well as dark dust nebulae. Nearby nebulae are usually elongated in a single direction, and this is preferentially parallel to the galactic equator. We may explain these distinctive formations by postulating that the nebulae are embedded in an external magnetic field, which controls their expansion. A nebula may spread out only along the magnetic lines of force, as noted above.

Shajn secured a striking correlation between the direction of elongation of the nebulosity and the direction of polarization of light from stars located in the same part of the sky. Especially strong correlation was found for directions of elongation highly inclined to the galactic equator; the nebulae involved were at quite high galactic latitudes. This situation is illustrated in the lower portion of Fig. 122, where the dashes denote magnitude and direction of polarization. The nebulae in the figure (mainly dark nebulae) extend over a large region of sky in Perseus, Taurus, and Auriga. So close is the relation between the directions of nebular elongation and polarization of the nearest stars, that it seemed possible to predict the directions of polarization for individual stars from the forms of the nebulae alone. Large deviations of the directions of nebular elongation and stellar polarization from the direction of the galactic equator testify to extensive local departures of the magnetic field from the regular field of the spiral arms.

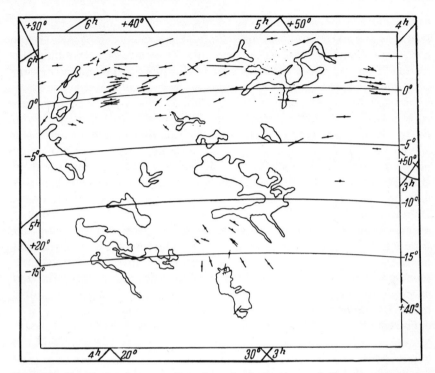

Fig. 122. Correlation between direction of polarization and direction of nebular elongation (Shajn 1955).

Radio astronomy can provide us with a method for measuring the intensity of the very weak interstellar magnetic fields. The method to which we refer is the analogue of a basic optical technique for deriving the magnetic fields present in cosmic objects; it utilizes the Zeeman effect.

Bolton and Wild (1957) first noted the possibility of using the 21-cm hydrogen radio line for this purpose. In the presence of a magnetic field the 21-cm line is split into the three components of a normal Zeeman triplet (Nafe and Nelson 1948); there are a π-component, at the "normal" field-free frequency ν_0 of the radio line, and two σ-components, at $\nu_0 \pm \varepsilon H/(4\pi mc)$. Here H denotes the longitudinal component of the field. The frequency interval $\Delta\nu$ between the two σ-components of the Zeeman triplet is $2.8 \times 10^6\,H$ c/s; thus if $H = 10^{-5}$ gauss, $\Delta\nu \approx 30$ c/s. Although this represents a very small amount of splitting, it can in principle be detected by radio-astronomical methods now in use. Bolton and Wild believe that 21-cm emission lines, which have been detected from many quarters of the Galaxy, will be of no assistance in this regard since they are too broad. However, as we shall see in Sec. 17, the radio spectrum of the source Cassiopeia A exhibits very narrow ($\delta\nu = 10$ kc/s) absorption lines in the 1420 Mc/s frequency region. The lines arise from separate clouds of interstellar gas; the velocity dispersion is very small (1.5 to 2 km/s) within each cloud. We may expect the intensity of the interstellar magnetic field to remain fairly constant throughout an individual cloud.

Nonetheless the observation is a difficult one even for lines as narrow as these; one must detect a line shift less than 1 percent of the line width. A problem of comparable difficulty has, however, been solved in optical astronomy. Bolton and Wild's proposal for measuring such small Zeeman shifts at radio wavelengths was motivated by the fully analogous method employed optically by Babcock (1953). It is known that the two σ-components of the Zeeman triplet are circularly polarized, but with opposite senses of rotation. The frequency of a narrow-band receiver is therefore set at the edge of the absorption line, where the profile attains maximum steepness; the polarization characteristic of the antenna is now switched continually, so that the receiver alternately accepts one, then the other, circularly polarized component. The power delivered to the output at the switching frequency is given, in terms of the antenna temperature T_a, by the expression

$$\Delta T = T_a \Delta\nu/\delta\nu,$$

where ΔT is the maximum decrease in antenna temperature in that portion of the radio spectrum where the absorption line is located. Since $\Delta\nu/\delta\nu = 2.8 \times 10^6\, H/10^4 \approx 3 \times 10^{-3}$, T_a must be quite high. At the present time it is necessary that $\Delta T \geq 1$ K deg for a reliable measurement; hence we must have $T_a \approx 1000°$K, which may in fact be attained with large radio telescopes. For example, a value $T_a \approx 2500°$K would be anticipated for the 250-ft Jodrell Bank paraboloid. Thus the effect might well be detected.

We may very soon look forward to an entirely new order of receiver sensitivity, through application of molecular amplifiers. These instruments can measure an increment $\Delta T \approx 0.1$ or even 0.01 K deg. It will not only be possible to employ the hydrogen absorption radio lines of discrete sources in a study of the interstellar magnetic field intensity; even emission lines from individual clouds of interstellar matter may be used. Individual clouds are readily observable at sufficiently high galactic latitudes. Their 21-cm lines are comparatively narrow, for the line width is a function only of the internal velocities.

The Bolton–Wild proposal of a radio technique for measuring the interstellar magnetic field intensity affords an opportunity for a substantial advance in our understanding of the nature of interstellar matter. It is particularly important to note that this method for determining the magnetic field strength is a *direct* one, in contradistinction to the various other methods outlined above.

To summarize our conclusions as to the interstellar magnetic field, the Galaxy possesses both a regular magnetic field, which extends along the spiral arms, and chaotic magnetic fields, which form a spherical system. The origin of the regular field and the interdependence between the regular and chaotic fields are not understood.

13. THE NONTHERMAL COMPONENT OF GALACTIC RADIO WAVES

Two independent papers appeared in 1950, suggesting a remarkably efficient mechanism for the generation of cosmic radio waves. The first paper, by Alfvén and Herlofson (1950), treated the problem of a mechanism for emission by radio stars (even though at the time hardly anything was known of the nature of radio stars!), while in the other and slightly later paper, Kiepenheuer (1950) discussed a mechanism for nonthermal galactic radio emission.

Alfvén and Herlofson believed—well-advisedly, as more recent investigation has shown—that the "radio stars" could not be objects of stellar character. They regarded the radio stars as consisting of rather extensive ($\approx 10^4$ AU) regions surrounding a star, regions apparently containing an intricately "tangled" magnetic field of fairly high intensity. Relativistic electrons (those whose kinetic energy $E \geq mc^2$) move through the field. Physicists had known for 40 years that under these conditions the radiation would possess a number of singular properties, which we shall discuss presently. It is essential for the frequency of the radiation from a relativistic electron in a magnetic field to exceed the cyclotron frequency $\varepsilon H/mc$. The phenomenon is observed in large synchrotrons, where visible light is radiated by an electron beam in a magnetic field. Alfvén and Herlofson estimate the energy E of the electron to be 1.5×10^8 ev if $H = 3 \times 10^{-4}$ gauss. One circumstance favoring the development of this comparatively high value for H would be the presence of an ample relative velocity between a star and an interstellar gas cloud. Thus the Alfvén–Herlofson position is that the "radio stars" are clouds of interstellar gas in rapid motion with respect to some type of star.

These views as to the nature of radio stars are very naïve ones, to be sure, and are not at all in accord with the views held today. Yet the Alfvén–Herlofson paper constituted the first relevant attempt to apply the theory of radiation from accelerated electrons in magnetic fields, a concept which has proved exceptionally fruitful.

The paper by Kiepenheuer, which appeared a few months after the Alfvén–Herlofson paper, suggested a new mechanism to explain galactic radio waves. Primary cosmic rays (more exactly, their electron components) are, according to Kiepenheuer's careful discussion, the sources of the nonequilibrium galactic radio emission. By employing the formulae for radiation from relativistic electrons, and by taking the concentration of relativistic electrons in interstellar space to be the same as that of heavy relativistic particles ($\approx 3 \times 10^{-11}$ cm^{-3}), Kiepenheuer found that the intensity of their radio emission would be of the same order as the observed intensity. It was assumed that the thickness of the radiating layer in the Galaxy is ≈ 300 pc, the intensity of the magnetic field is $\approx 10^{-6}$ gauss, and the energy of the relativistic electrons is $\approx 10^8$ ev. Kiepenheuer was therefore the first to show that nonequilibrium galactic radio emission might be regarded as a by-product of primary cosmic rays traveling through interstellar magnetic fields.

Beginning in 1951, a series of papers was published in the Soviet Union (Ginzburg 1951, 1953a; Getmantsev 1952; Ginzburg and Fradkin 1953), developing the Alfvén–Herlofson and Kiepenheuer proposals, and rendering them more precise. In the West, however, these ideas had meanwhile failed to attract the interest and attention which they surely deserved. The Soviet investigations had led to a refined theory for this synchrotron-radiation process; at the present time the theory is almost universally recognized. Therefore we shall now give a systematic exposition of this theory.

From classical physics, we know that an electron moves through a homogeneous magnetic field along a helical trajectory, spiraling around the lines of force. The motion is compounded of a progression along the lines of force, together with a rotation about them at the angular frequency of Larmor precession:

$$\omega_H = \frac{\varepsilon H}{mc};\qquad(13\text{-}1)$$

here H denotes the intensity of the component of the field normal to the instantaneous velocity of the electron. The electron radiates like a dipole with frequency ω_H.

But the radiation from the electron is entirely different at relativistic energies ($E > mc^2$), since radiation of electromagnetic waves by a relativistic particle is highly nonisotropic. Basically, the radiation from a relativistic electron is concentrated within a cone of vertex angle $\theta \approx mc^2/E$. Since $mc^2 \doteq 5 \times 10^5$ ev, we have, in the case of an electron of energy $E = 10^9$ ev, $\theta \approx 5 \times 10^{-4}$, or only $\approx 1'.5$! Thus if an electron moves along a trajectory, it will radiate in directions close to the direction of its instantaneous velocity.

Now suppose that an observer is located in the electron's orbital plane; he will perceive radiation only when the instantaneous velocity of the electron is directed toward him. As a result, the observer will record pulses of radiation, each of duration

$$\Delta t \approx (R\theta/c)(mc^2/E)^2.$$

The radius of curvature R of the orbit of an electron in a magnetic field is

$$R \approx c/\omega_H,$$

and in the relativistic case

$$\omega_H = (\varepsilon H/mc)\,(mc^2/E);$$

hence

$$\Delta t \approx \frac{mc}{\varepsilon H}\left(\frac{mc^2}{E}\right)^2.$$

These pulses recur with a frequency

$$\nu_H = \omega_H/2\pi.$$

This "interrupted" radiation may be regarded as a collection of harmonics of fundamental frequency ω_H. Actually this is tantamount to saying that the radiation will comprise all frequencies from zero to infinity, but with different weight; maximum radiation will occur at a frequency $\omega \approx 1/\Delta t$. Whereas a classical electron in a magnetic field radiates the single frequency $\varepsilon H/mc$ of Larmor precession, we now see that a relativistic electron will radiate a *continuous spectrum*. The problem is to find the distribution with respect to frequency of the energy of this radiation.

This problem had long ago been solved by Schott (1912), but his results had been quite forgotten. They were partially rediscovered by various authors, whose papers developed further the theory of radiation from relativistic particles in a magnetic field (Artsimovich and Pomeranchuk 1946; Ivanenko and Sokolov 1948; Vladimirsky 1948; Schwinger 1949; and others).

Our succeeding analysis is based on Vladimirsky's paper (1948). Schwinger (1949) has obtained practically the same results; it is his formulation which was used by Kiepenheuer (1950).

Vladimirsky shows that the energy radiated in 1 sec in the frequency interval $(\nu, \nu + d\nu)$ by a relativistic electron moving in a magnetic field is given by

$$P(\nu, E)\,d\nu = 2\pi P(\omega, E) = 16\varepsilon^3 H(mc^2)^{-1}p(\omega/\omega_m)$$
$$= 16(\varepsilon^2/c)\omega_H(\tfrac{1}{2}\omega/\omega_m)^{1/3}Y(u), \tag{13-2}$$

where

$$\omega \equiv \omega_H = (\varepsilon H/mc)\,(mc^2/E), \quad \omega_m = (\varepsilon H/mc)\,(E/mc^2)^2,$$
$$u = (\tfrac{1}{2}\omega/\omega_m)^{2/3}, \tag{13-3}$$

and H denotes the component of the magnetic field normal to the velocity of the electron. It is assumed that $E/(mc^2) > 1$. In limiting cases, the functions $p(\omega/\omega_m)$ and $Y(u) = p(\omega/\omega_m)2u^{3/2}u^{-1/2}$ reduce to the expressions:

$$p(\omega/\omega_m) = 0.256(\omega/\omega_m)^{1/3}, \qquad Y(u) = 0.256 \qquad [\omega/\omega_m \ll 1]; \quad (13\text{-}4a)$$

$$\left.\begin{array}{l} p(\omega/\omega_m) = \tfrac{1}{16}(\pi\omega/\omega_m)^{1/2} \exp[-2\omega/(3\omega_m)], \\[4pt] Y(u) = \tfrac{1}{16}(2\pi)^{1/2}u^{1/4} \exp[-\tfrac{4}{3}u^{3/2}] \end{array}\right\} \qquad [\omega/\omega_m \gg 1]. \quad (13\text{-}4b)$$

The function $p(\omega/\omega_m)$ attains a maximum at $\omega/\omega_m = 0.5$; its value is $p_{\max} = 0.10$. We conclude that a relativistic electron moving in a magnetic field radiates, basically, in a region of the spectrum near

$$\omega_m = \frac{\varepsilon H}{mc}\left(\frac{E}{mc^2}\right)^2.$$

The frequency of the radiation may therefore exceed the classical frequency $\varepsilon H/mc$ of Larmor precession by many orders of magnitude. Thus relativistic electrons of sufficiently great energy may radiate at quite high frequencies even in very weak magnetic fields. For example, $H = 10^{-5}$ gauss, a nonrelativistic electron can only radiate with the frequency $\nu_H = \varepsilon H/(2\pi mc) \approx 30$ c/s, about the same as for a very low-frequency sound wave. But in the same field, a relativistic electron, moving with an energy of 10^9 ev, will radiate with a frequency $\approx 1.2 \times 10^8$ c/s $= 120$ Mc/s! We have thus already reached the 2.5- to 3-m wavelength range, observable by radio astronomy.

It is clear from Eqs. (13-2) and (13-4) that if $\omega \ll \omega_m$ the spectral density of the radiation varies quite slowly, whereas if $\omega \gg \omega_m$ it falls off exponentially with ω. Vladimirsky has tabulated the function p at intermediate values ω/ω_m; we supply the graph as Fig. 123.

Let us regard the observed galactic radio emission from interstellar space as consisting of deceleration radiation by relativistic electrons in weak, randomly oriented magnetic fields. The intensity I_ν of the radiation will be

$$I_\nu = 2kT_b\lambda^{-2} = (1/4\pi) \int\int P(\nu, E)N(E, R)dEdR, \quad (13\text{-}5)$$

where $N(E, R)$ denotes the number of relativistic electrons per unit volume within the energy interval $(E, E + dE)$, and at distance R along the line of sight from the observer. The factor $1/4\pi$ in Eq. (13-5) arises from our assumption that the magnetic fields are randomly oriented, for the electron radiation may then be considered practically isotropic.

We shall also assume that the radio waves generated by the interstellar relativistic electrons are not absorbed by the interstellar medium, although, as pointed out in Sec. 9, absorption may occur in H II regions. Note that ionized gas clouds are strongly concentrated to the galactic plane. Since the absorption coefficient for

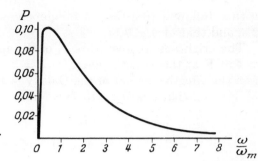

Fig. 123. The function $p(\omega/\omega_m)$.

radio emission by an ionized gas is proportional to the inverse square of the frequency [Eq. (9-32)], interstellar space is essentially transparent at sufficiently high frequencies. The discussion at the end of Sec. 9 shows that significant absorption of radiation distributed over the galactic plane may be expected only if $\nu < 50$ Mc/s. Therefore interstellar space must, at high galactic latitudes, be quite transparent at all radio frequencies, up to the very lowest.

We next assume that the energy spectrum $N(E)$ of the interstellar relativistic electrons is of the form

$$N(E) = KE^{-\gamma}. \tag{13-6}$$

This is substantiated by the energy distribution of primary cosmic rays (nuclei of various elements, mostly hydrogen) as observed at the Earth; a power dependence is exhibited. A final assumption is that the energy spectrum of relativistic electrons is constant throughout the interstellar medium.

One then arrives at the following expressions for the intensity:

$$I_\nu = 2kT_b\lambda^{-2} = (R/4\pi)\int_0^\infty P(\nu, E)N(E)dE$$

$$= \frac{3}{\pi}(2\pi)^{(1-\gamma)/2}\frac{\varepsilon^3 H}{mc^2}\left(\frac{2\varepsilon H}{m^3 c^5}\right)^{(\gamma-1)/2} Ru(\gamma)K\nu^{(1-\gamma)/2}$$

$$\approx 1.3 \times 10^{-22}(2.8 \times 10^8)^{(\gamma-1)/2}u(\gamma)KH^{(\gamma+1)/2}R\lambda^{(\gamma-1)/2}$$
$$\text{erg cm}^{-2}\text{ sec}^{-1}\text{ sterad}^{-1}. \tag{13-7}$$

The function

$$u(\gamma) = \int_0^\infty Y(u)u^{(3\gamma-5)/4}du$$

assumes the values 0.37, 0.163, 0.125, 0.087, 0.153 for values of γ of 1, 5/3, 2, 3, 7.

Now a dependence $T_b \propto \lambda^{2.8}$ is observed for the spherical component of cosmic radio waves, so that $I_\nu \propto \lambda^{0.8}$. If we are to identify the spherical component alone with radiation from interstellar electrons, it then follows [from the $\lambda^{(\gamma-1)/2}$ dependence in Eq. (13-7)] that $\gamma = 2.6$, and that $u(\gamma) \doteq 0.11$.

The brightness temperature T_b of the spherical component alone is $\approx 450°$K at the galactic poles, for $\lambda = 3$ m (see Sec. 4). In this direction the effective extent of the Galaxy is ≈ 7 kpc $\doteq 2 \times 10^{22}$ cm (Fig. 57, Sec. 5). Hence by Eq. (13-7) we have

$$I_\nu = 4.2 \times 10^7 H^{1.8} K$$

$$= 2kT_b\lambda^{-2} \approx 1.4 \times 10^{-18} \text{ erg cm}^{-2} \text{ sec}^{-1} \text{ sterad}^{-1}.$$

Taking $H = 10^{-5}$ gauss (Sec. 12), we obtain

$$K \approx 3.3 \times 10^{-17} \text{ cgs} \doteq 260 \text{ ev}^{1.6}/\text{cm}^3. \tag{13-8}$$

For the concentration of relativistic electrons of energy exceeding E_0 ev we have

$$N(E > E_0) = \int_{E_0}^{\infty} N(E) \, dE \approx (260/1.6)E_0^{-1.6} \text{ cm}^{-3}. \tag{13-9}$$

If $E_0 = 10^9$ ev, then $N(E > 10^9 \text{ ev}) \approx 4 \times 10^{-13} \text{ cm}^{-3}$. Again, the concentration of relativistic electrons of energy exceeding $E_0 = 3 \times 10^7$ ev will be $\approx 6.7 \times 10^{-11} \text{ cm}^{-3}$, which is of the same order as the concentration of heavy nuclei in the primary cosmic rays observed at the Earth. The cosmic rays are primarily protons of kinetic energy $> 5 \times 10^8$ ev.

We see, therefore, that a very small number of relativistic electrons moving through randomly oriented interstellar magnetic fields will not only ensure the production of radio waves of the observed intensity, but, fully as important, will yield the observed spectral composition. In his pioneering paper, Kiepenheuer (1950) had already concluded that an insignificant concentration of relativistic electrons in the interstellar medium would generate sufficient radiation to satisfy the observations. But it remained for Ginzburg, Getmantsev, and Fradkin to provide an adequate foundation for the theory, by demonstrating that the spectral composition of the radiation from magnetically decelerated relativistic electrons agrees with that of the cosmic radio waves observed.

Directly after S. B. Pikelner's discovery of a spherical system of rarefied intercloud matter and of the random magnetic fields associated with it, the author was able to identify the system with the spherical component of galactic radio waves (Shklovsky 1952*b*, 1953*a*). The very fact that a system of random interstellar magnetic fields appeared to coincide in space with a system of nonequilibrium radio sources argued strongly in favor of a magnetic-deceleration origin for the radio emission. Additional support for this identification has since been provided by extensive observations of this exceptional distribution of nonequilibrium galactic radio sources (see Secs. 4, 5, and 23).

A factor of great significance for the development of the above theory is the question of the electron component of the primary cosmic rays observed at the Earth. Electrons have not yet been detected in primary cosmic rays. The available experimental data indicate that the primary particles in the soft component (electrons, positrons, photons), of energy greater than 1.1×10^9 ev, comprise no more than 0.6 percent of all primary particles of energy greater than 10^9 ev (Critchfield, Ney, and Oleksa 1952). It might have been expected that conditions for observing the electron component of primary cosmic rays would be especially favorable near the geomagnetic poles, where low-energy cosmic rays might be incident. But recent observations have revealed a high-latitude cut-off in the cosmic rays, demonstrating that for some reason particles of energy below some limiting value will not reach the surface, even at high geomagnetic latitudes. For protons, the limiting kinetic energy is $\approx 5 \times 10^8$ ev; for electrons, $\approx 1.2 \times 10^9$ ev. The fraction 0.6 percent of the total particle concentration, 1.2×10^{-10} cm^{-3}, at high geomagnetic latitudes, does not contradict our previous value for the concentration of cosmic electrons having $E > 10^9$ ev, namely, 2×10^{-13} cm^{-3}. Furthermore, radio astronomy has shown that if we could increase the sensitivity of the techniques for recording primary cosmic electrons by only one order of magnitude, then electrons could in fact be detected in the primary component. Such observations are urgently required both for the theory of the origin of cosmic rays and for radio astronomy.

Ginzburg (1953a) has shown that relativistic electrons must be formed continuously in the interstellar medium by secondary processes—collisions of relativistic protons with atoms of the interstellar medium. Mesons are formed in such collisions, and decay into electrons and positrons. In each collision electrons and positrons acquire about 5 percent of the energy of the primary heavy relativistic particles. A relativistic positron moving in a magnetic field will radiate in the same way as a relativistic electron. Because of the extremely low density of the interstellar medium, positrons can survive for a considerable period of time.

If emission of radio waves in magnetic fields did not diminish the energy of the interstellar relativistic electrons and positrons which arise in this way, their concentration at the Earth would constitute several percent of the total concentration of primary cosmic rays of energy $> 10^9$ ev. Radio astronomy can thus explain why so few high-energy electrons are found in primary cosmic rays.

Although secondary processes in the interstellar medium must inevitably generate relativistic electrons, there is a more powerful source of these particles. It will be treated in detail in Chapter V.

The magnetic-deceleration theory of the spherical component of galactic radio waves can also be verified in the following way, a procedure of considerable interest. Getmantsev and Razin (1956) have considered whether, on the assumption that the radio waves arise from relativistic electrons moving in interstellar magnetic fields, the radiation might be polarized. We have pointed out that such electrons radiate within a cone of vertex angle $\theta \approx mc^2/E$, whose axis coincides with the direction of instantaneous velocity. This radiation is linearly polarized; the plane of vibration of the electric vector coincides with the osculating plane of the electron's trajectory. Since the magnetic field may be taken as homogeneous over a sufficiently large cell of turbulence in the interstellar gas, the radiation from such a cell will be almost 100-percent linearly polarized. The observer will detect radiation only from electrons moving nearly parallel to the magnetic field, so that the plane of polarization will be determined by the line of sight and a line normal to the direction of the magnetic field. Since the direction of electric vibration will differ for the electrons along the line of sight, a smoothing will occur: we may expect the radiation to be partially polarized, with, in general, a low degree of polarization.

Let radiation from n_0 "clouds" be incident on the antenna within the cone formed by its directional diagram, and let the radiation from the ith cloud be linearly polarized in some direction, with intensity I_i. If our coordinate axis is directed along the line of sight (along the antenna axis), then the degree ρ of polarization of the incident radiation is by definition

$$\rho = (I_x - I_y)/(I_x + I_y). \tag{13-10}$$

Calculations by Wild (1952) lead to the expressions

$$\langle \rho \rangle = \tfrac{1}{2}\left(\pi \sum_{i=1}^{N_0} I_i^2\right)^{1/2}\bigg/ \sum_{i=1}^{N_0} I_i,$$

$$\langle \rho^2 \rangle = \sum_{i=1}^{N_0} I_i^2 \bigg/ \left(\sum_{i=1}^{N_0} I_i\right)^2. \tag{13-11}$$

It may be shown that, under certain natural assumptions,

$$\langle \rho \rangle \approx l/(R\,\theta^{1/2}), \tag{13-12}$$

where l is the dimension of an element over which the magnetic field may be regarded as constant, R is the extent of the Galaxy in the direction in question, and θ is the antenna beamwidth. If we take, for example, $l = 10$ pc, $R = 5$ kpc, and $\theta = 10°$, then $\rho \doteq 0.5$ percent.

However, it is possible that l may be considerably greater, say 50 to 70 pc. Such a degree of polarization may be measured at meter wavelengths, if high-sensitivity techniques are employed.

Equation (13-12) assumes that the state of polarization remains invariant as radiation from an element of the interstellar medium travels toward the observer. This is not generally true, however. Ionized interstellar gas in a magnetic field represents a magnetically active medium for an electromagnetic wave propagated through it. The parameter $u = (\omega_H/\omega)^2$ governs the propagation; here $\omega_H = \varepsilon H/(mc)$ is the classical Larmor precession frequency. Under the conditions prevailing in the interstellar medium, $u = 10^{-12}$ or 10^{-13}, a very small quantity. The plane of polarization of a linearly polarized wave will rotate through an angle ψ as the wave traverses a line segment of length l (the Faraday effect). This angle is given by

$$\psi = \tfrac{1}{2}\, l\,(n_2 - n_1)\,\omega/c = 10^6\, N_\varepsilon\, H\, l\, \omega^{-2} \cos^2 \alpha, \qquad (13\text{-}13)$$

where n_1 and n_2 are the refractive indices for the extraordinary and ordinary rays, N_ε is the free-electron concentration, and α is the angle between the magnetic field and the direction of propagation of the wave (line of sight).

It is clear that radio waves arising from relativistic electrons in a single cloud will be polarized only if the angle ψ is small. As an example, let us take $l = 10$ pc, $\omega = 2\pi \times 10^8$, $H = 3 \times 10^{-6}$ gauss, and $\cos^2 \alpha \approx 1$; then we must have $N_\varepsilon \approx 6 \times 10^{-4}$ cm^{-3} in order that ψ be less than $\pi/20$, say. Highly rarefied interstellar gas has an atomic concentration $n_H \leq 0.1$ cm^{-3}. Hydrogen is neutral at great distances from the galactic plane, as we would expect (Ginzburg and Fradkin 1953). Under these conditions electrons will be formed through ionization of carbon atoms, whose abundance is several thousand times as small as that of hydrogen; as a result, $N_\varepsilon < 10^{-4}$ or 10^{-5} cm^{-3}. We may therefore anticipate the radiation from such clouds to be polarized.

But there is a complication. Equation (13-13) shows that the angle ψ depends strongly on the frequency of the radiation. If the radiation traverses a medium of strong rotatory power (such as the highly ionized intercloud medium near the galactic plane, where $N_\varepsilon \approx 0.1$ cm^{-3}), radiation received at different frequencies will be rotated through different angles. This differential-rotation effect, if prominent within the receiver bandwidth, will render the observed radiation unpolarized. This imposes certain conditions on the bandwidth, which it is desirable to make as narrow as possible. But by narrowing the bandwidth we must necessarily incur a loss in receiver sensitivity. It follows from Eq. (13-13) that the difference in the angles of rotation at two frequencies separated by $\Delta\omega$ is equal to

$$|\Delta\psi|_1 = 2 \times 10^6\, N_\varepsilon H l \omega^{-3} |\Delta\omega| \cos^2 \alpha. \qquad (13\text{-}14)$$

If we take $N_\varepsilon = 0.1$, but retain our earlier values for the other parameters entering into Eq. (13-14), we shall have $\Delta\omega/2\pi = 320$ kc/s, a bandwidth narrower by almost an order of magnitude than that customarily employed. Receiver sensitivity will in fact decline if such a bandwidth is used, but not seriously. It should be noted that the "angle of depolarization" will be even greater in practice, since the radio waves pass through n_1 clouds. Here

$$|\Delta\psi|_2 = |\Delta\psi|_1 \, n_1^{1/2}, \qquad (13\text{-}15)$$

with n_1 of the order of 10; thus once again the complication is not a vital one.

There can be no question but that the detection of this effect would be of paramount significance for the problems of cosmic radio waves. Apart from the fact that it would represent a decisive experimental demonstration of the magnetic-deceleration theory, we would gain a new and highly effective technique for studying the physical conditions in our Galaxy's magnetized corona of rarefied gas, which is almost inaccessible to optical investigation.

V. A. Razin (1956, 1957), in the Soviet Union, has essayed an experimental verification of the foregoing theory. The measurement of linearly polarized cosmic radio waves is by no means a simple experiment, for, as the antenna rotates about the normal to the wave front, a number of parasitic effects always set in, and these might be interpreted as polarization effects. Among these are effects due to asymmetry of the directional diagram, changes in the directional diagram itself and its impedance, and reception of strongly polarized radiation reflected from the Earth. The parasitic effects exceed the expected level of linearly polarized cosmic radio waves.

Razin circumvented these difficulties with an ingenious variant of the modulation method. The polarization effect depends on the receiver bandwidth [Eq. (13-14)]; hence modulation in bandwidth was introduced. Observations were made with two receivers, a wide-band ($\Delta\nu = 5$ Mc/s) and a narrow-band ($\Delta\nu = 0.2$ Mc/s), in parallel with a common heterodyne filter. The antenna was connected alternately to the two receiver inputs, at a frequency of 25 c/s.

The synphase antenna array of 36 half-wave dipoles was directed toward the zenith, and swept 360° in azimuth. The observations were conducted at 1.45 m, and also at 3.3 m but with modulation of the bandwidth of one receiver.

At 1.45 m, linear polarization was detected whenever the antenna was directed toward a point of galactic latitude $|b| > 5°$. The intensity of the linearly polarized component was 2 to 4 K deg, or 2 to 4 percent of the total intensity. At 3.3 m, the degree of polarization of cosmic radio waves was four or five times as small as at the shorter wavelength, in agreement with theory. Razin's valuable observations therefore fully support the synchrotron mechanism of cosmic radio emission. In view of the great importance such observations have for any consideration of the nature of cosmic radio waves, it would be expedient to repeat them with antennas of higher directivity.

We must emphasize, in conclusion, that the magnetic-deceleration theory for the nonequilibrium radio emission of the spherical component can satisfactorily explain all the observational data. Furthermore, new phenomena can be predicted from the theory, directing investigators toward the solution of new problems, not only in radio astronomy, but in such fields as the physics of cosmic rays. A number of fundamental problems in cosmic and terrestrial physics are closely associated; for example, the spherical distribution of the rarefied intercloud medium involves the delicate question of the polarization of cosmic radio waves, as well as the problem of the electron component of the cosmic rays observed at the Earth. Further relations, some quite unexpected, will become clear as we proceed. The theory has been developed to a point where we may almost assert, on the basis of both theory and observation, that nonthermal galactic radio emission constitutes a sound proof of the presence of randomly oriented magnetic fields in interstellar space.

As a result we may already form far-reaching astronomical conclusions on the basis of the radio data, as construed in the light of the synchrotron mechanism. We shall later adduce concrete examples which may serve to substantiate this position.

IV

Monochromatic Galactic
Radio Waves

14. THE 21-CENTIMETER HYDROGEN LINE

So far we have discussed cosmic radio waves whose emission spectra
are essentially continuous. Actually, the intensity of the radio emission in thermal equilibrium is given by

$$I_\nu = 2kT\nu^2 c^{-2}(1 - e^{-\tau_\nu}),$$

where $\tau_\nu \propto \nu^{-2}$ (to within a logarithmic factor). Throughout the enormous frequency interval where $\tau_\nu < 1$, $I_\nu = $ const; when $\tau_\nu > 1$, the
intensity decreases relatively slowly as ν^2. For nonequilibrium emission $I_\nu \propto \nu^{-n}$, where for different sources $1.2 > n > 0.2$, so that in this
case the emission also occurs over a very large frequency range. At
the present time the continuous radio spectrum of the strongest
sources has been investigated between 15 and 10 000 Mc/s.

Should we expect to find emission or absorption *lines* in the radio
spectra of cosmic objects? Transitions between very close levels of
several atoms and, particularly, molecules give rise to lines whose frequencies fall in the radio range. Recent years have seen the achievement of remarkable success in the development of molecular radio
spectroscopy. In astrophysics we make use of optical spectroscopy to
investigate the nature of the radiation from cosmic objects at almost
every step; hence we would expect that in radio astronomy a similar
role would be played by radio spectroscopy.

In 1944, when radio astronomy was still in its infancy, van de Hulst,
then a student in Holland and scientifically isolated because of the Nazi
occupation of his country, presented a paper at a colloquium at
Leiden in which he showed that transitions between the hyperfine
levels of the ground state 1^2S of the hydrogen atom produce a line of
wavelength about 21 cm. This line would, then, fall in the radio range.
Preliminary calculations led van de Hulst to believe that it would be
possible to observe this line in the radio spectrum of the Galaxy if it
were emitted by interstellar hydrogen.

Since international scientific ties were disrupted on account of the war, van de Hulst's brilliant idea did not spread immediately among astronomers and radiophysicists. An article containing an account of van de Hulst's paper was published in a rather inaccessible Dutch journal (Bakker and van de Hulst 1945). Reber and Greenstein (1947), in a review article, devoted a short paragraph to van de Hulst's suggestion.

The present author became interested in this idea in 1948. He investigated the problem and showed that the $\lambda = 21$ cm line of interstellar hydrogen should have been detectable in the galactic radio spectrum even with the equipment available at the time (Shklovsky 1949). This conclusion was based on calculations of the spontaneous transition probability between the hyperfine components of the ground state of the hydrogen atom. At the same time it was investigated whether emission or absorption lines of other interstellar atoms and molecules could also be observed.

The possibility of detecting interstellar hydrogen by radio was extremely exciting. Although hydrogen is the most abundant element in the universe, optical methods of investigation give very limited information about it. The hydrogen resonance lines of the Lyman series fall in the inaccessible ultraviolet region of the spectrum, and only in the solar spectrum have we been able to observe the emission lines Lα and Lβ, by means of rocket-borne instruments. As for the stars, nebulae, and interstellar gas, we are still far from being able to observe the Lyman series.

Until recent years only subordinate lines of hydrogen could be observed, mainly the Balmer series, and hence the observations gave only the population of excited levels. With this information and certain assumptions (for example, that the population of all levels is given by Boltzmann's law), it is possible to obtain a theoretical value for the number of neutral hydrogen atoms in the lowest quantum state. However, such assumptions are not always justified; in the majority of the interesting cases (for example, in the interstellar medium, nebulae, outer layers of the sun and stars) the Boltzmann distribution does not obtain since the matter and the radiation field in these objects are far from thermodynamic equilibrium.

In the case of H II regions we can assume from the start that observations of the brightness of the Balmer lines give the concentration of protons rather than that of hydrogen atoms. However, in H I regions, where hydrogen practically never is raised above the second quantum level, optical observations are impossible (and we must remember that most of the interstellar medium is in the form of H I regions).

The observation of the 21-cm line of hydrogen opened the possibility of studying the physical characteristics of the interstellar medium, such as its density and temperature. Such observations are not affected by the interstellar absorption caused by cosmic dust, and hence it is possible to investigate remote portions of our stellar system whose emission is very difficult to detect optically. Since radiophysical methods permit measurement of the emitted frequency with much higher precision than optical methods, it is possible to make detailed studies of the line profiles. From them we can derive exhaustive information about the motions of the emitting atoms of interstellar hydrogen, whether they be thermal or macroscopic (for example, the motion of clouds of interstellar gas, or the motion of the interstellar medium due to galactic rotation).

To the preceding remarks we should add that such observations can be carried out at any time of the day and in almost any weather, while optical observations of H II regions demand ideal conditions. Therefore it is natural that the problem of detecting monochromatic galactic radio waves attracted immediate attention and was solved relatively quickly.

We shall now review the theory of the formation of the 21-cm line. It is well known that the ground state of hydrogen, 1^2S, splits into two very close levels on account of its hyperfine structure, due to the mutual interaction between the intrinsic magnetic moments of the proton and the electron. The separation between the two levels is very small because the magnetic moment of the proton is small (2.7 nuclear magnetons, that is, 1838 times as small as the Bohr magneton, which is the intrinsic magnetic moment of the electron). Each hyperfine level will be characterized, in addition to the usual quantum numbers, by the quantum number $F = J + I$, where I is the spin of the nucleus, $J = L + S$ the total angular momentum, L the orbital angular momentum, and S the spin of the electron. For the ground state of hydrogen $L = 0$, $S = \frac{1}{2}$, $J = \frac{1}{2}$, $I = \frac{1}{2}$, and for the two hyperfine levels F can be either unity or zero.

The magnitude of the additional energy due to the hyperfine structure can be computed from a formula given by Fermi and Bethe in the *Handbuch der Physik*:

$$W = \frac{h\nu_0}{n^3}\left[\frac{F(F + 1) - I(I + 1) - J(J + 1)}{J(J + 1)\,(2L + 1)} \right], \qquad (14\text{-}1)$$

where n is the principal quantum number, $\nu_0 = g(I)\alpha^2 cR$, R is the Rydberg constant expressed in inverse centimeters, α is the fine structure constant, and $g(I)$ the Landé factor for a proton. According to Eq. (14-1), the difference in energy between the components of the hyperfine structure of the ground state of hydrogen is

$$h\nu = (8/3)h\nu_0. \qquad (14\text{-}2)$$

The size of the splitting $h\nu$ can be measured directly by means of the magneto-resonance methods of Nafe, Nelson, and Rabi (1947; Nafe and Nelson 1948); of Nagle, Julian, and Zacharias (1947); or of Prodell and Kusch (1950). According to the latter authors,

$$\nu = 1420.4058 \pm 0.0003 \text{ Mc/s},$$

or $\nu_0 = 532.65$ Mc/s, a value which differs by only 0.3 percent from the one given by Eq. (14-2).

From ν_0 we can compute the hyperfine splitting of other levels of the hydrogen atom. The results of the computations for $n = 1$ and 2 are shown in Table 17. Notice, however, that levels with $n > 1$ do not offer any astrophysical interest, since the overwhelming majority of the hydrogen atoms in the interstellar medium are in the ground state; the radio lines arising from excited states would therefore be several orders of magnitude weaker than the $\nu = 1420.4$ Mc/s line (we shall refer again to this question in Sec. 18).

Transitions between two hyperfine levels are forbidden, since they do not involve a change in the orbital quantum number L. This selection rule, however, applies only to the case of electric dipole radiation, and hence transitions between the hyperfine-structure components may occur owing to magnetic dipole radiation.

According to Condon and Shortley (1935) and Pasternack (1940), the probability of such a transition is given by the expression

$$A_m(A, B) = \frac{4\pi^2 \varepsilon^2 h}{3m^2 c^2} \frac{\sigma^3}{2F_A + 1} S(A, B)$$

$$= 2.696 \times 10^{-11} \frac{\sigma^3}{g_A} S(A, B), \tag{14-3}$$

where the frequency ν is given in reciprocal centimeters as a wave number σ, A and B represent the initial and final states of the atom, $g_A = 2F_A + 1$ is the statistical weight of the initial state (in our case $F = 1$ and $g_A = 3$), and $S(A, B)$ is a dimensionless parameter, the "line strength." According to the basic principles of quantum mechanics,

$$S_m(A, B) = \sum_{A, B} (A \,|\, \mathbf{M} \,|\, B)^2, \tag{14-4}$$

where the summation extends over all levels differing in the magnetic

TABLE 17. Hyperfine splitting in hydrogen.

Level	ν (Mc/s)	Level	ν (Mc/s)
$1^2S_{1/2}$	1420.4	$2^2P_{1/2}$	59.2
$2^2S_{1/2}$	177	$2^2P_{3/2}$	23.7

quantum number M. The quantity $\mathbf{M}$, the total magnetic moment operator expressed in Bohr magnetons, is given by the relation

$$|\mathbf{M}| = L + 2S + \gamma I, \tag{14-5}$$

where L and S are the orbital and spin quantum numbers, and γ is the ratio of the proton and electron magnetic moments. The magnetic moment of the hydrogen nucleus is about one-thousandth that of the electron. Therefore, in Eq. (14-5), the term γI can be neglected in comparison with L and $2S$, as was first pointed out by Purcell (Wild 1952).

In our case, the initial level A and the final level B differ only in the quantum number F; the quantum numbers J, L, and S remain the same. Since the interaction with the magnetic moment of the nucleus is very small, the eigenfunctions of A and B are almost identical and hence all the summands in Eq. (14-4) are practically equal. Their magnitude is determined by the number of Zeeman components of the levels A and B, which, in turn, depends only on the quantum numbers M (each level splits into $2F + 1$ states, differing in the value of the magnetic quantum number M, while in the absence of an external magnetic field the energy of all states is the same). Thus, while the value of each summand in Eq. (14-4) depends on the quantum numbers J, L, and S, the number of terms depends only on F. In the case of pure Russell–Saunders coupling, which in our case is valid to sufficient accuracy, the line strength is given by the formula

$$\begin{aligned} &S_m \left\{ (F + 1) JLS; FJLS \right\} \\ &= \frac{(2F + 1)\left[S(S + 1) - L(L + 1) + 3J(J + 1) \right]^2}{4J(J + 1)}. \end{aligned} \tag{14-6}$$

Setting $L = 0$, $S = \frac{1}{2}$, $J = \frac{1}{2}$, $F = 0$ we get $S_m = 3$; substituting this value of S_m in Eq. (14-3) we obtain

$$A_{21} \equiv A_m(A, B) = 2.85 \times 10^{-15} \text{ sec}^{-3}. \tag{14-7}$$

It follows, then, that the mean lifetime of the "excited" hydrogen atom (that is, of the atom in state $F = 1$) is

$$\tau_{21} = A_{21}^{-1} \approx 3.5 \times 10^{14} \text{ sec}$$

or 11 million years, as compared with lifetimes of 10^{-8} sec for allowed optical transitions!

The reasons for such a long lifetime are the low frequency of the line (since $A_{21} \propto \nu^3$) and the fact that it arises from a forbidden transition (the transition probabilities for magnetic-dipole radiation are of the order of 10^5 times as small as those for electric-dipole radiation, regardless of frequency).

We must keep in mind that even under the most favorable condi-

tions the hydrogen gas in the interstellar medium will, in the overwhelming majority of cases, make a transition from the higher to the
lower hyperfine state without the emission of radiation, as a result of
a collision of the second kind. The effective cross section for such collisions is rather high, since there is a large probability for electron
exchange. Even if the effective cross section were $\sigma \approx 10^{-16}$ cm^2, that
is, of the order of the kinetic cross section (and it could be considerably greater), the time between collisions would be

$$\tau_{\text{coll}} \approx (nv\sigma)^{-1}.$$

If we assume that the concentration of hydrogen atoms in the interstellar medium is $n = 0.1$ cm^{-3} and that $v = 10^5$ cm/s, then
$\tau_{\text{coll}} = 10^{12}$ sec. In regions where $n \approx 10$ cm^{-3}, $\tau_{\text{coll}} \approx 10^{10}$ sec, or
30 000 times less than the mean lifetime of the hydrogen atom in the
upper hyperfine state.

According to the latest calculations of Purcell and Field (1956),
when neutral hydrogen atoms in the $1^2S_{1/2}$ state collide, exchange of
electron spin takes place. If this is the case, the effective cross section
of the state with $F = 1$ is very large. If $T = 1°$K, then $\sigma_{\text{H}-\text{H}} = 19.6 \times
10^{-15}$ cm^2, which is 200 times as great as $\pi a_0{}^2$. The temperature de-
pendence is $\sigma \propto T^{-0.27}$, and hence at $T = 100°$K, $\sigma = 5.65 \times 10^{-15}$
cm^2, a value considerably larger than that given by kinetic theory.

Thus in the interstellar medium the basic elementary processes
determining the population of the hyperfine levels are collisions of
the first and second kinds. We must satisfy the equilibrium condition
that the number of collisions of the first kind per unit volume and
time equals the number of collisions of the second kind. Then the
population of the two hyperfine levels will be given by Boltzmann's
formula,

$$n_2/n_1 = (g_2/g_1)e^{-h\nu/kT}, \tag{14-8}$$

where T is the kinetic temperature of the interstellar gas. Even for
very low temperatures, $h\nu \ll kT$, and hence

$$n_2/n_1 \approx g_2/g_1 = 3. \tag{14-9}$$

Since the initial state of the transition giving rise to the 21-cm line
is maintained in equilibrium by collisions, the energy emitted in this
line depends on the internal energy of the interstellar gas. In other
words, the monochromatic galactic radio emission is thermal and we
can estimate its expected intensity.

It would seem a fantastic idea to try to detect emission with such
a small Einstein coefficient, especially when most of the transitions
from the upper to the lower states are radiationless. Yet this is not so.
We can write the equation of transfer for the 21-cm line as follows:

$$dI_\nu/ds = -k_\nu'I_\nu + \epsilon_\nu, \tag{14-10}$$

where s is the distance in the line of sight from the observer, I_ν the intensity of the emission, ϵ_ν the energy emitted per unit volume and unit time in the frequency interval $d\nu$ about ν, and $k_\nu' = k_\nu[1 - (g_1/g_2)(n_1/n_2)]$ the absorption coefficient (including the negative absorptions). Since in our case

$$\frac{n_2}{n_1} = \frac{g_2}{g_1}e^{-h\nu/kT},$$

where T is the kinetic temperature of the interstellar gas, the factor which takes into account the negative absorptions becomes

$$\left(1 - \frac{g_2 n_2}{g_1 n_1}\right) = (1 - e^{-h\nu/kT}) \approx \frac{h\nu}{kT}.$$

As the natural width of the 21-cm line is extremely small, the absorption coefficient k_ν' will be determined mainly by Doppler effect. Consequently,

$$k_\nu' = \frac{2\pi^{3/2}\varepsilon^2}{2\pi mc}\frac{1}{\Delta\nu_D}\, n_1 f_{12} e^{-(\Delta\nu/\Delta\nu_D)^2}(1 - e^{-h\nu/kT}), \tag{14-11}$$

where f_{12} is the oscillator strength:

$$f_{12} = \frac{g_2}{g_1}A_{21}\frac{mc^3}{8\pi^2\nu^2\varepsilon^2}.$$

The emission per unit volume is given by the expression

$$\epsilon_\nu = \frac{1}{4\pi^{3/2}}n_2 A_{21} h\nu \frac{1}{\Delta\nu_D}e^{-(\Delta\nu/\Delta\nu_D)^2}, \tag{14-12}$$

where the Doppler half-width is $\Delta\nu_D = \nu\langle v\rangle/c$. From Eq. (14-12) we can surmise that the radial velocities of the hydrogen atoms, due to either thermal or macroscopic motions, follow a Gaussian distribution.

Solving Eq. (14-10) we find

$$I(\nu) = \int_0^\infty \epsilon(\nu)\exp\left[\int_0^s k'(\nu)ds\right]ds. \tag{14-13}$$

Also, according to Kirchhoff's law,

$$\epsilon(\nu) = \frac{2\nu^2}{c^2}kT\cdot k'(\nu). \tag{14-14}$$

Expressing the intensity I_ν in terms of the brightness temperature $T_b(\nu) = I(\nu)c^2/2k\nu^2$, we have

$$T_b(\nu) = \int_0^\infty Tk'(\nu)\exp\left[-\int_0^s k'(\nu)ds\right]ds = \int_0^{\tau_\nu'} Te^{-\tau_\nu}\,d\tau_\nu, \tag{14-15}$$

where $\tau_\nu = \int_0^s k'(\nu)d\nu$ is the optical depth, and $\tau_\nu{}^1$ the optical depth of the whole Galaxy in a given direction. If the kinetic temperature of the interstellar gas is constant, then

$$T_b(\nu) = T(1 - e^{-\tau_\nu{}^1}). \tag{14-16}$$

Equation (14-16) confirms that the monochromatic radio emission from hydrogen is of thermal origin.

Our analysis of the problem has been rather idealized, since we did not take into account the presence of a continuous spectrum. If we do, together with the coefficients $\epsilon(\nu)$ and $k'(\nu)$ we should introduce corresponding coefficients $\epsilon_1(\nu)$ and $k_1'(\nu)$ for the continuum. Then the equation of transfer (14-10) can be rewritten as

$$\frac{dI_\nu}{ds} = \epsilon(\nu) + \epsilon_1(\nu) - [k'(\nu) + k_1'(\nu)]I_\nu. \tag{14-17}$$

Solving Eq. (14-17) we have

$$T_b = \frac{c^2}{2k\nu^2} \int_0^\infty [\epsilon(\nu) + \epsilon_1(\nu)] \exp\left\{-\int_0^s [k'(\nu) + k_1'(\nu)]ds\right\} ds, \tag{14-18}$$

and for the region in the continuum adjacent to the radio line, where $\epsilon(\nu)$ and $k'(\nu)$ are equal to zero,

$$T_b = \frac{c^2}{2k\nu^2} \int_0^\infty \epsilon_1(\nu) \exp\left[-\int_0^s k_1'(\nu)ds\right] ds. \tag{14-19}$$

Since the Galaxy can be considered transparent for decimeter radio waves in the continuum, Eq. (14-19) becomes

$$T_b = \frac{c^2}{2k\nu^2} \int \epsilon_1(\nu)ds. \tag{14-20}$$

Furthermore, in the region of the 21-cm line, $k'(\nu) \gg k_1'(\nu)$. Therefore Eq. (14-18) can be rewritten thus:

$$T_b(\nu) = \frac{c^2}{2k\nu^2} \int_0^\infty [\epsilon(\nu) + \epsilon_1(\nu)]e^{-\tau_\nu} ds$$

$$= \frac{c^2}{2k\nu^2} \int_0^\infty \epsilon(\nu)e^{-\tau_\nu} ds + \frac{c^2}{2k\nu^2} \int_0^\infty \epsilon_1(\nu)e^{-\tau_\nu} ds, \tag{14-21}$$

or, making use of the theorem of the mean for the second term on the right-hand side of Eq. (14-21), we get, with the help of Eqs. (14-16) and (14-20),

$$T_b(\nu) = T[1 - \exp(-\tau_\nu{}^1)] + T_b \exp(-\langle\tau_\nu\rangle). \tag{14-22}$$

The difference in brightness temperature between different points on the line profile and the continuum will be equal to

$$\Delta T = T[1 - \exp(-\tau_\nu{}^1)] + T_b[1 - \exp(-\langle\tau_\nu\rangle)]. \quad (14\text{-}23)$$

Since the brightness temperature in the continuum near $\lambda = 21$ cm is considerably lower than the kinetic temperature of H I regions (that is, $T_b \ll T$) even in the direction of the galactic center, and since $\tau_\nu{}^1$ and $\langle\tau_\nu\rangle$ are of the same order of magnitude $(\langle\tau_\nu\rangle \approx \frac{1}{2}\tau_\nu{}^1)$, then $\Delta T > 0$; that is, the line will appear in emission. In the opposite case we would observe an absorption line.

In the direction of the galactic center, and also of the anticenter, where the differential rotation of the Galaxy does not influence the profile, the Doppler width of the radio line is determined by the random velocities of the interstellar gas and by the thermal velocities of the emitting hydrogen atoms (in the case of relatively high kinetic temperatures). Let us estimate the brightness temperature in the center of the 21-cm line under different assumptions about the kinetic temperature of the interstellar gas. Assume that the total extent of the Galaxy is 20 kpc in the direction of the center and 4 kpc in the direction of the anticenter. For the mean concentration of hydrogen atoms along the line of sight we shall take 1 cm^{-3} and for the mean velocity of the interstellar gas 7 km/s, following O. A. Melnikov. Then the Doppler half-width is

$$\Delta\nu_\mathrm{D} = \nu\langle v\rangle/c = 3.3 \times 10^4 \text{ sec}^{-1}.$$

Calculations show that for any reasonable value of the kinetic temperature of the interstellar gas the brightness temperature in the center of the line should exceed 20°K.

Obviously the bandwidth Δf of the receiver should be smaller than the half-width of the 21-cm line. In the opposite case the brightness temperature "spreads" over a wider band and, since the sensitivity of the receiver is proportional to $(\Delta f)^{1/2}$, its value will be decreased approximately $(\Delta f)^{1/2}$ times.

The limiting sensitivity of the receiver is given by the well-known formula

$$\Delta T \approx NT_0(t\Delta f)^{-1/2}, \quad (14\text{-}24)$$

where t is the characteristic time of the recording device. If we take $\Delta f = 3 \times 10^4$ c/s, $t = 100$ sec, $N = 10$, $T_0 = 300$°K, then $\Delta T = 2.7$ K deg, which is considerably smaller than the expected brightness temperature in the region of the 21-cm line.

Increasing sensitivity makes it possible to narrow down the bandwidth of the receiver, thus allowing $T_b(\nu)$ to be determined, and thereby the line profile. This fact, as we shall see below, is of great significance.

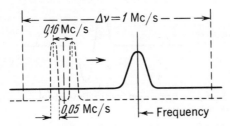

Fig. 124. Explanation of the method for observing the hydrogen radio line used by Christiansen and Hindman (1952).

All the preceding considerations were basically known by 1950 and efforts were being made in several countries to design and construct equipment capable of detecting the hydrogen radio line. The problem was attacked simultaneously in Holland, the United States, and Australia.

The efforts of the investigators were crowned with complete success. The 21-cm line was first observed in the United States by Ewen and Purcell (1951) on 25 March 1951; in May of the same year, it had already been observed by Muller and Oort (1951) in Holland, and within two months by Christiansen and Hindman (1952) in Australia.

In the United States, as well as in Holland and Australia, a frequency-modulation method was used. The basic features of the method are the following. The receivers are essentially double superheterodynes with narrow passbands which are switched back and forth between two frequencies. Both frequencies can be slowly swept together through a passband of a few megacycles per second centered at $\nu = 1420.4$ Mc/s. In this way the receiver measures the difference in intensity between two neighboring regions in the spectrum.

Figure 124 shows, schematically, the arrangement used by Christiansen and Hindman. The two passbands are $\Delta f_1 = 160$ kc/s apart; the bandwidth is 50 kc/s. If both bands are outside the spectral region where the 21-cm line lies, the receiver registers the difference between the cosmic radio waves at two frequencies in the continuum. Since Δf_1 is small, the intensities at the two frequencies are practically the same and the recording equipment will show no signal. As the passband moves by an amount $\Delta\nu$, the leading band reaches the frequency region where the radio line lies, and the records will show the difference between the emission in the continuum and that due to the line. When the leading band moves beyond the region of the 21-cm line, this region will be reached by the trailing band and again the difference between the line and the continuum will be recorded, this time in the opposite direction.

Figure 125 shows a schematic record; it does not actually correspond to the profile of the radio line because in the Australian equip-

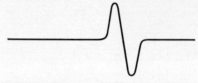

Fig. 125. A record of the hydrogen radio line obtained with the frequency-modulation method (schematic).

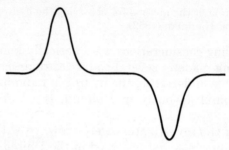

Fig. 126. Transformation of the hydrogen radio line for sufficiently great Δf_1 (schematic).

ment Δf_1 was not sufficiently large. The profile sought can, however, be easily derived from the records after some analysis. If, on the other hand, Δf_1 considerably exceeds the width of the hydrogen line, the record will resemble Fig. 126. In this case an undistorted profile is obtained when each band passes through the region of the 21-cm line. Figure 127 illustrates an actual, typical record, obtained by the Australian observers.

15. SPIRAL STRUCTURE AND THE DYNAMICS OF THE GALAXY

As we have already mentioned, the 21-cm line was first detected in 1951 by Ewen and Purcell, who used a small, pyramidal antenna. These investigators, however, were not able to observe much detail in the monochromatic galactic radio waves.

The first serious work in this direction was that of Christiansen and Hindman. Their observations were carried out with a section of an equatorially mounted paraboloidal reflector of surface area about 25 m²; the beamwidth was approximately 2°. The profile of the line was measured for all the observable sky at intervals of 5° in right ascension and of 1° to 2° in declination. The isophotes in Fig. 128 were constructed on the basis of the results of these observations.

A comparison of the 21-cm isophotes with continuum isophotes (Fig. 25, for example) shows a different intensity distribution. In Fig. 128 we see that the intensity of the monochromatic emission in the direction of the anticenter is somewhat greater than in the direction

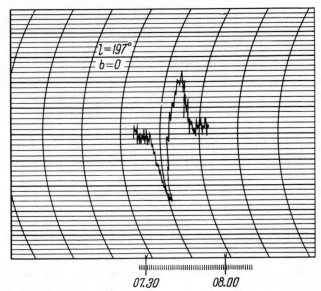

$l = 197°$
$b = 0$

07.30   08.00

Fig. 127. An actual record of the 21-cm hydrogen radio line (Christiansen and Hindman 1952).

of the center, while the radio waves in the continuous spectrum concentrate toward the galactic center. Furthermore, the monochromatic radio waves are considerably less concentrated toward the galactic plane than the thermal radio waves in the continuum.

Such differences between the intensity distribution at 21-cm and monochromatic radio waves are explained by the fact that the interstellar gas is opaque to 21-cm radiation, while continuous radiation (of wavelength less than 3 to 5 m) passes through the interstellar medium practically without absorption. Calculations show that the interstellar gas remains opaque for the monochromatic radiation only in the case of low kinetic temperatures. This important conclusion is

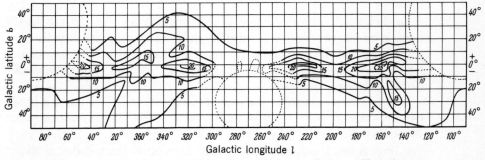

Fig. 128. Isophotes of hydrogen radio emission from the Galaxy (Christiansen and Hindman 1952).

in agreement with the assumption that the kinetic temperature of H I regions is considerably lower than that of H II regions, as we shall see below.

From measurements of the half-width of the 21-cm line in the directions of the center and the anticenter, which are free from the effects of differential galactic rotation, it has been possible to estimate the peculiar velocities of the hydrogen clouds at about 8 km/s.

As we have already pointed out, the Galaxy is opaque to 21-cm radiation in the directions of both the center and the anticenter. In other directions, because of differential rotation and Doppler effect, the Galaxy becomes "transparent" to the monochromatic radiation, making it possible to observe interstellar hydrogen to considerably larger distances.

In several directions the radio-line profiles split into two components. In such cases we are observing two galactic spiral arms, one relatively close to us, the other at some greater distance. The splitting of the profiles proves the presence of large relative radial velocities between clouds of interstellar gas located in the two arms, due to differential galactic rotation.

Another interesting point is the detailed intensity distribution of the 21-cm line. For example, in the isophotes shown in Fig. 128 there is a bright strip at $l = 150°$, extending almost to $b = -40°$. The cause of this interesting feature will be given in Sec. 16.

Thus, the very first investigations of the hydrogen radio line in the Galaxy had already given very valuable results of general astronomical significance. Further observations developed and extended this pioneer work.

The first systematic investigation of the interstellar medium by means of the 21-cm line was made in Holland by van de Hulst, Muller, and Oort (1954). This investigation was started in 1952, that is, shortly after Muller first detected the hydrogen line.

The Dutch group used as antenna a reflector from a German radar receiver of the "Great Würzburg" type, 7.5 m in diameter and 1.7 m in focal length. The mounting of the antenna was azimuthal. At the focus there were a half-wave dipole and a reflecting disk; the receiver was connected to the feed by means of a coaxial cable. The beamwidth was $1°.9$ in the horizontal direction, and $2°.7$ in the vertical direction (for $\lambda = 21$ cm).

The receiver operated by the switching method. Two narrow bands 648 kc/s apart were switched at a rate of 430 c/s and could be swept through a region of the spectrum 3 Mc/s wide containing the hydrogen line. A block diagram of a revised version of the receiver is shown in Fig. 129.

The original receiver used double frequency conversion. The first

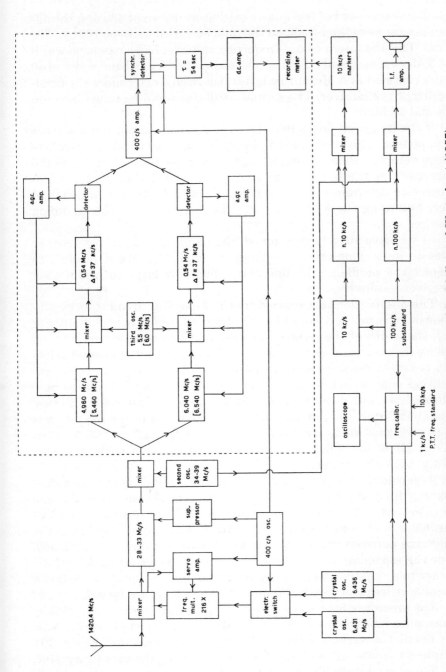

Fig. 129. Block diagram of the receiver used by Muller and Westerhout (1957).

oscillator consisted of two quartz oscillators driven at 430 c/s; the difference between their frequencies could be varied between 100 and 700 kc/s. The first intermediate-frequency amplifier had a passband width of 3 Mc/s centered at 30 Mc/s. The second had a passband width of 40 kc/s centered at 5 Mc/s; this latter bandwidth determines the bandwidth of the receiver. The second oscillator could be tuned between 24 and 27 Mc/s.

If the hydrogen line is detected, the receiver output has a 430-c/s component. This component is amplified and converted in a synchronous detector; then it goes through a low-pass filter which limits the passband of the indicator to frequencies lower than 0.01 c/s and therefore determines the time constant of the indicator, which is 24 sec. Following the filter there is a d.c. amplifier and the recording device.

As we have already mentioned, the bandwidth Δf of the receiver was 40 kc/s. This value is sufficiently small to make it possible to detect the profile of the line and sufficiently large to insure good receiver sensitivity.

The observations are limited by the following circumstances. Obviously, the smallest detail in the observed profile is of the order of Δf, the width of the receiver passband. If two narrow bands sweep along the spectrum with a speed of 1 Mc/s per hour (a smaller velocity is very difficult to achieve since it would require very large stabilizing equipment), then it takes a band of 40 kc/s 2 min to pass through a point. Practice has shown that a time interval of 100 sec does not distort the profile. Since the noise factor of the receiver was 6 (a very good value for this frequency range) and $T_0 = 290°$ K, then $\Delta T = 2.2$ K deg. The actual sensitivity of the receiver was, apparently, somewhat lower.

As a rule the observational method was the following. The antenna was directed toward a definite region of the sky and every 2.5 min it corrected its horizontal coordinates. From these observations a line profile was obtained. When the width of the line was less than the distance between the bands (648 kc/s), two contours were obtained, one corresponding to a positive deviation of the recording device, the other to a negative one. The record had a linear intensity scale and a nonlinear frequency scale. Figure 130 shows a typical record.

The procedure for reducing the data, although conceptually simple, was quite laborious. Van de Hulst, Muller, and Oort (1954) describe in detail the various steps in the reduction procedure. Figure 131 shows 54 profiles of the 21-cm line every 5° along the galactic equator.

The observed profiles result from the solar motion combined with the motions of all the interstellar hydrogen atoms in the line of sight which contribute to the 21-cm radiation.

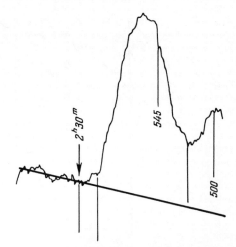

Fig. 130. A sample line profile (van de Hulst, Muller, and Oort 1954).

To take into account the solar motion with respect to the inter-stellar medium, the Dutch investigators used the standard value for the velocity: $v_\odot = 20$ km/s in the direction $\alpha = 18^h00^m$, $\delta = +30°$ (1900).

The motions of the interstellar hydrogen atoms are the following: (*a*) thermal motions, determined by the kinetic temperature of the H I regions (hydrogen, mostly neutral); (*b*) random motions of the interstellar gas clouds; (*c*) galactic rotation.

According to Oort's basic theory of galactic rotation, the angular velocity ω at a point in the Galaxy a distance R from its center is given by the formula

$$\omega(R) = \omega(R_0) + (R - R_0)\omega'(R_0) = \omega(R_0) - (R - R_0)2A/R_0, \quad (15\text{-}1)$$

where R_0 is the distance from the sun to the galactic center and A is Oort's constant. This expression is valid if $R - R_0$ is a small quantity. The relative radial velocity Δv_r due to galactic rotation can be obtained from a formula first given by Bottlinger (1933):

$$\Delta v_r = R_0[\omega(R) - \omega(R_0)]\sin(l - l_0), \quad (15\text{-}2)$$

where l_0 is the galactic longitude of the galactic center. This expression does not depend on the smallness of $R - R_0$ but it does assume that the galactic rotation is circularly symmetric (that is, that the angular velocity at any point depends only on its distance to the center). From Eq. (15-2) it follows that the radial velocity due to differential galactic rotation vanishes for $l - l_0 = 0°$ or $180°$, that is, for the galactic center and anticenter.

For $l - l_0 < 90°$ the line of sight goes through regions of both positive and negative radial velocity. The largest radial velocity will occur around point D, closest to the galactic center (Fig. 132), while at

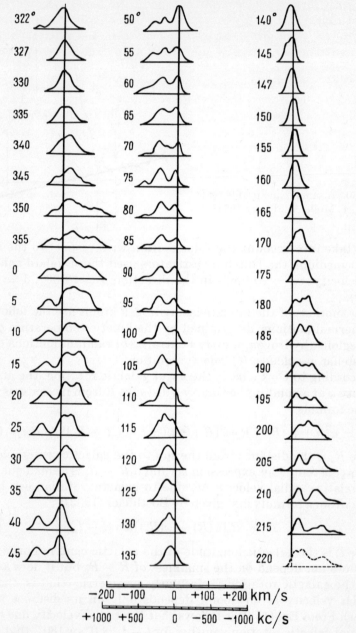

Fig. 131. Profiles of the 21-cm hydrogen radio line at 54 longitudes (van de Hulst, Muller, and Oort 1954).

sufficiently large distances from this point the radial velocity remains almost constant. Let us assume now that in the region around point D there is a spiral arm, where the concentration of interstellar gas is increased. Then an intensity maximum will be observed in the direction l, and the line profile will show a sharp cut-off on the high-frequency side. Such a situation is exemplified in Fig. 131 for l between 15° and 30°. By measuring the frequency at which this cut-off occurs it is possible to determine the maximum radial velocity in the direction l. The distance of the emitting region to the galactic center follows immediately (see Fig. 132):

$$CD = R_0 \sin (l - l_0) = R_0 \sin l',$$

if R_0 and ω_0 are known. There are many determinations of the value of R_0, and the method based on the analysis of the hydrogen radio line allows an independent check on the value of this important quantity. According to Eq. (15-1),

$$\omega(R) - \omega(R_0) = -2A (R - R_0)/R_0.$$

If we substitute this value of $[\omega(R) - \omega(R_0)]$ into Eq. (15-2), it follows that

$$\Delta v_{max} = 2AR_0(1 - \sin l') \sin l'. \tag{15-3}$$

The quantity Δv_{max} is known from the analysis of the line profile and hence we can determine AR_0 from Eq. (15-3). The weighted mean value of AR_0 for the four profiles between $l = 15°$ and $l = 30°$ is $AR_0 = 161$ km/s. Taking $A = 19.5$ km sec^{-1} kpc^{-1}, we have $R_0 = 8.26$ kpc, in very good agreement with the best determinations. (For example, Baade's result, based on an analysis of the RR Lyrae variables in Sagittarius, gives $R_0 = 8.16$ kpc.)

Thus, we have shown that it is possible to estimate the angular velocity of rotation at different distances from the galactic center. This can be done when there is a spiral arm in the region closest to

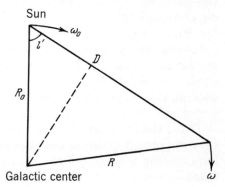

Fig. 132. Geometric method for reducing the records (van de Hulst, Muller, and Oort 1954).

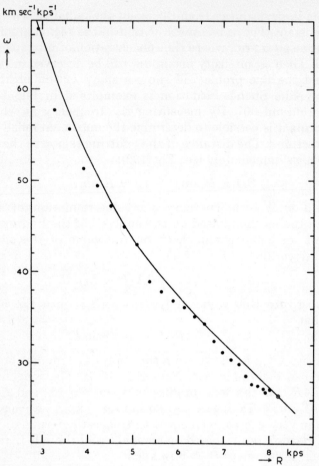

Fig. 133. Angular velocity of the Galaxy (km sec^{-1} kpc^{-1}) as a function of the distance from the center (kpc), according to the radio observations (Kwee, Muller, and Westerhout 1954).

the galactic center along the line of sight. Figure 133 shows the dependence of the angular velocity upon distance to the galactic center; Fig. 134 gives a similar graph for the linear rotational velocity.

It is necessary to stress once more that the preceding work is based on the assumption that all macroscopic galactic motions are radially symmetric. Analysis of the line profiles leads us to believe that, in general, this assumption is justified. At the present time the observed systematic deviations from radial symmetry are negligible.

Thus, the analysis of the profiles of the 21-cm line give new and very important observational data which help the development and improve the accuracy of the theory of galactic rotation. It has also made it possible to trace the galactic rotation to regions at large distances from the sun which are almost inaccessible to optical astronomy because of interstellar absorption. And we must also keep in mind that we have so far only described the first pioneering work in this direction.

As soon as $\omega(R)$, $\omega(R_0)$, and R_0 are known, we can determine the distance to the emitting cloud of interstellar gas with the help of Eq. (15-2) and the measured value of Δv_r. In this way it is possible to localize in space the different maxima and minima of the 21-cm line profiles at different galactic longitudes. When this is done it becomes apparent that the clouds of interstellar hydrogen are concentrated in a few relatively narrow but very extended formations which we naturally identify with the spiral arms of the Galaxy. In the space between the arms the hydrogen concentration is at least one order of magnitude smaller than in the arms.

The study of the spiral structure of our own stellar system with the methods of optical astronomy is very difficult because of interstellar absorption. The first reliable determinations of the galactic spiral structure based on the distribution of H II regions, emitting the inter-stellar Hα line, and also on the distribution of O-associations, was done during 1952 and 1953 by Morgan and his associates (Morgan, Whitford, and Code 1953; Morgan, Sharpless, and Osterbrock 1952). In 1952 the present author (Shklovsky 1952a) found evidence of spiral structure from an analysis of the galactic thermal radio waves in the continuum.

According to the optical observations, the sun is in one of the spiral arms. This arm has received the name of "Orion." The arm which ex-tends through the well-known galactic clusters h and χ Persei has received the name of "Perseus." Finally, the first spiral arm in the

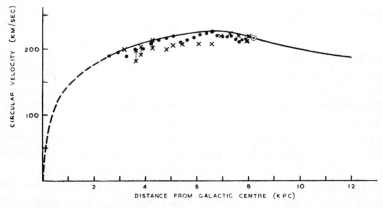

Fig. 134. Circular velocity (km/s) as a function of distance from the center (Kwee, Muller, and Westerhout 1954).

direction of the galactic center is known as the "Sagittarius" arm.

The analysis of the profiles of the hydrogen radio line gives clear indication of the presence of spiral structure in the Galaxy. For example, in Fig. 131 between $l = 40°$ and $l = 115°$ there is a strongly pronounced maximum at $\Delta v = 0$. This shows that a considerable mass of interstellar hydrogen along this direction is concentrated in the immediate vicinity of the sun. An entirely different picture is observed in the region $170° < l < 220°$. Here the first maximum in the profile occurs at $\Delta v = +25$ km/s, indicating that the density of interstellar hydrogen in this direction is very small in the neighborhood of the sun and increases considerably at distances of 500 to 1000 pc. From the preceding data we can surmise that the sun is located near the inner boundary of a spiral arm. An analogous conclusion was drawn from the analysis of the brightness distribution of the thermal radio waves (Chapter III). The Perseus arm is clearly seen from the line profiles for $50° < l < 115°$ (Fig. 131).

The analysis of the 21-cm profiles led to the discovery of a new spiral arm, completely inaccessible to optical observations. The profiles for $45° < l < 345°$ show a strongly pronounced maximum for negative values of Δv. This maximum corresponds to an outer arm, which received the name of "distant arm," and which may be an extension of the Perseus arm. The distant arm has a considerably larger extent in the direction perpendicular to the galactic plane than in others. Its thickness (between points where the hydrogen density has decreased by a factor of two), according to measurements made at galactic latitudes $b = 0°$, $10°$, $20°$, and $30°$, is about 800 pc, that is, approximately three times the thickness of the closer arms.

Between $l = 60°$ and $l = 120°$ appears a new, very weak arm (that is, of relatively small density). It extends considerably beyond the sun and its thickness is comparable to that of the distant arm. It is possible that this very diffuse arm is at the extreme periphery of our stellar system. It is interesting to note that the spiral arms lie in planes slightly inclined to each other.

From an analysis of the spiral structure of the Galaxy based on the hydrogen line profiles it follows that our stellar system belongs to a type intermediate between Sa and Sb; its spiral structure is probably similar to that of NGC 488 and NGC 4594.

Kwee, Muller, and Westerhout (1954) investigated the density distribution of interstellar hydrogen for those regions of the Galaxy for which $R < R_0$. The equipment used for these observations incorporated the latest improvements. The noise factor of the receiver was reduced and the time constant of the recording device increased to 54 sec. As a result the sensitivity of the receiver was increased to $\Delta T = 1.0$ K deg. The distance between the two narrow bands was increased from 648 to 1080 kc/s, thus avoiding superposed records even for the very wide profiles. With this equipment they obtained 21-cm profiles every 2°.5 between $l = 320°$ and $l = 45°$ at $b = -1°.5$. The records were obtained with a fixed reflector at fixed frequencies. Such a technique made it possible to determine the thickness of the emitting region at different points on the profile.

Figure 135 presents several 21-cm profiles for a region near the galactic center. In these profiles we notice the presence of weak wings extending on the negative radial-velocity side which may be caused by a mass of hydrogen in the inner part of the Galaxy ($R < 3$ kpc) in

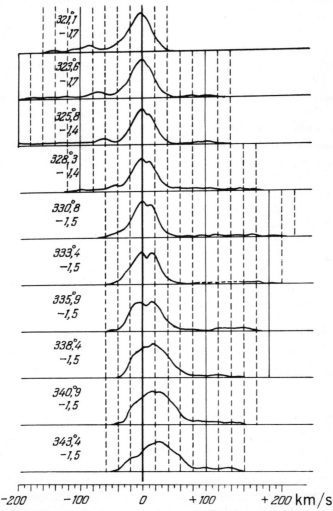

Fig. 135. Profiles of the 21-cm hydrogen radio line in the vicinity of the galactic center (Kwee, Muller, and Westerhout 1954).

rapid turbulent motion. The mean value of any component of the velocity of this motion (for example, the component along the line of sight) is about 50 km/s. The width of this layer is estimated by Kwee *et al.* (1954) to be 240 pc on the basis of observations at fixed frequencies. The analysis of the line profiles permits the study of some of the detailed spiral structure in the inner regions of the Galaxy (Fig. 136). The points in Fig. 136 give the position of the O-associations, according to Morgan; clearly these groups of hot stars in the direction of Sagittarius agree with the position of one of the inner spiral arms discovered from the analysis of the 21-cm profiles. We should point out, however, that the data on the spiral structure and the dynamics of the inner regions of the Galaxy remain to the present time much more scanty than corresponding data for the outer regions. Further observations would be very welcome.

Undoubtedly, the analysis of the 21-cm line profiles puts in the hands of the investigator the most powerful tool for studying the spiral structure of the Galaxy. As new and bigger reflectors are built, and as the sensitivity of the receivers is increased, the data will become more and more complete.

The analysis of the 21-cm line of hydrogen makes possible not only the study of the dynamics and the spiral structure of the Galaxy, but also of many important physical characteristics of the interstellar gas —in particular, its density and temperature.

The optical depth τ_ν of interstellar hydrogen at any frequency ν can be written as

$$\tau_\nu = \int_0^\infty k'(v)ds = N_1(v)A_{21}\frac{g_2}{g_1}\frac{c^3}{8\pi\nu^3}\frac{h\nu}{kT}, \tag{15-4}$$

where $N_1(v)$ is the number of hydrogen atoms in the ground state in a column of unit cross-sectional area, having a radial velocity between v and $v + dv$. Taking into account the statistical weights of the upper and lower hyperfine levels ($g_2 = 3$, $g_1 = 1$), we find that $N_1 = N/4$, where N is the total number of hydrogen atoms. Inserting numerical values into Eq. (15-4) we get

$$N(v) = 1.83 \times 10^{13}\, T\tau_\nu.$$

The intensity of the radio waves of frequency ν in a direction where the optical depth is τ_ν is given by the expression

$$I(\nu) = 2\nu^2 kTc^{-2}(1 - e^{-\tau_\nu}) = I_0(1 - e^{-\tau_\nu}). \tag{15-5}$$

To determine I_0 and the absorption coefficient from the observations, two groups of 21-cm profiles showing the Orion arm were compared. The first group of profiles corresponds to $80° < l < 100°$ and the second to $40° < l < 45°$. In the first group, the line of sight cuts the

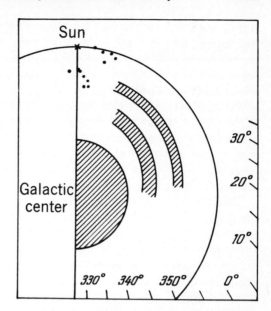

Fig. 136. Sketch of the spiral
structure in the inner parts
of the Galaxy (Kwee, Muller,
and Westerhout 1954).

arm at a large angle (that is, it goes through the arm); in the second
case, the line of sight goes along the arm.

For $l = 45°$ the maximum radial velocity due to differential galactic
rotation is 7 km/s, and it goes to zero at a distance of 4.2 kpc. In a first
approximation we can assume that the differential galactic rotation is
negligible. If we let τ_4 be the optical depth at the center of the 21-cm
line (that is, at $\Delta v = 0$) for $l = 45°$, one can assume that, in the
mean, τ_4 is formed at a distance of 4.2 kpc, where the density of
hydrogen is the same as in the spiral arm, and that differential galactic rotation is absent. For the region $80° < l < 100°$ let us call the
optical depth τ_3. Neglecting the differential galactic rotation, the
effective thickness of the arm in this direction can be estimated as
600 pc. Considering, as a first approximation, that the arm is homogeneous, we find that $\tau_4/\tau_3 = 7$. The maximum intensities of the Orion
arm for $80° < l < 100°$ and $40° < l < 45°$ are, respectively, $I_3 = 51$
and $I_4 = 105$ (in arbitrary units). Hence it follows that

$$\frac{I_3}{I_4} = \frac{1 - e^{-\tau_3}}{1 - e^{-\tau_4}} = 0.49;$$

and, since $\tau_4 = 7\tau_3$,

$$\tau_3 = 0.65, \quad \tau_4 = 4.6.$$

From the values of I_4 and τ_4 we find that $I_0 = 106$. It follows that
absorption coefficient is about 0.91 kpc^{-1}.

Analogous arguments based on other line profiles give values of I_0
and of the absorption coefficient of the interstellar medium which are

approximately the same. In the mean the absorption coefficient appears to be 0.95 kpc^{-1}. We must keep in mind, however, that the observations were made with a bandwidth of 40 kc/s, which corresponds to a resolution of radial velocity of 8.4 km/s. If we assume that the scatter in radial velocities (due to the random motions of clouds of interstellar gas) in one coordinate is 8 km/s, it is easy to see that if the observations were performed with a receiver of infinitely narrow passband the absorption coefficient would be larger, about 1.20 kpc^{-1}. This means that, neglecting differential galactic rotation, the intensity of the radio waves decreases by a factor of e when traversing a segment of spiral arm 830 pc long.

The absolute magnitude of the intensity of the thermal radio waves (and, therefore, of the kinetic temperature of the H I regions) can be obtained by calibrating the equipment. Measurements give, for H I regions, $T = 125°$K.

Equation (15-5) involves the implicit assumption that the kinetic temperature of the interstellar gas is the same everywhere along the line of sight. In the opposite case, that is, if the interstellar medium consists of several optically thin layers of different temperature, we can derive an expression for I_ν similar to Eq. (15-5), but where the temperature would be the harmonic mean of the temperatures of the different layers of interstellar gas in the line of sight.

The question of the kinetic temperature of H I regions has been taken up by several authors. Spitzer and Savedoff (1950) computed the kinetic temperature from the condition of local equilibrium between the energy gained per unit time as a consequence of the photoionization of atoms and the energy lost per unit time as a consequence of inelastic collisions.

In the H I regions only atoms with ionization potentials lower than that of hydrogen can be ionized, such as C and Si, since all the emission beyond the Lyman series limit is absorbed by hydrogen. Computations give a very low value for the temperature of H I regions (from 20° to 200°K, depending on the values adopted for the effective cross sections of the inelastic collisions, the dilution coefficient, and the cosmic abundances of the elements). However, Kahn has recently shown that collisions between clouds are an important source of internal energy. As a result of these collisions there is a transformation of mechanical energy into thermal energy of the interstellar gas which raises its temperature to very high values, of the order of 3000° to 4000°K. These heated regions will subsequently cool down as a result of inelastic collisions between electrons and C II, Si II, and Fe II ions (these ions have excited states with very low excitation potentials). Seaton (1955), on the basis of the cross sections that he computed for such inelastic collisions, investigated the cooling time of those regions of interstellar gas heated by the mechanism described above.

According to Seaton, a cloud of interstellar gas heated to $5000°K$ can cool down to a temperature of a few tens of degrees Kelvin in a time of the order of 10^7 yr. From the condition that the harmonic mean of the kinetic temperature shall be $125°K$, as suggested by observations of the 21-cm line, we can write

$$\frac{1}{t_1} \int_0^{t_1} \frac{1}{T}dt = \frac{1}{125}, \qquad (15\text{-}6)$$

and from Eq. (15-6) we can compute both the time between collisions (which is equal to the cooling time) and the arithmetic mean of the kinetic temperature. Let us notice that in Eq. (15-6) the mean was taken for only one cloud. Obviously we need the mean of the aggregate of clouds in the line of sight, all at different stages of cooling.

Table 18 shows the results of the calculations. In the first column are listed the initial kinetic temperatures attained after the collision between two clouds; in the second column, the cooling time t; in the third, the final kinetic temperature; in the fourth, the arithmetic mean of the temperature. Since the time between collisions of clouds of interstellar gas is of the order of 10^7 yr (as is easily shown), the results of Table 18 should correspond to reality. We must emphasize, however, that the question of the kinetic temperature of H I regions has not yet been finally settled, since up to the present time no computations have been made which take into account satisfactorily the effects of thermal conduction and of dissipation due to turbulent motions in the gas, which are complicated by the presence of interstellar magnetic fields. Further 21-cm observations will undoubtedly give new and important data on the temperature of different clouds of interstellar gas which will help the development of a more precise theory.

Together with the determination of the kinetic temperature of H I regions, the analysis of the observed 21-cm line profiles makes possi-

TABLE 18. Cooling of interstellar clouds.

T_0 ($°K$)	Cooling time (10^6 yr)	T_{final} ($°K$)	$\langle T \rangle$ ($°K$)
5000	15.8	22	1870
4000	12.6	24	1420
3000	9.6	27	1000
2000	6.9	33	600
1000	4.3	43	300

ble the investigation of the density of the interstellar gas. In this respect we must keep in mind that the clouds of interstellar gas are concentrated near the galactic plane in a layer of negligible thickness, about 200 to 300 pc. This means that, with a directional diagram of about 2° to 3°, the interstellar gas at a distance greater than about 7000 pc will not "fill" the main lobe of the antenna. To take this effect into account special corrections are necessary.

From the observed 21-cm line profiles and Eq. (15-5) it is possible to determine $\tau(v)$ in every direction. Then we can write down the relation

$$N_{\text{H}} = \int N(v)dv = \int n_{\text{H}}ds = 2.02 \times 10^{15} \int\limits_{-\infty}^{+\infty} \tau(v)dv.$$

Van de Hulst, Muller, and Oort (1954) computed N_{H} along the galactic equator, for longitudes $30° < l < 320°$. The magnitude of N_{H} lies between 0.87×10^{22} cm^{-2} and 2.3×10^{22} cm^{-2}. From these numbers the mean concentration of hydrogen atoms was found after computing the length of the emitting columns of unit cross-sectional area.

In the directions of the galactic center and anticenter, where owing to the absence of differential galactic rotation the optical depth in the center of the 21-cm line is large, N_{H} cannot be determined with adequate accuracy. The mean value of the hydrogen concentration in different directions was about 0.70 cm^{-3}.

On the basis of their analysis of the 21-cm line profiles, van de Hulst, Muller, and Oort, drew a chart showing the density distribution of hydrogen in the Galaxy (Fig. 137), which clearly shows the spiral structure of the outer parts of our stellar system. The "dead cone" in Fig. 137 with its axis in the direction of the anticenter ($l \approx 147°$) indicates the region where N_{H} cannot be reliably determined because of the great optical depth.

Future investigations, similar to those of the Dutch group but performed with larger antennas, will undoubtedly add more detail and precision to the chart shown in Fig. 137. However, such charts are already an excellent achievement of contemporary astronomy, an achievement undreamt of a few years ago.

Finally we should point out that the analysis of the 21-cm profiles provides a new possibility for studying the random motions of clouds of interstellar gas. Optical methods for investigating these motions (for example, the analysis of the interstellar lines of ionized calcium) give proof that the distribution of random cloud velocities cannot be represented by a Gaussian curve, since there is a considerable excess of large velocities. Blaauw (1952) represented the empirically determined distribution of radial velocities by means of the expression

$$\varphi(v) = (2\eta)^{-1} e^{-|v|/\eta}.$$

According to Spitzer (1948), $\eta = 9$ km/s.

To determine η from the 21-cm profiles, the Dutch investigators studied the region $65° < l < 130°$. In these directions the radial velocities due to differential galactic rotation are negative and there-

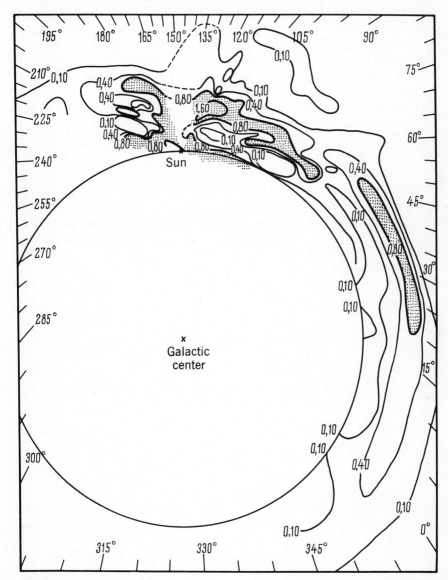

Fig. 137. Contour map of hydrogen density in the galactic plane as derived from 21-cm observations (van de Hulst, Muller, and Oort 1954).

fore the positive wing of the contours must be due solely to the random velocities of the clouds. The value derived is $\eta = 8.5$ km/s, in agreement with the optical data. Another method for determining the random velocities is based on the analysis of the 21-cm profiles in the region of the anticenter. Estimates of the optical depth at the center of the line in the direction of the anticenter give the value $\tau_0 = 2.5$; the best agreement with the observed line profile for the accepted law of velocity distribution is then obtained by adopting $\eta = 8.3$ km/s.

Westerhout (1957a) has published an extensive catalogue of 21-cm line profiles for a region along the galactic equator 20° wide, from $l = 318°$ to $l = 220°$. Profiles were derived for 694 points in the sky. The data were obtained with the 7.5-m reflector described above. On the basis of this catalogue, Westerhout investigated in more detail the distribution of neutral hydrogen in the outer parts of the Galaxy, and built a three-dimensional model of this distribution. In particular he

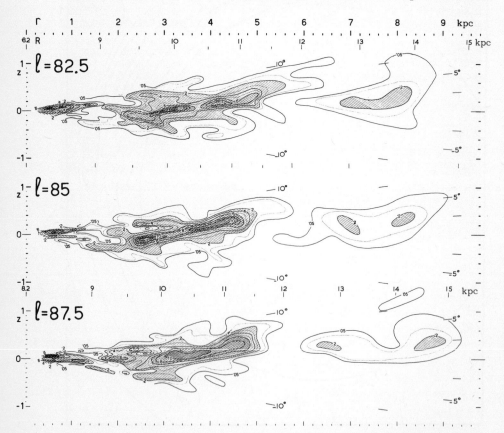

Fig. 138. Contour maps of hydrogen density in three meridional sections; r, distance from the sun; R, distance from the galactic center (Westerhout 1957a).

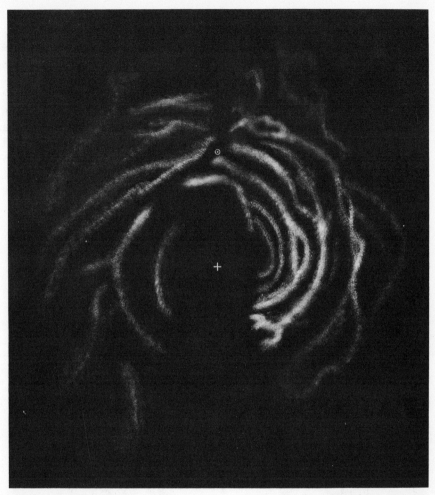

Fig. 139. Spiral structure in the Galaxy as seen from the North Galactic Pole (Oort, Kerr, and Westerhout 1958).

constructed curves joining points of equal density of interstellar hydrogen in planes passing through the sun and perpendicular to the galactic plane, for different galactic longitudes. Figure 138 shows three such "meridional sections," for $l \approx 85°$, out of a total of about seventy. From these, Westerhout derived the spiral structure of the Galaxy as it would be seen from the North Galactic Pole.

Figure 139 presents the results of an extension of this investigation, as reported recently by Oort, Kerr, and Westerhout (1958). The Orion and Perseus arms stand out quite clearly, and one more arm can be noticed, the "outer" arm, at a distance of 12 or 15 kpc from the sun. In this drawing, Westerhout has greatly exaggerated the density contrasts in order to emphasize the structure of individual arms, as well as interarm detail. He does not regard all positions of separate features as definitive, since clouds of hydrogen occupying a common site in the Galaxy but differing in velocity will appear as distinct features in the diagram.

Simultaneously with the investigation described above, Schmidt (1957a) continued the study of the spiral structure of the inner parts of the Galaxy begun by Kwee, Muller, and Westerhout (1954). As a result of an analysis of the observational data for several hundred regions of the sky near the galactic equator a chart has been drawn giving the projection of the maximum density of interstellar hydrogen on the galactic plane (Fig. 140). This chart shows considerably more

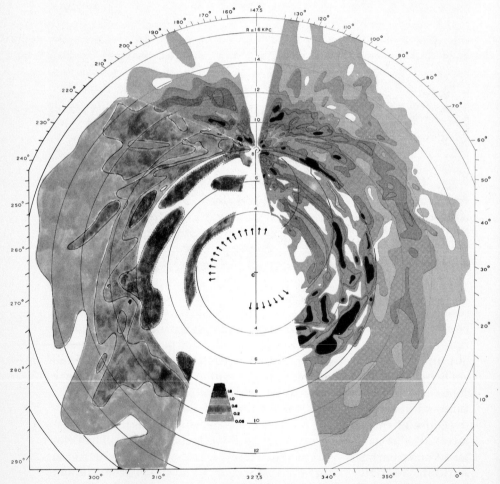

Fig. 140. Contour map of neutral-hydrogen density in the galactic system (Oort, Kerr, and Westerhout 1958).

detail than Fig. 137; in particular, it depicts very clearly the spiral structure of the inner parts of our stellar system. In the region $R < 8.2$ kpc it is possible to differentiate four or five spiral arms.

Thus, even the first investigations of the 21-cm line, though carried out with modest instruments, gave extremely valuable information about the physical properties of the interstellar medium and the dynamics and structure of the Galaxy. It is beyond doubt that, in the near future, the use of more powerful instruments will considerably augment this information.

16. FURTHER OBSERVATIONAL RESULTS

Along with the study of the general characteristics of the Galaxy (such as its spiral structure and its rotation), the analysis of 21-cm line profiles is very useful in the investigation of local details. A considerable amount of work in this direction has been done at the Harvard College Observatory, with the 24-ft radio telescope erected at the Agassiz Station at the end of 1953. The beamwidth of the antenna is about 1°.7. The receiver was built to operate by the usual frequency-modulation method; its passband can be set at 5, 15, 50, or 200 kc/s. Usually the observations were carried out with a bandwidth of 15 kc/s, which gives a probable error in the radial velocity of ± 3 km/s. The minimum signal ΔT that could be detected in the first investigations was about 5 K deg.

The investigations at Harvard followed a well-prepared program. Exactly as in Holland, they established close technical cooperation between astronomers and radiophysicists. Such cooperation and division of labor is undoubtedly the best approach. The radio telescope that analyzes monochromatic cosmic radio waves is an astronomical instrument of a new type. Naturally it belongs at an astronomical observatory and must be operated by astronomers and physicists. The observational program and the reduction and analysis of the results can only be performed by trained astronomers, who must also be responsible for the important task of comparing the optical and radio observations. Only under such conditions can work progress, but unfortunately such conditions are not realized everywhere.

One of the first projects carried out at Harvard was a study of the complex dark nebula in the region of Taurus and Orion (Lilley 1955). Lilley divided the region into five sections perpendicular to the galactic equator, between $l = 141°$ and $l = 153°$, extending over a 30° interval in galactic latitude. A remarkable peculiarity noticeable in all five sections is a sharp asymmetry between the northern and southern galactic hemispheres. Figure 141 shows the maximum intensity in the profile (in units of ΔT) as a function of galactic latitude b, for $l = 147°$. The same graph depicts the run of the interstel-

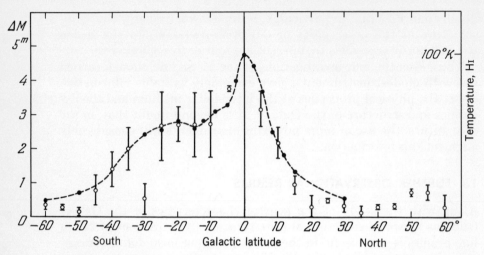

Fig. 141. Maximum intensity of the 21-cm line (black dots, broken curve) and interstellar absorption (open circles, vertical bars) as functions of galactic latitude in the vicinity of the dark nebulae in Taurus and Orion (Lilley 1955).

lar absorption in this region computed by Hubble from an analysis of distant galaxies. The general correlation between the distribution of monochromatic 21-cm radiation and light absorption is quite obvious.

Thus, it appears that in interstellar space gas clouds are associated with dust clouds. In support of this assumption Lilley used the following additional data. At the fixed galactic latitude $b = -15°$ he obtained 21-cm line profiles for the interval in longitude $110° \leq l \leq 200°$. In the region covered by the observations the interstellar light absorption changes considerably, and there are three zones of high absorption. Figure 142 shows schematically the three "zones of avoidance" (above) and the dependence of the maximum intensity of the radio-line profile on l (below). It can clearly be seen that the three tongue-shaped zones of avoidance in Orion, Taurus, and Perseus correspond to three maxima in the intensity of the 21-cm radiation.

Figure 143 shows the relation between the optical depth τ_{gas} of the monochromatic radio waves and the optical depth τ_{dust} of the cosmic dust. The first is determined from the relation

$$Te^{-\tau_{gas}} = T - \Delta T_{max},$$

and the second is given by the equation

$$\tau_{dust} = \Delta m/1.086,$$

where Δm is the interstellar light absorption. From Fig. 143 we can see that there is a general correlation between τ_{gas} and τ_{dust}. The dispersion of the points is apparently due to observational errors (mainly

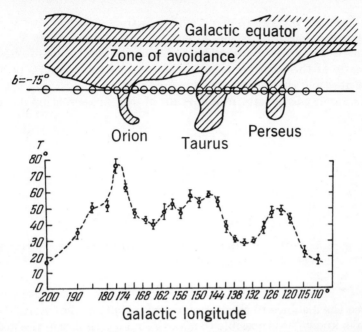

Fig. 142. The "zone of avoidance" for galaxies (above), and the intensity of the hydrogen radio line as a function of galactic longitude (Lilley 1955).

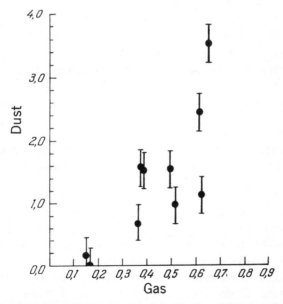

Fig. 143. The optical depth for the 21-cm radio line (H I) as a function of the optical depth for cosmic dust (Lilley 1955).

optical). From the values of τ_{gas} and τ_{dust} thus obtained we can compute the number of hydrogen atoms N_{H} and of dust grains N_{dust} in a column of unit cross-sectional area. For the computation of N_{dust} use was made of the existing data on the radii of the particles of interstellar dust and their light-absorbing properties.

The mean gas density in a column of unit cross-sectional area of length l_1 is

$$\langle \rho_{\text{gas}} \rangle = N_{\text{H}} m_{\text{H}} / l_1,$$

and for dust

$$\langle \rho_{\text{dust}} \rangle = \tfrac{4}{3} a_0{}^3 \delta N_{\text{dust}} / l_1,$$

where a_0 is the radius of the particle and δ its density. It follows that

$$\langle \rho_{\text{gas}} \rangle / \langle \rho_{\text{dust}} \rangle = 1.5 \times 10^{-11} N_{\text{H}} / N_{\text{dust}}$$

(with $a_0 = 3 \times 10^{-5}$ cm and $\delta = 1$). According to the observations, $N_{\text{H}} / N_{\text{dust}} = 0.66 \times 10^{13}$, and hence

$$\langle \rho_{\text{gas}} \rangle / \langle \rho_{\text{dust}} \rangle = 100.$$

Since the distance to the complex dark nebula in the Taurus–Orion region is known, it is possible to make a rough estimate of the mass of neutral hydrogen in it. It turns out that if a characteristic dimension of the nebula is 65 pc (according to Greenstein), the mass of neutral hydrogen is about 20 000 $\mathfrak{M}_\odot$, and the mass of dust is 100 times smaller.

Although these estimates are only very approximate, they are sufficient to show that the masses of diffuse nebulae are very high: they exceed stellar masses by factors of 10^3 to 10^4. This conclusion has great cosmological significance. Shajn and Gaze (1952), on the basis of an analysis of ionized nebulae observed in the Hα line of hydrogen, first proposed that the masses of diffuse nebulae attain very large values.

Note that the radio method of determining nebular masses is considerably more accurate than the optical. Optical observations provide us with the emission measure from the nebula,

$$\mathcal{E} = \int N_\varepsilon{}^2 ds.$$

However, the magnitude of $\mathcal{E}$ is very sensitive to the density distribution of the gas in the nebula along the line of sight, and thus, in the final analysis, to the structure of the nebula, which may be very complicated. To determine the mass we make the usual assumption of uniform density distribution, which, of course, may introduce considerable error. Therefore, the optical method makes it possible to determine only the upper limit of the mass of gas in an ionized nebula. On the other hand, the analysis of the 21-cm line profiles gives directly the number of atoms of neutral hydrogen in a column of unit cross-sectional area, that is, $\int n_H ds$. This quantity is not dependent on the structure of the nebula.

The basic result of Lilley's observations was the establishment of the close dependence between interstellar gas and dust, which is particularly pronounced at moderate and high latitudes. Roughly speaking, the interstellar gas and dust are strongly mixed and their relative proportion remains more or less constant for different aggregates of the interstellar medium. We should point out that Christiansen and Hindman (1952) were the first to notice that the 21-cm isophotes are related to regions of increased absorption. In their isophotes (Fig. 128) we can clearly see a "tongue" of increased radio-wave intensity at $l = 150°$, between $b = -5°$ and $b = -30°$. The Australian authors related this tongue to the anomalously high light absorption in this region of the sky, due to interstellar dust. They observed a similar situation in Ophiuchus.

Heeschen (1955), also at Harvard, arrived at the same general conclusion. He investigated a region near the galactic center, and detected a general positive correlation between the number of hydrogen atoms in a column of unit cross-sectional area and the interstellar light absorption in the direction of the well-known complex dark nebula in Ophiuchus. Nevertheless, Heeschen did not detect a detailed correlation between the variations of τ_{gas} and τ_{dust} in small regions. This shows that the dependence between gas and dust is of a more complex character than we would deduce from Lilley's results.

Bok, Lawrence, and Menon (1955) investigated the two darkest concentrations of dust in the nebula in Taurus, one exceptionally dark region near ρ Ophiuchi, and several areas of relatively low absorption. If gas and dust were found in the same proportion in all nebulae we would expect the intensity of the radio line from dark regions to be at least 50 percent greater than that from less dark areas. However, no systematic difference has been observed between "very dark" and "not so dark" regions. Van de Hulst, Muller, and Oort (1954), who made observations at high galactic latitudes, have confirmed this result.

According to the American authors, these results can be explained in two ways. First, we could assume that in the darkest regions a considerable fraction of the hydrogen is in molecular form. In such a case the intensity of the radio line in these regions would not increase, even though the ratio ρ_{gas}/ρ_{dust} is kept constant. Second, as van de Hulst, Muller, and Oort also point out, in these dark nebulae the kinetic temperature may be relatively low, as suggested by the work of Kahn and Seaton on H I regions. However, according to the American authors, this second explanation is inadmissible since, if it were true, the 21-cm profiles from "dark" regions and from comparison regions would be different, contrary to observations.

Therefore we need further observations to clarify the character of the dependence between interstellar gas and dust. There is no doubt about the existence of a general positive correlation between them but its details remain unexplained.

For detailed investigations it is imperative to use instruments of high resolving power. In April 1956, Harvard dedicated a 60-ft radio telescope, which should give a new impulse to the study of monochromatic radio waves.

Together with the study of the relation between interstellar gas and dust, the Harvard group has carried out important investigations on the large-scale motions of interstellar gas in several regions of our stellar system.

Menon's work (1955) on the Orion region offers considerable interest. This region contains the well-known Orion nebula, an association of hot stars, the Great Arc of Barnard, an extended maximum of thermal radio waves (Bolton, Westfold, Stanley, and Slee 1954), and a number of other interesting features. Its relatively high galactic latitude ($-10°$ to $-25°$) makes it possible to observe this region by itself, without a background of radio waves from other parts of the Galaxy. Galactic rotation can also be neglected, thus facilitating the investigation of the inner motions of the interstellar medium.

Menon's work consisted in obtaining 21-cm profiles every $3°$ in the region $160° < l < 184°$, $-25° < b < -10°$. He detected considerable variations in the profiles even for neighboring regions. The results of the observations can be interpreted to mean that a mass of interstellar gas is expanding radially with a velocity of 19 km/s from some point 500 pc from us, and at the same time it is receding with a velocity of 5 km/s. These results have great significance in the problem of stellar formation in "O-associations," first introduced by Ambartsumian, but we shall not go here into the details of this important and interesting cosmological problem.

Lawrence (1956) made the first attempt to observe the 21-cm line arising from separate clouds of interstellar gas. It is well known that the interstellar absorption lines of ionized calcium and neutral sodium observed in the spectra of some stars show a splitting into several components. Each of these components originates in a definite cloud of interstellar matter, and all these clouds move with high velocities with respect to each other. Therefore the wavelengths of the optical absorption lines arising from separate clouds will be somewhat different, and this effect will give the line the appearance of being split into several components.

Lawrence investigated the 21-cm profiles in a region where there are stars whose spectra show split interstellar lines. He chose stars at relatively high galactic latitudes so that there would be only a few clouds along the line of sight; otherwise the picture would have been too complex.

For two stars Lawrence found good agreement between the radial velocities computed both from the maxima in the 21-cm profile and from the interstellar absorption lines. Lawrence's results are shown in Table 19. The sixth column gives the radial velocity derived from the maximum in the 21-cm profile in the direction of the given star; the seventh column shows the radial velocities of the different components of the Ca II lines, arranged in order of decreasing intensity of the components. For the star HD 219 188 the radial velocity derived from the radio observations is very close to the velocity shown by the most intense component of the Ca II lines. For the other star the velocity of the "radio cloud" is almost equal to the velocity of the component of the Ca II lines second in intensity.

TABLE 19. Radial velocities of interstellar clouds.

Star	l	b	Spectral type	Photographic magnitude	v (21-cm) (km/s)	v (Ca II) (km/s)
HD 219 188	52°	−51°	B2	6.9	− 9	− 7
						+18
						−29?
HD 215 733	54	−37	B2	7.7	−10	−26
						−11
						−44
						−57

We should point out that considering the relatively low resolving power of the radio telescope used by Lawrence ($\approx 1°.7$) it is difficult to compare his results with optical observations. In the radio case the emission is received from all the clouds located within a rather large solid angle. In the optical case the observed absorption lines are due only to those clouds that are projected on the star. It is easy, then, to understand why for two other stars investigated by Lawrence no agreement was found between the radial velocities determined by radio and optical methods; this lack of correlation, however, does not diminish the value of the results shown in Table 19.

Independent confirmation of the basic conclusion resulting from these observations, namely, that to different clouds of interstellar calcium correspond different clouds of neutral hydrogen, comes from the analysis of the intensity distribution of hydrogen emission at different frequencies in the neighborhood of the star HD 219 188. Figure

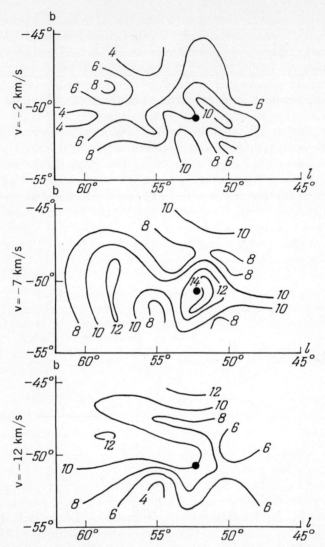

Fig. 144. Distribution of hydrogen emission in the vicinity of HD 219188 at three frequencies (Lawrence 1956).

144 shows this distribution for three frequencies, which correspond to radial velocities of −2, −7, and −12 km/s. The black dot shows the position of the star; clearly, a cloud whose radial velocity is − 7 km/s is projected against the star. It is very interesting to see how strongly the intensity distribution changes for different radial velocities. Glancing at these drawings we can get an idea of the complexity of the structure of the interstellar medium and of the magnitude of the random velocities of its different elements.

It is easy to imagine how this picture would look if it were observed with the 250-ft reflector that will soon be in use! A wealth of detail will be detected and the correspondence with the optically observed interstellar lines will probably be established beyond doubt. Furthermore, detailed observations with a large radio telescope will substantially enrich our knowledge of the structure and dynamics of the interstellar medium.

In the U.S.S.R., the Crimean expedition of the Physics Institute of the Academy of Sciences put into use, at the end of 1955, equipment for the observation of the 21-cm line, operated by B. M. Chikhachev and R. L. Sorochenko. The bandwidths of the receiver are 5 kc/s and 20 kc/s. The dimensions of the antenna are 18 m $\times$ 8 m, and it has an azimuthal mounting (Fig. 145). With the 20-kc/s bandwidth the sensitivity of the equipment is $\Delta T = 2$ K deg. With this equipment it is expected that a varied program will be carried out.

Of special interest is the projected investigation of the intensity of the 21-cm radiation from the region of the galactic pole. We have already mentioned that Pikelner arrived at the conclusion that there is very rarefied interstellar gas between denser clouds which constitute the spherical system, something like a gaseous halo encircling

Fig. 145. The radio telescope with which the Crimean expedition of the Physics Institute, U.S.S.R. Academy of Sciences, observed the hydrogen radio line.

our Galaxy. If, as we may expect, this gas is not ionized, it will emit a measurable amount of 21-cm radiation. Let us assume, for example, that the mean concentration of hydrogen atoms between clouds is 3×10^{-2} cm^{-3}, and that its extent in the direction of the galactic pole is about 5000 pc; furthermore, assume that the dispersion in the velocities of the "intercloud" random motions is large, of the order of 50 km/s. Then, according to the formulas of Sec. 15, we find that the brightness temperature in the region of the hydrogen radio line is about 4° K, high enough to be measured. The line width should be about 250 kc/s.

Another interesting problem, requiring a prolonged series of observations, is the investigation of galactic rotation away from the galactic plane, say at $b = 15°$ or $20°$. From these observations we could deduce new and interesting information about the dynamics of our stellar system.

There is one more worthwhile project: an attempt to detect interstellar hydrogen in globular clusters. We usually assume that globular clusters are devoid of interstellar gas, but actually there are no observational data on the subject and hence our assumption should be carefully examined. Globular clusters are very convenient objects for radio-astronomical observations; their angular diameters are of the order of a few tens of minutes of arc, and their linear extents about 100 pc. Considering the small dispersion in the random velocities within the cluster (1 to 2 km/s), the radio line should be quite narrow: $\Delta \nu_D \approx 10$ kc/s. Since the radial velocities of the clusters are high, the 21-cm line would not be lost in the galactic background. If the concentration of interstellar gas in the cluster is as high as 0.1 cm^{-3}, the 21-cm line would be sufficiently intense to be detected by a sensitive receiver with a 10-kc/s passband. We assume, of course, that the directional diagram of the antenna is smaller than the angular diameter of the cluster.

17. THE 21-CENTIMETER LINE IN ABSORPTION

From a general analysis of the mode of formation of the 21-cm line we would expect that under certain conditions the line may be observed in absorption. The most interesting results in this connection were obtained at the Naval Research Laboratory at Washington with their 50-ft reflector.

In principle, the interstellar 21-cm absorption line arises in the same way as the optically observed interstellar lines. The latter (mainly the H and K lines of Ca II, the sodium D lines, and several molecular lines) are observed in the spectra of distant hot stars; in a similar fashion, the 21-cm line has been observed in the radio spectra of the brightest discrete sources.

The analysis of the interstellar optical absorption lines makes it possible to determine the distance to the stars in whose spectra these lines are observed; in exactly the same way we can determine the distances to radio sources from observations of the 21-cm line in absorption. Vitkevich (1952) first suggested this method of distance determination, and Hagen, Lilley, and McClain (1955) made the first observations to this effect. At about the same time the 21-cm absorption line in the spectra of discrete sources was studied by Williams and Davies (1954) in England, and by several Australian scientists.

Let us suppose that a radio telescope, capable of detecting 21-cm radiation, is first directed toward a discrete source and later to a point in the sky in the vicinity of the source. The line profiles obtained in the two cases will show marked differences. In the second case a typical emission profile will be observed, while in the first case the profile of the radio line will appear in absorption superposed on the spectrum of the discrete source. In other words, the presence of a discrete source in the lobe of the antenna distorts the line profile considerably.

For simplicity, let us assume that there is no interstellar hydrogen present at the source and that the kinetic temperature and optical depth of the hydrogen remain constant within the limits of the antenna pattern. Then the observed profile in the direction of the source will be given by the expression

$$T(\nu) = T_a e^{-\tau(\nu)} + T_k(1 - e^{-\tau(\nu)}), \qquad (17\text{-}1)$$

where T_a is the antenna temperature of the source in the spectral region near $\nu = 1420.4$ Mc/s, but not including the 21-cm line, and T_k is the kinetic temperature of the H I region. The first term on the right-hand side of Eq. (17-1) gives the absorption of the radio waves from the source by the interstellar gas present between the source and the observer; the second term gives the emission from the interstellar gas.

Let us assume that the receiver has a second passband, a "comparison band," placed very near the main one but in the spectral region where the absorption by the interstellar hydrogen is no longer present (at the Naval Research Laboratory the two bands were placed 3.25 Mc/s apart). With this arrangement the antenna temperature observed in the comparison band is almost equal to T_a. From the observations in the comparison and signal bands it is possible to obtain a record of the difference in antenna temperatures,

$$\Delta T' = (T_k - T_a)(1 - e^{-\tau(\nu)}). \qquad (17\text{-}2)$$

Figure 146 shows schematically how the radiometer works. In case a there is no interstellar hydrogen present, and $\Delta T = T_s - T_c = 0$. In

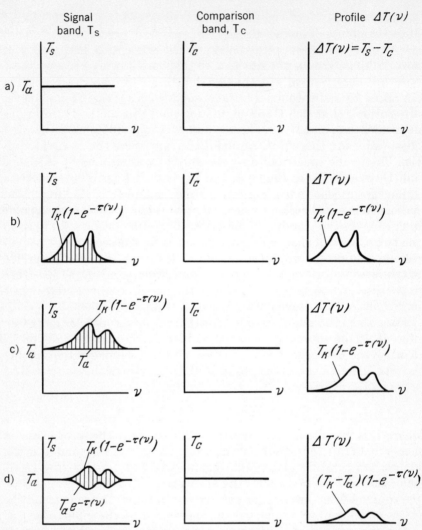

Fig. 146. Operation of the radiometer for detecting the hydrogen radio line in absorption (Hagen, Lilley, and McClain 1955).

case b the antenna is directed to a region of the sky where there is no discrete source present, and hence ΔT gives the emission profile. In case c the interstellar gas is behind the source, and ΔT again gives an emission profile. Finally, in case d the interstellar hydrogen is between the source and the antenna and the observed profile shows the effects of absorption.

It is safe to assume that the line profile in a region contiguous to the source (about 1° away, say) is similar to the profile that would be observed at the point where the source is found, if the source were not there ("expected" profile). This is equivalent to the assumption that the kinetic temperature and the optical depth of the interstellar gas in adjacent regions of the sky are the same. Therefore the profile away from the source is given by the relation

$$\Delta T = T(1 - e^{-\tau(\nu)}). \qquad (17\text{-}3)$$

From Eqs. (17-2) and (17-3) we obtain

$$T = \frac{\Delta T(\nu) \cdot T_a}{\Delta T(\nu) - \Delta T'(\nu)}, \quad \tau(\nu) = -\ln\left[1 - \frac{\Delta T(\nu) - \Delta T'(\nu)}{T_a}\right]; \qquad (17\text{-}4)$$

all the quantities on the right-hand side of Eq. (17-4) can be determined from observation.

Figure 147 displays schematically two profiles, one with the antenna pointing toward Cassiopeia A, and the other toward an adjacent region of the sky. Figure 148 shows the original record of the first profile. According to Eqs. (17-2) and (17-3) we would expect both profiles to show identical, or, at any rate, similar, dependence on frequency (to the accuracy of the surface of the reflector). However, as we can see from Fig. 147, no similarity is observed. Furthermore, according to Eq. (17-2), the maximum depth of the absorption line should be $T_a - T_k$. When Cassiopeia A was observed with the 50-ft reflector at $\lambda = 21$ cm, $T_a = 115°$, $T_k \approx 100°\text{K}$. Therefore, according to Eq. (17-3),

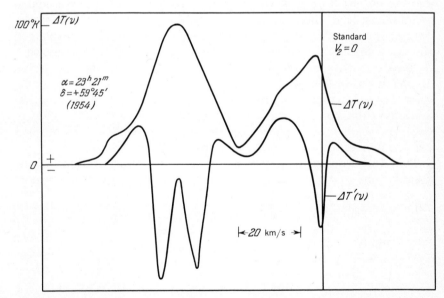

Fig. 147. Profiles of the hydrogen radio line with the telescope directed toward Cassiopeia A (lower profile) and toward an adjacent region of the sky (upper profile) (Hagen, Lilley, and McClain 1955).

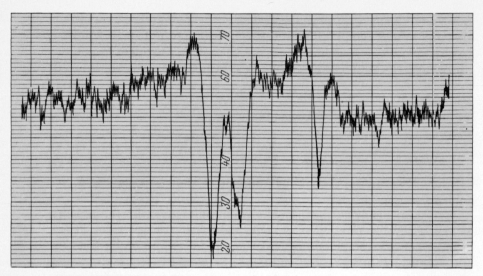

Fig. 148. The original tracing corresponding to the lower profile in Fig. 147 (Hagen, Lilley, and McClain 1955).

ΔT should be less than 15° K, but in the actual observations the depth of the absorption line reached 80° K.

In order to explain the strong discrepancies between the expected and the observed profiles, the American authors made the reasonable assumption that there exist relatively dense clouds of interstellar hydrogen moving about in a rather rarefied medium. They explained the three deep minima observed in the radio spectrum of Cassiopeia A by means of three condensations of interstellar hydrogen projected on the discrete source. Table 20 presents the data characterizing these condensations. The dispersion in the velocities of the internal motions (third column) was computed from the width of the absorption peaks after taking into account the passband of the receiver (which, incidentally, was quite narrow: 5 kc/s, or 1 km/s).

The American investigators made such an interpretation for the observations of the spectrum of Cassiopeia A in the 21-cm region.

TABLE 20. The absorption spectrum of Cassiopeia A.

Radial velocity of absorbing clouds (km/s)	Maximum optical depth	Random velocity (km/s)
− 1.0	0.96	1.7
−38.5	1.47	2.3
−48.6	2.04	1.8

They placed one of the clouds in the nearby Orion arm and the other two in the more distant Perseus arm. Then it follows that Cassiopeia A is at least 3000 pc away, and that its diameter is 5 or 6 pc.

Muller (1957), with the 25-m reflector at Dwingeloo, has recently completed a series of observations similar to those described above. Figure 149 shows the expected and observed profiles of the 21-cm line in absorption for Cassiopeia A. The heavy curve shows the true absorption spectrum. In general, Muller's results agree with those of the American authors. Naturally, his original records show considerably more detail than the schematic curves of Fig. 149. In the region of the Orion arm he detected four absorption lines, all quite narrow, while the emission profile of this arm was rather broad. It is significant that four maxima appear in the expected profile at the frequencies of the four absorption lines. This may show that the clouds producing the absorption belong to a large aggregate of clouds in the arm. The dispersion in the velocities of the inner motions in each cloud would be about 1 or 2 km/s, which is considerably lower than the velocity dispersion of the clouds in the Orion arm. The second trough in the spectrum may be due to the Perseus arm.

It is interesting to note that, as is apparent from Fig. 149, there is not a single interstellar cloud between the Orion and Perseus arms.

Muller's results confirm the idea of the American authors that the distance to Cassiopeia A is about 3 kpc. Recent optical observations of Minkowski fully agree with these results.

Muller also investigated absorption effects in the Crab Nebula (Fig. 150). A double absorption line is observed. We can assume that the part of the expected (emission) profile corresponding to negative radial velocities is formed in regions farther away from the sun than the Crab Nebula. At least for the time being, there is no need to make other assumptions in order to explain Fig. 150.

Williams and Davies (1954) made an interesting attempt to estimate the distance to Cygnus A from an analysis of the radio line in absorption. Figure 151 is a schematic picture of the distribution of interstellar hydrogen in the directions of Cygnus A and Cassiopeia A, according to Oort, van de Hulst, and Muller; the Orion and Perseus arms are also shown. Figure 152 shows the profile of the hydrogen line in the direction of Cygnus A. The two maxima in this profile correspond to the Orion and Perseus arms.

The intensity of the discrete source was measured at the frequencies corresponding to the two maxima in Fig. 152 and to points 500 kc/s on either side. The ratio $(I_b - I_a)/I_b$ for each maximum describes the absorption observed in each spiral arm. Since in the direction of Cygnus A the second arm is at a distance of 9500 pc, we can infer that the discrete source is behind the arm, and thus essentially beyond the

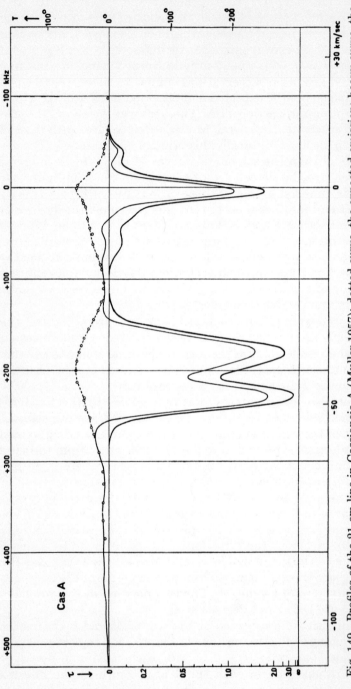

Fig. 149. Profiles of the 21-cm line in Cassiopeia A (Muller 1957): dotted curve, the expected emission; heavy curve, the true absorption; thin curve, the observed profile.

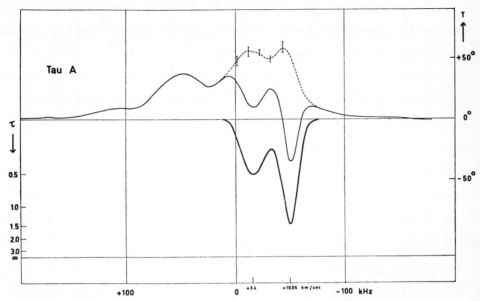

Fig. 150. Profiles of the 21-cm line in the Crab Nebula (Muller 1957); same notation as for Fig. 149.

edge of the Galaxy. Radio observations therefore confirm the extragalactic nature of this remarkable source. Williams and Davies, in a similar way, estimated the distance to Cassiopeia A, and reached the conclusion that this source is nearer than the Perseus arm. However, the antenna used by the English group was smaller (beamwidth $\approx 1°.5$) and the passband of the receiver wider than those used by the Americans. With respect to Cygnus A it is somewhat unexpected to find

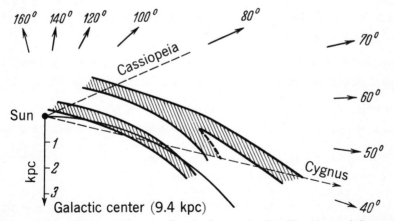

Fig. 151. Distribution of interstellar hydrogen in the directions of Cygnus A and Cassiopeia A (Williams and Davies 1954).

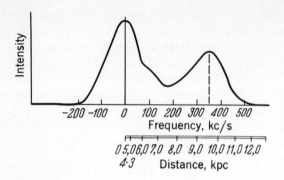

Fig. 152. Profile of the hydrogen radio line in the Cygnus A region (Williams and Davies 1954).

absorption due to the Perseus arm. Since the galactic latitude of this source is $b = 4°.9$, at the distance of the Perseus arm (9.5 kpc) the line of sight passes 800 to 900 pc above the galactic plane, where relatively dense clouds of interstellar gas may be absent.

McClain (1955) has studied the absorption spectrum of the well-known source Sagittarius A (which is in the direction of the galactic center) with the 50-ft reflector of the Naval Research Laboratory, by the same method as described above. The question of the nature of this source was for a long time the subject of animated discussion. The coordinates of Sagittarius A are the same as those of the galactic center. On this basis McGee, Slee, and Stanley (1955) assumed that this source was the nucleus of the Galaxy at radio frequencies. However, further investigations revealed a peculiarity in its spectrum. Observations at high frequencies seemed to indicate that we were dealing with thermal emission from an H II region. Recent measurements at relatively low frequencies, obtained with antennas of high directivity, have finally thrown light on the question. Shain's observations at 19.7 Mc/s and Mills's observations at 3.5 Mc/s (both obtained with cross-shaped antennas) show absorption effects in the central region of Sagittarius A. From this fact we can conclude that there is actually an H II region projected on the radio source, and causing the absorption. Optically, this H II region is very difficult to observe because of the absorption caused by interstellar dust.

Westerhout's isophotes at 1390 Mc/s clearly show that even at such high frequencies the emission from the entire source is not thermal. The strong concentration of isophotes near the galactic nucleus can only be interpreted as the result of an enormous increase of radio emission at the galactic center. The existence of a galactic "radio nucleus" can be considered an established fact.

It is to be expected that the study of the absorption spectrum of Sagittarius A at 21 cm will also help elucidate the nature of this source.

Figure 153 shows the "expected" and the observed profiles of the

21-cm line for Sagittarius A. The expected profile shows a sharp depression at $\Delta v = +5.5$ km/s, a feature observed for all 21-cm line profiles in this region of the sky. Apparently it can be explained by the presence of clouds of neutral hydrogen of relatively low kinetic temperature ($T \approx 40°$K) located quite near the sun.

By applying Eq. (17-4) to the observed and expected profiles we can compute the optical depth τ of the interstellar gas located between the observer and the source (Fig. 154). This quantity is zero for velocities less than -16 km/s or greater than $+32$ km/s. This shows that the interstellar gas radiating outside this velocity interval and forming the wings of the observed profile is located behind the source. Thus the gas producing the absorption, and therefore located between the source and the sun, has velocities lying in the narrow range between about $+30$ and -16 km/s.

In order to estimate the distance to this source it is necessary to establish a relation between the mean radial velocities of the clouds of interstellar hydrogen and their distances. In the case of Sagittarius A it is impossible to take advantage of differential galactic rotation

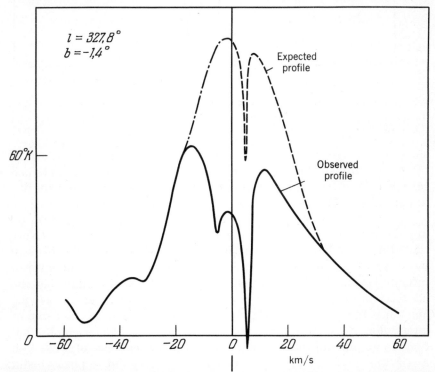

Fig. 153. Expected and observed profiles of the hydrogen radio line for the source Sagittarius A (McClain 1955).

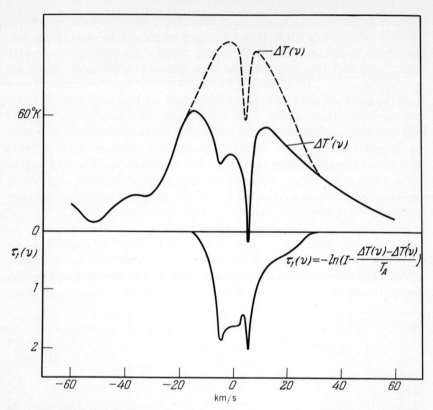

Fig. 154. Profile of the 21-cm line (above) in the direction of Sagittarius A, and the computed optical depth of interstellar hydrogen between the observer and the source (McClain 1955).

since this is zero in the direction of the center (or the anticenter). In order to establish the required relation, 21-cm profiles were obtained for $l \approx 327°.8$ (the galactic longitude of Sagittarius A) and for different galactic latitudes: $\pm 0°.5$, $\pm 1°.0$, $\pm 1°.5$, $\pm 2°.5$, $\pm 4°$.

For corresponding points of the profile the dependence of T_a on angular distance from the source ($b = 1°.4$) was determined. (Note that the source lies almost exactly on the galactic plane.) In this way it was possible to determine the "angular half-width" of the layer of interstellar gas for different points on the profile. Assuming that the interstellar gas is distributed in a plane layer of thickness $z = 240$ pc, and since $\tan(b/2) = z/2d$ (where d is the distance to the cloud of interstellar gas we are interested in), it is possible from such observations to get an empirical curve showing the dependence of $(z/d)^{-1}$ on Δv (in km/s) (Fig. 155). The two curves correspond to positive and negative velocities. The scale on the ordinate axis on the right gives directly the distance in kiloparsecs, if $z = 240$ pc.

Since Fig. 154 shows that the absorption takes place for -16 km/s $< \Delta v < +32$ km/s, then by means of Fig. 155 we can conclude that the distance to the source is between 3 and 4 kpc, from the positive curve, and between 2 and 5 kpc, from the negative curve. Hence we can safely assume that the distance to Sagittarius A is about 3 or 4 kpc. Davies and Williams (1955) obtained analogous results.

Recently, however, the hydrogen emission at the galactic center was investigated by van Woerden, Rougoor, and Oort (1957). They detected the innermost spiral arms of the Galaxy at a distance $R < 2$ kpc from the center, moving toward the sun with velocities of 50 to 100 km/s. One of these arms produces very strong absorption on Sagittarius A; from this fact we can conclude that the source is indeed located at the center of the Galaxy.

Westerhout (1958), on the basis of his own isophotes of the galactic center at 1390 Mc/s and of Mills's observations at 85 Mc/s, has given what appears to be the definitive interpretation of the source located at the galactic nucleus. His basic conclusions are the following. At the center of the Galaxy there is a nonthermal source of extent $2° \times 1°$ (300×150 pc) whose maximum brightness temperature at $\nu = 1390$ Mc/s is $25°$K, and at 85 Mc/s is $36\,000°$K. Inside it there is a thermal source of extent $0°.55 \times 0°.25$ (80×35 pc) whose maximum brightness temperature at $\nu = 1390$ Mc/s is $500°$K; at 85 Mc/s the source appears in absorption and, according to Mills's isophotes, the brightness temperature of the observed minimum is $28\,000°$ K. If the thermal source identified with Sagittarius A were not at the center of the Galaxy but between the sun and the center, then, because of its

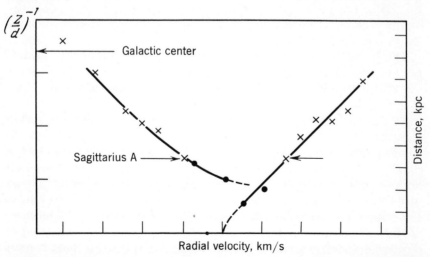

Fig. 155. Dependence of $(z/d)^{-1}$ on Δv for the source Sagittarius A (McClain 1955).

large optical depth, it would completely shield the nonthermal radiation emitted by the galactic nucleus. In such a case the brightness temperature at 85 Mc/s would be considerably lower than the value observed.

Thus, the latest radio observations have revealed a very interesting H II region at the nucleus of the Galaxy. According to Westerhout's estimate, the total mass of ionized hydrogen is $2.5 \times 10^5 \, \mathfrak{M}_\odot$ and the concentration is 85 cm^{-3}. This discovery has the remarkable significance of being the first peculiarity detected at the galactic center.

Westerhout concludes that in the inner regions of the Galaxy ($R < 3$ kpc) the interstellar hydrogen is mainly in the neutral state. From the latest Dutch observations at 21 cm it appears that this hydrogen is moving away from the galactic center in all directions with a velocity of from 200 km/s at the innermost regions to 50 km/s at $R = 3$ kpc. H II regions can first be detected at $R = 3$ kpc, and most of the ionized hydrogen is found between 3.5 and 4.0 kpc from the center.

Such a peculiar distribution of interstellar hydrogen can be explained in the following fashion. In some way ionized gas is continuously formed at the galactic nucleus. As it flows out from the nucleus the protons and electrons in this gas recombine within a few hundred thousand years; it can be shown that this process takes place within 50 pc from the nucleus, in full agreement with the observed dimensions of the H II region located at the center. As the gas continues to expand it becomes neutral. While the gas is in motion it does not condense to form stars, but between 3.5 and 4 kpc the conditions for stellar formation become particularly favorable; large quantities of interstellar gas form young, hot stars which ionize the surrounding gas to form H II regions.

The preceding is only a preliminary interpretation of the observational data. Many questions remain open—above all, where does the gas at the center of the Galaxy come from? The yearly "production" of gas at the nucleus is estimated to be $\approx 1 \, \mathfrak{M}_\odot$.

At the present time it is very difficult to answer this important question. In our opinion the only possible source of gas lies in the old population II stars, very abundant at the galactic center. The gas could originate from planetary nebulae, but if this were the case their number in a small region around the nucleus should be $\approx 10^6$! In any case the number of stars in this region would be of the order of several billion and the stellar density 10^5 times that in the neighborhood of the sun. These conclusions are so foreign to contemporary astrophysics and stellar astronomy that we shall not discuss them further in this book.

We have to face the fact that there is some outstanding peculiarity at the galactic center. To radio astronomy belongs the merit of having discovered this peculiarity, whose study will, in our opinion, become one of the central problems of astrophysics and cosmology.

Is there any other special feature at the galactic nucleus? Apparently there is. Even with the shortest exposures Baade has observed a bright disk $2''.5 \times 1''.5$ (≈ 6 pc) at the center of M 31, that is, considerably smaller than in our Galaxy. In this connection we should point out that the radio emission per unit volume is significantly smaller for M 31 than for our stellar system.

A special group of galaxies is composed of objects with very bright emission at the nucleus. They were observed by Seyfert as far back as 1943. The emitting regions have extents up to 100 pc and the emission lines are extremely wide, corresponding to random velocities up to 8000 km/s! Two such galaxies—NGC 1275 and 1068—are radio sources: could it be that the emission from the nuclei of our own Galaxy and from Seyfert's galaxies has the same nature? The near future may provide an answer to this question.

As we have seen, the analysis of the 21-cm line in absorption has already produced a series of significant results. When larger antennas come into operation this method of investigation will acquire even greater importance.

18. THE RADIO SPECTRUM OF THE GALAXY

In the preceding sections of this chapter we have discussed in detail the emission and absorption of the 21-cm line of hydrogen by different cosmic objects. Can we expect to find other emission or absorption lines in the radio spectra of these objects?

In the hydrogen atom a small splitting takes place not only for the ground state but also for excited states. Figure 156 shows the experimentally determined fine and hyperfine structure of the second quantum state of hydrogen. There are three fine-structure levels, $2^2P_{3/2}$, $2^2S_{1/2}$, and $2^2P_{1/2}$, each of which in turn splits into two hyperfine components. Therefore, the radio line $2^2P_{3/2}$–$2^2S_{1/2}$ should consist of three very close components, whose intensity ratios can be computed theoretically. According to Wild (1952), the center of gravity of this line should fall at the frequency $\nu = 9847 \pm 5$ Mc/s.

The transition $2^2P_{3/2}$–$2^2S_{1/2}$ is permitted. The transition probability for dipole radiation is given by the formula

$$A_{21} = \frac{64\pi^4\nu^3}{3hc^3g_2} S_{21}. \tag{18-1}$$

For the hydrogen atom,

$$S_{21} = \frac{9n^2}{16J} (4J^2 - 1)(n^2 - l^2)a_0^2 l^2,$$

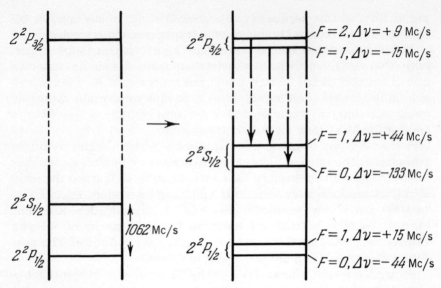

Fig. 156. Diagram of the fine structure (left) and hyperfine structure (right) of the second quantum state of the hydrogen atom.

where a_0 is the Bohr radius, g_2 the statistical weight of the upper level, and n, l, J are the usual quantum numbers. It follows, then, that the Einstein coefficient for this line is $A_{21} = 6.5 \times 10^{-7}$ sec^{-1}. The line width, taking into account the width of the level, is quite large, of the order of 100 to 200 Mc/s. The permitted transition between the levels $2^2S_{1/2}$–$2^2P_{1/2}$ should give a line of frequency of about 1000 Mc/s. Its transition probability, however, is considerably smaller than that of the $\nu = 9847$ Mc/s line.

Can these hydrogen lines be observed in the radio spectrum of cosmic objects? In particular, we are interested in the interstellar gas and in gaseous nebulae. We would expect to find these lines in H II regions, as opposed to the 21-cm line, which we expect to be emitted or absorbed only by H I regions, since only in H II regions can the Lyman α radiation field be sufficiently strong. It was assumed earlier that this field would be accompanied by a significant overpopulation of the levels $2^2P_{1/2, 3/2}$. However, computations based on the macroscopic motions in H II regions and their associated Doppler effect show that the population of these levels is sharply reduced. We can show that the number of hydrogen atoms in the $2^2P_{3/2}$ level is quite insufficient to produce the emission line $\nu = 9847$ Mc/s, especially if we take into account its very large natural width.

As for the level $2^2S_{1/2}$, its population could be quite high. This is the initial level for the emission line at $\nu \approx 1000$ Mc/s and for the absorption line at $\nu = 9847$ Mc/s. The number of hydrogen atoms in this level can be estimated from the following considerations. According to A. Ya. Kipper, the optical emission in the continuum observed in ionized gaseous nebulae can be accounted for by the transition $2^2S_{1/2}$–$1^2S_{1/2}$, accompanied by the emission of two photons. The sum of the energies of these two photons must be equal to the energy of a Lα photon, but the frequency of each photon can take any value not greater than $\nu_{L\alpha}$.

According to Kipper, the total (that is, integrated over frequency) Einstein coefficient for such a double process is about 10 sec^{-1}. On the other hand, the intensity of the continuous spectrum of ionized gaseous nebulae is comparable with the intensity of their Hα line. For the weakest gaseous nebulae, of emission measure about 1000, the intensity of the Hα line is approximately 10^{-4} erg sec^{-1} cm^{-2} sterad^{-1}. From this quantity it is a simple matter to estimate $N(2^2S_{1/2})$, the number of hydrogen atoms in the $2^2S_{1/2}$ level, in a column of unit cross-sectional area passing through the nebula:

$$N(2^2S_{1/2}) \approx \frac{4\pi I_{H\alpha}}{h\langle\nu\rangle \cdot A_{21}} \approx 2 \times 10^7 \text{ cm}^{-2}.$$

For the brightest emission objects (emission measure $\approx 10^7$), such as the central part of the Orion nebula, $N(2^2S_{1/2}) \approx 10^{11}$ cm^{-2}. However, in this case it is not possible to observe any emission or absorption line of measurable intensity: $N(2^2S_{1/2})$ turns out to be too small. For example, the optical depth of the line $\nu = 9847$ Mc/s, taking into account its large natural width, is of the order of 10^{-7}.

So far our analysis has been concerned with permitted lines between the fine-structure levels of the second quantum state of hydrogen lines whose transition probabilities are quite large. It is clear, therefore, that there is no possibility of observing forbidden lines arising from transitions between hyperfine components, since the transition probabilities of these lines are about 10^5 times as small as those for permitted lines.

Wild (1952) suggested that the $\nu = 9847$ Mc/s line may be observed in absorption in the solar radio spectrum, but we do not find his arguments sufficiently convincing. However, problems concerned with the radio spectrum of the sun are beyond the scope of this book.

Therefore, hydrogen—the most abundant element in the universe—produces only one observable radio line ($\lambda = 21$ cm) under the conditions present in the interstellar gas or in gaseous nebulae. What can be said about other elements? First of all, the cosmic abundance of the element has to be relatively high. This requirement would exclude everything except deuterium, helium, carbon, nitrogen, and oxygen. In addition, the spin of the nucleus of the most abundant isotopes of helium, carbon, and oxygen is zero, and therefore these atoms do not show any hyperfine structure. Shklovsky (1952c) has shown that, if the abundance of deuterium atoms in the interstellar medium is the same as on the Earth (D/H $= 2 \times 10^{-4}$), a weak absorption radio line of deuterium may be detected with the most sensitive equipment available. This line would arise as a transition between the hyperfine components of the ground state, that is, in the same way as the 21-cm line arises in hydrogen. The spin of the deuterium nucleus is equal to 1. Therefore for the upper state of the transition $F = \frac{3}{2}$, and for the lower state $F = \frac{1}{2}$.

The magnitude of the splitting of the ground state of deuterium due to the interaction between the spins of the nucleus and the electron has been measured directly several times. According to Kusch (1955),

$$\nu_{\mathrm{D}} = 327.384\ 30 \text{ Mc/s} \pm 30 \text{ c/s}; \tag{18-2}$$

hence the wavelength of the line is $\lambda = 91.6$ cm. The transition probability can be computed from Eqs. (14-3) and (14-6); since, in this case, $L = 0$, $S = \frac{1}{2}$, $J = \frac{1}{2}$, and $I = 1$, we obtain

$$A_{21} = 6.6 \times 10^{-17} \text{ sec}^{-1}, \tag{18-3}$$

which means that the lifetime of the deuterium atom in the upper hyperfine state is 550 million years. However, as we shall see below, this is no record.

Let us estimate now the optical depth τ_{D} at the center of the deuterium radio line. According to the equations in Sec. 14 we can express τ_{D} in terms of τ_{H}, the optical depth at the center of the 21-cm line:

$$\tau_{\mathrm{D}} = \left(\frac{g_2}{g_1 + g_2} \right)_{\mathrm{D}} \left(\frac{g_1 + g_2}{g_2} \right)_{\mathrm{H}} \frac{\nu_{\mathrm{D}}}{\nu_{\mathrm{H}}} \frac{N_{\mathrm{D}}}{N_{\mathrm{H}}} \tau_{\mathrm{H}} \approx 0.4 \frac{N_{\mathrm{D}}}{N_{\mathrm{H}}} \tau_{\mathrm{H}}. \tag{18-4}$$

If $N_{\mathrm{D}}/N_{\mathrm{H}} = 2 \times 10^{-4}$ (as on the Earth), $\tau_{\mathrm{D}} \approx 10^{-4}\tau_{\mathrm{H}}$. In the direction of the center of the Galaxy $\tau_{\mathrm{H}} \approx 10$ and hence τ_{D} should be of the order of 10^{-3}.

As has been shown above, we may expect the presence of a weak absorption line of deuterium in the radio spectrum of the Galaxy. This would account for the fact that the brightness temperature of galactic radio waves in the continuum in the direction of the center is about 300°K, a value higher than the harmonic mean of the kinetic temperature of H I regions (see Sec. 14).

The detection of the deuterium line in the radio spectrum of the Galaxy would have great astronomical and cosmological significance. The discovery of the radio line of hydrogen gave astronomy a powerful method for investigating the physical conditions in the interstellar medium as well as the structure and dynamics of our stellar system; all this was possible because the 21-cm line is very intense. The deuterium line, however, is very weak and hence could not be used in the study of the same problem. On the other hand, the discovery of the deuterium line is of special interest because it would shed light on the question of the isotopic composition of the interstellar gas.

If the interstellar gas is formed by material ejected from stars (and this is the point of view adopted by some astronomers), then it should not contain any appreciable amount of deuterium. This is because even at relatively low temperatures, about 2×10^6 °K, the deuterium present in stellar cores undergoes thermonuclear reactions and is almost completely burnt up. It has been computed that the equilibrium concentration of deuterium which has gone through a thermonuclear process is of the order of 10^{-17} that of hydrogen. Note also that at the present time there are no data revealing the presence of deuterium in the solar atmosphere. The most thorough investigation in this difficult field seems to be the recent work of Severny (1956), from which we can conclude that if any deuterium is present in the solar atmosphere its concentration is several times as small as on the Earth. On the other hand, the interstellar gas or, at least, a significant part of it may have not originated from stars, but may actually be the "relics" of the primordial matter that was present at the time of the formation of the Galaxy. In this case the concentration of deuterium in the interstellar medium cannot really be estimated.

Thus, the discovery of the 91.6-cm line would give important clues about the origin of the interstellar gas and the early history of our stellar system, about which we still know so little.

As far as we know, the first attempt to observe the interstellar radio line of deuterium was made in Canada. The experimental arrangement was similar to that usually employed for the observation of the 21-cm line. Although the time constant of the recording equipment was very long (several hours), the deuterium line was not detected. We should point out, however, that the observers made the mistake of investigating the region of the anticenter. Obviously, the Canadian group was guided by observations of the hydrogen line, whose brightness in the region of the anticenter is higher than in the opposite direction. They did not take into account that the Galaxy is opaque to the hydrogen line, but transparent to the deuterium line, so that in the case of deuterium the optical depth (and, consequently, the strength of the line) in the direction of the center is considerably higher than toward the anticenter.

Getmantsev, Stankevich, and Troitsky (1955) made an attempt to detect the deuterium line in absorption in the direction of the center. After taking the mean of a large number of records, they did not detect any positive effect. However, future observations may not confirm this result.

Stanley and Price (1956) arrived at the same negative results.

Stankevich (1958) repeated the attempt to detect the deuterium line in absorption. The frequency-modulated radiometer had a noise factor of about 5 and its time constant could be set at 6 or 12 min. As antenna he used an 8-m paraboloid.

To an accuracy of about 0.3 to 0.4 K deg in antenna temperature, no real depression in the spectrum was detected in a frequency interval of 200 kc/s centered at $\nu = 327.384$ Mc/s.

From these observations it follows that the abundance ratio of deuterium to hydrogen in the interstellar medium is $<10^{-3}$.

Thus, the question of the presence of a deuterium line in the radio spectrum of the Galaxy still remains open, and careful investigations in this field are needed.

The best possibility for observing the deuterium line in absorption seems to be in the spectrum of strong sources, such as Cassiopeia A. In such cases the optical depth at the center of the deuterium line would be $\tau_\mathrm{D} \approx 10^{-4}$ (if we assume an abundance $N_\mathrm{D}/N_\mathrm{H} = 2 \times 10^{-4}$); the depth of the absorption would be

$$\Delta T = T_\mathrm{a}\tau_\mathrm{D},$$

where T_a is the antenna temperature. With a sufficiently large reflector (about 25 m in diameter), T_a may be quite large, of the order of several thousand degrees, and ΔT may become measurable.

We shall now discuss the possibility of observing other radio lines.

It is well established that the nitrogen atom has spin unity. This leads to a splitting of its ground state $^4S_{3/2}$ into three hyperfine components with $F = \frac{5}{2}, \frac{3}{2}$, and $\frac{1}{2}$. Therefore we can expect two radio lines, corresponding to the transitions $(\frac{5}{2}-\frac{3}{2})$ and $(\frac{3}{2}-\frac{1}{2})$. Under the assumption of pure Russell–Saunders coupling, the relativistic magnetic interactions for the ground state of nitrogen vanish. However, owing to small departures from Russell–Saunders coupling, weak magnetic interactions will occur.

Recently Heald and Beringer (1954) have obtained the magnetic-resonance spectrum of N^{14}. As a result of these experiments they give for the constant A of magnetic interaction the value

$$A = 10.45 \pm 0.02 \text{ Mc/s.} \tag{18-5}$$

The constant B of electric quadrupole interaction, due to the presence of a quadrupole moment in N^{14}, was shown to be less than 0.03 Mc/s.

According to Lew (1953) the hyperfine-structure splitting of any level is given exactly by the expression

$$W = \frac{AC}{2} + B\,\frac{\frac{3}{8}\,C(C+1) - \frac{1}{2}I(I+1)\,J(J+1)}{2(2I-1)\,J\,(2J-1)}, \tag{18-6}$$

where $C = F(F + 1) - J(J + 1) - I(I + 1)$, and A and B are the constants referred to above. Comparing Eqs. (18-6) and (14-1) we see that $A = \nu_0$.

From Eq. (18-6) it follows that the separation between two hyperfine components whose quantum numbers are, respectively, $F + 1$ and F is

$$\Delta\nu = A(F + 1).$$

Taking $F + 1 = \frac{5}{2}$ and $\frac{3}{2}$ as the initial levels, we can compute the wavelengths of the two nitrogen radio lines. They are:

$$\Delta\nu_1 = 26.12 \pm 0.05 \text{ Mc/s,} \quad \lambda_1 = 11.48 \text{ m;}$$
$$\Delta\nu_2 = 15.67 \pm 0.03 \text{ Mc/s,} \quad \lambda_2 = 19.14 \text{ m.}$$
(18-7)

The magnitude of S_{21} can be computed from Eq. (14-6). With $F = \frac{5}{2}$, $J = \frac{3}{2}$, $L = 0$, $S = \frac{3}{2}$, we have $S_{21} = 60$; if $F = \frac{3}{2}$, $S_{21} = 30$. The magnitude of S_{21} for the nitrogen radio lines is unexpectedly high owing to a favorable combination of the quantum numbers.

For the 11.48-m line the Einstein coefficient is $A_{21} = 3.5 \times 10^{-19}$ sec^{-1}, and for the 19.14-m line, $A_{21} = 3.5 \times 10^{-20}$ sec^{-1}. In other words, the lifetime of nitrogen in the $F = \frac{3}{2}$ state is about 10^{12} yr.

The cosmic abundance of nitrogen is 3×10^{-4} that of hydrogen. In the direction of the galactic center, assuming a kinetic temperature for the interstellar gas of 120°K, a mean radial velocity of 8 km/s, and a column of unit cross-sectional area of length $l = 10$ kpc, we can easily compute that $\tau_{\nu_1} \approx 10^{-3}$, $\tau_{\nu_2} \approx 2 \times 10^{-4}$.

Observational evidence for the presence of the nitrogen lines seems to be the very large value of the brightness temperature in this region of the continuous galactic radio spectrum. In the direction of the galactic center, at $\lambda \approx 11.5$ m, $T_b = 100\,000$°K. Therefore the power of the incident radiation is several orders of magnitude greater than the power of the equipment's noise. The latter is determined by the magnitude of NT_0, where $T_0 \approx 300$°K, and N in this frequency range is quite small (of the order of 2 or 3). Under these circumstances, the fluctuations of the noise level will oscillate around the mean value of the antenna temperature (and not around the proper noise of the receiver, as happens for small values of the antenna temperature T_a).

According to Eq. (14-6), for $T_a \gg T$ and $\tau_\nu \ll 1$, the depth of the line can be determined from the relation

$$\Delta T = T_a \tau. \tag{18-8}$$

In order to detect the line, we must have $\Delta T > (\Delta T)_{\text{fluct}}$, or

$$\tau > (\Delta f \cdot t)^{-1/2}, \tag{18-9}$$

where Δf is the bandwidth of the receiver and t its time constant. The condition (18-9) can be rewritten thus:

$$t > (\Delta f)^{-1}\tau^{-2}. \tag{18-10}$$

For observations in the region of the galactic center the effective optical depth at the center of the 11.48-m line can be taken approximately as half the computed value, since the sources of the general galactic radio waves are only weakly concentrated toward the center. The bandwidth Δf of the receiver should be of the order of the Doppler width of the nitrogen line, that is, about 1 kc/s. Therefore, in order to detect the 11.48-m line the time constant should be greater than 2 hr. The observation of the nitrogen radio line in absorption in the spectra of intense discrete sources is considerably more difficult than in the direction of the galactic center, since for those the optical depth is significantly smaller.

We should add that we have not studied the absorption of radio waves in this frequency range in H II regions. At such long wavelengths the radiation from these sources has considerably decreased, and ΔT would be very small. Furthermore, a highly directional antenna would be required for these observations; an arrangement similar to the Mills cross would serve the purpose.

Thus, even though the time constant t can today be extended to the order of 1 or 2 hr, the observation of the nitrogen radio lines is still quite difficult. It is not, however, beyond the possibilities of present-day radio astronomy and we hope that work in this direction will be carried out in the near future.

In 1948 we had already expressed the opinion that molecular lines may be expected in the radio spectrum of the Galaxy. It is well known that molecules are present in the interstellar medium, together with atoms and dust. These molecules show in the spectra of early-type stars as weak interstellar absorption lines. At the present time CH, CH_2, and CN are known to exist in the interstellar material. Knowing the equivalent widths of these lines and the distances to the stars in whose spectra they are observed, we can compute the product Nf, where N is the number of molecules in a column of unit cross-sectional area between the star and the observer and f is the oscillator strength. Unfortunately, this last quantity is not known for the molecules mentioned above with sufficient accuracy to permit a reliable estimate of the concentration of interstellar molecules.

Under the conditions present in interstellar space, namely, very low radiation density and infrequent collisions, the overwhelming majority of the molecules are in the lowest electronic, vibrational, and rotational state. Therefore interstellar molecules produce narrow lines, rather than bands, in stellar spectra. In the case of diatomic molecules, for which the projection of the orbital angular momentum of the electron on the molecular axis (in molecular spectroscopy this number is denoted by the letter Λ) is different from zero, the rotational terms are doubly degenerate.

Changing the circular motion of the electron around the molecular axis to the "reverse" direction gives rise to a new state, whose energy to a first approximation is the same as if the electron's motion were "direct." If the molecule did not rotate, that is, if the molecular axis were fixed, the degeneracy would not be removed, that is, the two states corresponding to opposite senses of rotation of the electron would have identical energies. Actually, however, the degeneracy is removed by the weak interaction between the rotation of the molecule and the motion of the electron, which splits each rotational term into two. This phenomenon receives the name of "Λ-doublet" and is observed by analysis of the fine structure of molecular electronic spectra. Shklovsky (1949) showed that transitions between the components of Λ-doublets of several interstellar molecules may give lines in the galactic radio spectrum. These transitions are not forbidden by any of the known selection rules in molecular spectra.

The projection of the orbital angular momenta of the electrons on the molecular axis differs from zero only for very few molecules in the ground state. To this group of molecules belong CH, OH, and SiH, whose ground state is $^2\Pi$ (that is, $\Lambda = 1$). As we have already mentioned, the molecule CH has been detected in the interstellar medium, but OH and SiH so far have not. According to Dunham (1940), the equivalent width of the CH line at $\lambda = 4300.3$ A ($X^2\Pi$–$A^2\Delta$) varies, for the six stars investigated, between the limits 0.003 and 0.014 A. For weak lines (in the linear part of the curve of growth) the equivalent width W, the oscillator strength f, and the number of molecules N in a column of unit cross-sectional area are related by the well-known formula

$$N = \frac{mc^2}{\pi\varepsilon^2\lambda^2 f}W = 1.13 \times 10^{20}\frac{W}{f\lambda^2}\,\text{cm}^{-2},$$

where W and λ are measured in angstroms. According to Spitzer and Bates (1951), the oscillator strength f for $\lambda\,4300.3$ is equal to 0.06. Taking into account the distance to the stars in whose spectra the interstellar molecular line is observed, we obtain for the mean concentration of CH in the interstellar medium the value

$$n_{\text{CH}} = 10^{-9}\,\text{cm}^{-3}.$$

On the other hand, according to Herzberg (1939), $f = 0.002$ for this line, and then $n_{\text{CH}} = 3 \times 10^{-8}\,\text{cm}^{-3}$. In a later paper Spitzer (1954) gives the value $n_{\text{CH}} = 2 \times 10^{-9}\,\text{cm}^{-3}$, but does not offer any arguments in support of this value. Therefore we can say that the concentration of CH has not yet been accurately determined and further work in this direction is needed. For future work we shall adopt the value $n_{\text{CH}} = 2 \times 10^{-9} \text{ cm}^{-3}$.

In 1948, when we suggested that transitions between the components of Λ-doublets may give rise to observable radio lines, there were no experimental data available, since up to that time no investigations in this field of molecular radio spectroscopy had been made. Sanders, Schawlow, Dousmanis, and Townes (1953) were able to detect two lines of OH corresponding to transitions between the components of the Λ-doublet for the fourth and fifth rotational levels of the lowest vibrational and electronic state of the molecule. We should point out that such investigations of free radicals present very difficult experimental problems, on account of their large chemical activity. According to Sanders *et al.*, the strong absorption lines were obtained by passing the radiation through a gas tube 1 m long, kept at a pressure of 0.1 mm, and filled with a gas whose relative concentration of hydroxyl was several tenths percent. From these conditions we find that the number of absorbing molecules was about 10^{15} cm^{-2}, and that the optical depth of the absorbing gas was of order unity.

Let us assume that the mean concentration of interstellar OH is the same as that of CH, that is $\langle n_{\text{OH}} \rangle = 2 \times 10^{-9} \text{ cm}^{-3}$ (so far optical methods have not revealed the presence of OH in interstellar space). Then in a column of unit cross-sectional area and 10 kpc long, lying in the galactic plane, there will be $2 \times 10^{-9} \times 3 \times 10^{22} = 6 \times 10^{13}$ molecules. Under these conditions the optical depth of the Galaxy for monochromatic radio waves would be of the order of 0.1. Of course we should not exclude the possibility that the number of OH molecules in the column may be considerably larger. The fact that the resonance band of OH (at about 3066 A) has not yet been observed can be explained on the basis of a relatively small oscillator strength.

However, as we have repeatedly stressed, in the interstellar medium the molecules are in the lowest vibrational and rotational states, while Sanders *et al.* investigated lines arising from transitions between the components of Λ-doublets for high rotational levels. Therefore we have to compute the frequencies of the radio lines arising from transitions between the components of the Λ-doublet of the lowest rotational level of OH, CH, and SiH. Although OH and SiH have not yet been detected, their presence in interstellar space is quite probable. For example, during a collision between one O atom and one H atom a molecule of OH may be formed because of the favorable shape of one of the potential curves (the presence of a shallow minimum).

Shklovsky (1953c) has computed the wavelengths for the most interesting transitions between the components of the Λ-doublets of the ground states of interstellar molecules.

It is well known that the coupling between the different angular momenta of a molecule (the electronic Λ, the spin S, and the rotational R) can assume various forms. A similar situation is encountered in the atomic case, where we can have the extreme cases of L–S coupling, j–j coupling, and all intermediate cases. For molecules we can also distinguish two extreme coupling schemes, type a and type b, according to the classification proposed by Hund. In case a the coupling energy between the spin and the orbital angular momenta is large compared with the energy difference between rotational terms, and in case b it is small. Figure 157 shows schematically the position of the rotational terms for both types of coupling. In the case under discussion, the orbital angular momentum $\Lambda = 2$, and the spin $S = \frac{1}{2}$; J is the total angular momentum, and $\Omega = \Lambda + S$ (for type a coupling). For the same electronic state there may be a continuous transition from type a to type b coupling as the rotational quantum number R increases. Each of the levels in Fig. 157 actually splits into the two components of the Λ-doublet.

According to Van Vleck (1929), in the general case of intermediate coupling (transition case between type a and type b) the splitting of the rotational levels into a Λ-doublet is given by

$$\Delta \nu = [(\tfrac{1}{2}p + q)(\pm 1 + 2X^{-1} - YX^{-1})$$
$$+ 2qX^{-1}(J - \tfrac{1}{2})(J + \tfrac{3}{2})](J + \tfrac{1}{2}). \tag{18-11}$$

Here p and q are the Λ-doublet parameters, computed from the analysis of the fine structure of the optical spectra of the molecules; $J =$

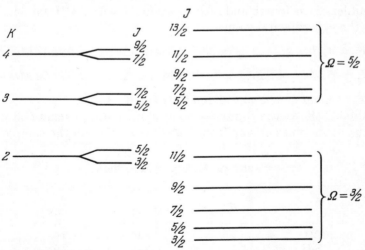

Fig. 157. Splitting of the rotational terms of a molecule in cases b and a respectively, for $\Lambda = 2$, $S = \frac{1}{2}$.

$K \pm \frac{1}{2}$; $Y = A/B_v$, where A is the multiplet splitting and B_v is the rotational constant; and X is given by the relation

$$X = +[Y(Y - 4) \pm 4(J + \tfrac{1}{2})^2]^{1/2}. \tag{18-12}$$

The plus sign in Eq. (18-11) corresponds to T_2 terms, and the minus sign to T_1 terms. T_2 terms arise from the following cases: type a coupling, regular doublets of $^2\Pi_{3/2}$; type b coupling, $J = K - \frac{1}{2}$ and inverted doublets of $^2\Pi_{1/2}$. T_1 terms arise from: type a coupling, regular doublets of $^2\Pi_{1/2}$; type b coupling, $J = K + \frac{1}{2}$ and inverted doublets of $^2\Pi_{3/2}$. "Regular" and "inverted" have the same meaning as in atomic spectra.

For OH the ground state is $^2\Pi_i$, that is, a series of rotational terms. For the molecules CH and SiH the ground state is $^2\Pi_{1/2}$ (regular doublets). Table 21 lists the parameters required in the computation of the

TABLE 21. Basic parameters of some free radicals.

	A (cm^{-1})	B_v (cm^{-1})	$Y = A/B_v$	p (cm^{-1})	q (cm^{-1})
CH	+ 28.4	14.207	+ 2.0	+0.032	+0.037
SiH	+144	7.4	+ 19.5	+0.049	+0.0154
OH	−138	18.47	− 7.45	+0.231	−0.038

Λ-doublets [the data are taken from Mulliken and Christy (1931)]. To check the accuracy of $\Delta\nu(J)$ computed from Eq. (18-11) we can calculate by means of this equation the separation of the components of the Λ-doublet of the fourth and fifth rotational levels of OH, for which we have the experimental values:

$$K = 4; J = \tfrac{9}{2}: \quad \Delta\nu_{\text{exp}} = 23\,822 \text{ Mc/s}, \quad \Delta\nu_{\text{theor}} = 23\,600 \text{ Mc/s};$$

$$K = 5; J = \tfrac{11}{2}: \quad \Delta\nu_{\text{exp}} = 36\,989 \text{ Mc/s}, \quad \Delta\nu_{\text{theor}} = 37\,060 \text{ Mc/s}.$$

The agreement can be considered fully satisfactory.

Table 22 shows the computed values of the frequencies and wavelengths of the lines arising from transitions between the components

TABLE 22. Molecular radio lines.

	J	$\Delta\nu$ (Mc/s)	λ (cm)
CH	$\frac{1}{2}$	3180	9.45
SiH	$\frac{1}{2}$	2395	12.5
OH	$\frac{3}{2}$	1640	18.3

of the Λ-doublets of the lowest rotational level of OH, CH, and SiH. The accuracy of the computed $\Delta\nu$ may be estimated as ± 10 Mc/s. Townes (1954) has computed $\nu = 1668$ Mc/s for the OH radio line, but in his paper he does not give any details of the calculation. The OH line, because of the spin of the nucleus of the hydrogen atom, splits into four components. However, we can expect to observe only the two components corresponding to the transitions $\Delta F = 0$, where $F = J + I$ (I is the spin of the proton). The spacing between the components of the 18.3-cm line should be about 1 Mc/s. The splitting due to hyperfine structure should be observable also for the CH and SiH lines.

Besides those lines shown in Table 22, other radio lines may also be produced by the same interstellar molecules.

The rotational level $^2\Pi_{3/2}$ ($K = 1$) in CH, whose energy is 28 cm^{-1} greater than that of the deepest level $^2\Pi_{1/2}$, is metastable. If the lifetime of the CH molecule in the $^2\Pi_{3/2}$ state is not less than the time between collisions of CH with hydrogen atoms, then the population of the $^2\Pi_{1/2}$ and $^2\Pi_{3/2}$ levels will be given by Boltzmann's relation, according to which the population of the $^2\Pi_{3/2}$ level will be twice the population of the $^2\Pi_{1/2}$ level. The transition between the components of the Λ-doublet of the $^2\Pi_{3/2}$ level gives a line at $\nu = 7025$ Mc/s ($\lambda = 4.27$ cm). In a similar way, in the OH molecule the transition between the components of the Λ-doublet of the level $^2\Pi_{1/2}$ gives a line at $\nu = 4350$ Mc/s ($\lambda = 6.9$ cm), and in SiH a line at $\nu = 360$ Mc/s ($\lambda = 83.5$ cm).

So far we have analyzed only hydride molecules. The interstellar gas also contains CN, but, since its ground state is a Σ state, the Λ-doublets for the rotational terms are absent. In the case of NO, which has not yet been detected in the interstellar medium but whose presence there can by no means be excluded, the ground state is $^2\Pi_{1/2}$. Transitions between the components of the Λ-doublet of the lowest rotational level of the ground state of this molecule give a line whose frequency is about 450 Mc/s ($\lambda = 67$ cm).

Thus, from theoretical considerations we can compute the frequencies of several lines due to interstellar molecules. The detection of these lines would give astronomy a new tool for investigating the nature of the interstellar medium, which is still far from clear. This problem has a deep cosmological significance because of its connection with the origin of cosmic dust in our stellar system.

It is now opportune to ask: is it possible to observe the molecular radio lines with present-day techniques? What are the possibilities for the near future?

As we pointed out before, the lower limit for the number of CH molecules (in the lowest rotational level of $^2\Pi_{1/2}$) in a column of unit cross-sectional area lying in the plane of the Galaxy and directed toward its center can be taken as $N = 6 \times 10^{13}$ cm^{-2}. For OH we can assume the same value of N. Silicon seems to be one of the more abundant elements in the universe and therefore the presence of SiH in interstellar space is to be expected. The emission by transitions between the components of a Λ-doublet is of dipole type and, to judge from the experimental data of Sanders *et al.* (1953), the oscillator strengths of these transitions are quite large. We can write down the Einstein coefficient for spontaneous emission as

$$A_{21} = \frac{8\pi\varepsilon^2 \nu^2}{mc^2} \frac{g_2}{g_1} f_{12}.$$

Since the lifetime of the molecule in the upper state is of the same order as the time between collisions of the molecule with a hydrogen atom (in clouds of interstellar gas), the distribution of molecules between the two levels of the Λ-doublet of the lowest rotational level will roughly follow Boltzmann's law. The optical depth of the column referred to above is

$$\tau = NK_\nu' \approx 10^{-2}.$$

According to Eq. (14-23), the brightness temperature at the center of the 18.3-cm line will be of the order of 1° to 2°K. Therefore, the emission line of OH should be observable with the sensitivity of present-day receivers built for monochromatic radio waves. This would require, however, the use of large reflectors, since the effective width of the emitting gas layer around the galactic plane is about 1° to 2°. The search for the OH line may be, therefore, one of the next tasks of radio astronomy.

In the case of the CH line ($\lambda = 9.45$ cm), the lifetime of the upper state is shorter than the time between collisions of CH with hydrogen atoms. Therefore, the relative population of the two levels of the Λ-doublet will no longer be given by Boltzmann's law, since the upper level will be depleted by collisions. In such a case the brightness temperature at the center of the 9.45-cm line can be estimated from the following considerations. Almost every collision between an H atom and a CH molecule in the upper level of the Λ-doublet will produce a quantum with the frequency of the radio line. The number of collisions per unit volume is given by

$$Z_{\text{coll}} = n_{\text{CH}} n_{\text{H}} v\sigma.$$

Therefore the intensity of the radiation is

$$I = n_{\text{CH}} n_{\text{H}} \frac{l\beta v\sigma}{4\pi} \frac{h\nu}{\Delta\nu_{\text{D}}} = n_{\text{CH}} n_{\text{H}} \frac{l\beta v\sigma hc}{4\pi \, v_{\text{CH}}},$$

where $\beta = 0.1$ is the fraction of the line of sight passing through a cloud of interstellar gas, and

$$\Delta \nu_D = \nu_{CH} \nu / c.$$

Taking $n_H = 10$ cm^{-3}, $n_{CH} = 2 \times 10^{-8}$ cm^{-3}, $\sigma = 10^{-16}$ cm^2, $\nu = 2 \times 10^5$ cm/s, $\nu_{CH} = 5 \times 10^4$ cm/s, and $l = 3 \times 10^{22}$ cm, we find that

$$I = 4 \times 10^{-18} \text{ erg sec}^{-1} \text{ cm}^{-2} (\text{c/s})^{-1} \text{ sterad}^{-1}.$$

The brightness temperature at the center of the 9.45-cm line is given by the relation

$$T_b = I\lambda^2/2k;$$

it is about 2 °K, which is within the limits of detection by present-day equipment. Therefore, an attempt to observe this radio line may be successful.

In this respect we must keep in mind the following. In several dark nebulae (of the type discussed in Sec. 15) the concentration of molecules may be considerably higher than that adopted so far. Several authors are of the opinion that a considerable fraction of the hydrogen in these dark nebulae is in molecular form. If this is so, the concentration of OH, CH, and so on may be exceptionally high. It would be very interesting to try to observe molecular radio lines from such objects, and in particular from the complex dark nebulae in Taurus. For these observations a very large reflector would be required with sufficient resolving power.

So far we have only discussed the possibilities of observing the interstellar molecular radio lines in emission. Conditions would not be any less favorable for observing the same lines in absorption in the spectra of the more intense discrete sources. In this case the depth of the absorption at the center of the line is given by

$$T_a(1 - e^{-\tau}) \approx -T_a\tau,$$

where τ is the optical depth.

For the sources Taurus A and Cygnus A the extent of the absorbing column is about 1000 pc. The optical depth of this column at the center of the OH and CH lines is about 0.01 to 0.02. For Cassiopeia A, the strongest source, it would be somewhat smaller. If, for example, the 250-ft paraboloid is used for such observations, the antenna temperature of the strong sources at $\lambda = 18$ cm would be of the order of 1000° to 2000°K. Hence the expected depth of the absorption line would be about 10° to 20°K, which is easily observable.

Barrett and Lilley (1957) made an attempt to detect the 18-cm line of OH. The observations were carried out with the 50-ft reflector of the Naval Research Laboratory. They covered regions in the neighborhood of the center and the anticenter and also investigated the absorption spectrum of Cassiopeia A, in a frequency band 14 Mc/s wide, centered at $\nu = 1567$ Mc/s. Their observations gave negative results, indicating that the abundance of OH is OH/H $< 10^{-5}$.

And thus we have outlined a new age in radio astronomy, which we may call "radio astrospectroscopy." Undoubtedly, further developments along these lines will enrich very considerably our knowledge of the universe.

V

Radio Astronomy and the Origin of Cosmic Rays

19. RADIO EMISSION BY THE EJECTA OF SUPERNOVAE

Taurus A was the first discrete source to be identified with an optical object. The source represents the celebrated Crab Nebula, the remnants of a supernova explosion in the year 1054; it is one of the most intense of all sources. Its spectrum has been carefully studied throughout the immense frequency range of nine octaves—from 18.3 to 9400 Mc/s. The spectrum exhibits an interesting property: the radiation flux (per unit frequency interval) is a much more slowly decreasing function of frequency than for other discrete sources of nonequilibrium radiation (Fig. 69, Sec. 6).

Here we must again emphasize that the radio emission from the Crab Nebula is of a nonequilibrium character: it is *nonthermal*. Thus, in the central portion of the Crab, the brightness temperature approaches 4×10^6 °K at $\lambda = 3$ m (see Sec. 6), a value considerably higher than the kinetic temperature in the hottest H II regions. The spectrum of the Crab also displays a nonequilibrium character. For an optically thin layer of ionized gas, the intensity of the radio emission is independent of frequency (Sec. 9). It had long been believed that the Taurus A radio spectrum possessed this very property. But high-frequency observations have convincingly demonstrated that the radio flux from Taurus A in fact diminishes, if slowly, as frequency increases. Thermal emission cannot be responsible for this type of spectrum. We note that the flux observed at low frequencies may be considerably less than the actual flux because of absorption by the interstellar medium and, in particular, by the nebula itself, which undoubtedly contains much ionized gas.

It is highly suggestive that the site of a great cosmic catastrophe, the explosion of a supernova, should be occupied by a powerful discrete source of radio waves. It points to the play of important processes which must have become effective at the time of the supernova outburst, and which must surely be intimately related to the genesis of this remarkable phenomenon. The question naturally arises whether

all peculiar nebulae resulting from supernova outbursts are radio
sources. Might we even regard all galactic discrete sources of nonequilibrium radio emission as remnants of supernova explosions?

The supernova phenomenon is of course one of the rarest in the
Galaxy. By a statistical analysis of other galaxies, Zwicky (1942) concluded that, on the average, a supernova occurs once every 300 to
500 yr in each galaxy. At least three outbursts have been observed in
our stellar system over the past thousand years: in 1054, in 1572
(Tycho Brahé's star), and in 1604 (Kepler's star); the attempts to detect discrete radio sources at the positions of these supernovae have
been of outstanding interest.

Hanbury Brown and Hazard (1952a) sought to detect a discrete
source at the site of the 1572 supernova, using the 218-ft fixed reflector of the University of Manchester. Tycho's records had designated
a point in Cassiopeia at which the supernova had flared up. Now the
most powerful discrete source in the sky, Cassiopeia A, is located
$\approx 10°$ from this point. It is a difficult matter to detect a comparatively weak source near a very intense one, for radiation from the
latter may enter the side lobes of the directional diagram, becoming
confused with any weak source that may be present. The observations
did reveal a discrete source at $\alpha = 0^h 21^m 49^s \pm 2^s$, $\delta = +64°15' \pm 35'$
(1950), whose flux at 1.89 m was $F_\nu = 1.7 \times 10^{-24}$ w m^{-2} (c/s)$^{-1}$, some
ten times as small as the flux from the Crab Nebula. Tycho's visual
position of the supernova reduces to $\alpha = 0^h 22^m 2^s$, $\delta = +63°52'$ (1950),
to an accuracy of $\pm 2' = 8^s$. A true discrete source had indeed been
observed, rather than side-lobe radiation from Cassiopeia A; special
interferometric observations showed the radiation from the new
source and from Cassiopeia A to be entirely non-coherent.

Analysis of Chinese chronicles has yielded a maximum brightness
of $-6^m.5$ for the 1054 supernova, as compared with -4^m for the 1572
outburst. The apparent stellar magnitude of the Crab Nebula, the
remnants of the 1054 outburst, is $\approx 9^m$, whereas no emission objects
brighter than 15^m have been detected at the position of Tycho's star.
On this basis, Hanbury Brown and Hazard concluded that the radio
flux from the ejecta of supernovae is proportional to the optical flux
from the supernovae at the epoch of maximum light, and is not correlated with the optical radiation from the nebular remnants of the
explosion.

The discovery of Hanbury Brown and Hazard was confirmed by a
Cambridge group (Shakeshaft *et al.* 1955). The coordinates of Object
No. 34 of the Cambridge catalogue are $\alpha = 0^h 22^m 46^s \pm 5^s$, $\delta = 63°57'$
$\pm 5'$ (1950). The flux at 3.7 m was $F_\nu = 2.86 \times 10^{-24}$ w m^{-2} (c/s)$^{-1}$; the
angular diameter was $< 9'$. It is beyond doubt that the source does in
fact represent the ejecta of the 1572 supernova. The agreement in
declination is perfect, and it is readily possible to account for the

small discrepancy in right ascension (44^s at $\delta = +64°$) through errors in the interferometry, errors in Tycho's visual position, distortions induced by the powerful Cassiopeia A, and the finite angular dimensions of the source, which are still unknown. According to the measurements at $\lambda = 1.89$ and 3.70 m, the spectrum of the supernova ejecta in this frequency range is of the form $F_\nu \propto \lambda^{0.8}$. This is a typical spectrum of nonequilibrium radiation.

In 1955 a fairly powerful source of radio waves was detected at the position of Kepler's star, the 1604 supernova; it is listed in the Cambridge catalogue as No. 1485. The coordinates are $\alpha = 17^h26^m24^s \pm 15^s$, $\delta = -21°22' \pm 15'$ (1950). The radiation flux is equal to 1.36×10^{-24} w m^{-2} (c/s)$^{-1}$ at 3.7 m, 13 times as small as for the Crab Nebula; the brightness of the supernova at its 1604 maximum was estimated at $-2^m.5$. Hence the relation noted by Hanbury Brown and Hazard is fulfilled in the present case as well: the greater the brightness of a supernova at maximum, the greater is the radio flux from its ejecta.

In the same year, Mills, Little, and Sheridan (1956b) secured a reliable record of the radio flux from the 1604 supernova, using the Sydney cross at $\lambda = 3.5$ m. The Sydney position of the source identified with this supernova is $\alpha = 17^h27^m42^s \pm 6^s$, $\delta = -21°30' \pm 3'$ (1950); the flux $F_\nu = 0.38 \times 10^{-24}$ w m^{-2} (c/s)$^{-1}$.

At Cambridge, Baldwin and Edge (1957) have investigated the radio emission by the ejecta of the 1572 and 1604 supernovae. The observations were carried out with the large Cambridge interferometer at $\lambda = 1.9$ m. They found no exact correspondence between the position of the radio source and the most probable coordinates of the 1572 supernova: in right ascension the source was located 35 to 40^s following. Nevertheless no doubt can be entertained that this source has in fact been correctly identified with Tycho's supernova. In particular, recent observations of the position of the radio source carried out at Palomar Mountain have revealed a faint nebula, the remnants of the supernova outburst. The nebula is receding from the site of the outburst at a considerable velocity.

Baldwin and Edge give an observed position for Kepler's 1604 supernova of $\alpha = 17^h27^m47^s \pm 5^s$, $\delta = -21°16' \pm 10'$ (1950); this agrees very well with the position of a faint nebula which Baade (1943) had detected at the site of Kepler's star.

The Cambridge measurements yield a radio flux for the 1572 and 1604 supernovae of 1.0 and 0.5×10^{-24} w m^{-2} (c/s)$^{-1}$ respectively. One important observation was a determination of the angular diameter of the radio source corresponding to the 1572 supernova. The value was surprisingly great: $5'.4$ between half-intensity points. The angular diameter of the source corresponding to the 1604 supernova was $<1'$.

We have seen that powerful discrete sources of radio waves have been detected at every site of a galactic supernova. One may assert that the formation of such a source is a regular consequence of *every* supernova explosion. The observed angular dimensions of the sources corresponding to the 1054 and 1572 supernovae indicate that the radio emission is localized in expanding nebulosity formed upon explosion of the star. We have at once the important result that these radio sources must be relatively young and unstable formations; they have come into existence very recently, and cannot long persist.

Thus the Cassiopeia source did not exist until 1572, nor did Taurus A exist until 1054. On the other hand, the nebulae formed upon explosion of the supernovae are expanding with great velocity. A gradual retardation will set in; eventually the nebulosity will dissipate through the interstellar gas. Oort (1946) calculates that a nebula of mass $\approx 10 \mathfrak{M}_\odot$ and the velocity of expansion ≈ 1000 km/s will be retarded by a factor of two in velocity in ≈ 5000 yr. We shall see presently that the process actually runs its course far more rapidly, since the mass of the envelope ejected at the explosion is considerably less than $10 \mathfrak{M}_\odot$. It is natural to expect that the dissipation of the nebular remnants of the supernova will be accompanied by a gradual disintegration, or fragmentation, together with the disappearance of the associated discrete source; older sources of this type, and the corresponding nebulae, would be objects of considerable angular dimensions.

Such sources are in fact observed. A typical sample is the extended source of nonequilibrium radio emission identified with the peculiar nebula IC 443 + S 40 (photograph: Fig. 94, Sec. 8). The characteristic ringlike form of this nebula, its fine filamentary structure, the great dispersion in the velocity of the internal motions (up to 100 km/s)—all these point to its formation as the result of an old supernova explosion. Another object, of even greater extent and very low radio surface brightness, was observed in Aquarius with the 218-ft Manchester radio telescope; it is associated with a faint nebula which is morphologically very like IC 443 + S 40. Finally, an extended radio source has recently been detected near the prominent group of filamentary nebulae in Cygnus (Fig. 95). It has long been recognized (for example, Oort 1946) that this nebulosity probably constitutes the ejecta of a supernova which appeared 10 000 to 20 000 yr ago.

Thus there are strong grounds for believing that the discrete sources of Class I are peculiar nebulae formed upon the explosion of supernovae in our Galaxy. (We recall that the galactic sources of thermal radio emission, the ionized gaseous nebulae, have been classified in a special group, rather than with the Class I sources; see the end of Sec. 8.)

All the aspects of the supernova phenomenon therefore assume exceptional interest for radio astronomy: the frequency at which outbursts occur, the mode of formation at the outburst, the time required for dissipation into the interstellar medium, and, of course, the physical processes in the nebulae. On the other hand, if the ejecta of supernovae are indeed powerful radio sources, then radio-astronomical data can materially enrich our understanding of the nature of supernovae.

One of the first questions to arise is whether we may identify some of the most intense radio sources with the ejecta of galactic supernovae; in particular, may the most intense discrete source, Cassiopeia A, be regarded in this way?

Cassiopeia A is a galactic object, as follows from both optical and radio observations (Secs. 8 and 17). Furthermore, several facts indicate that the radio nebula is of comparatively recent formation. Baade and Minkowski find the proper motion of its filaments to be $0''.1$ to $0''.3$ per year, so that the radio nebula is expanding at a like velocity; the angular radius of $\approx 200''$ suggests that it was formed no later than 700 to 2000 yr ago. A particularly important argument for the "youth" of Cassiopeia A is its unusually high surface brightness; for since galactic radio nebulae unremittingly expand and dissipate into interstellar space, it is quite clear that their surface brightness must fall rapidly as they evolve. We shall examine this conclusion carefully in the following section.

May we then regard the radio nebula Cassiopeia A as the remnants of a supernova outburst? The Crab Nebula is the only optical object which has been decisively established to be the ejecta of a supernova. Baade and Minkowski (1954a) did not agree to such an interpretation for Cassiopeia A, since its structure does not resemble that of the Crab. They believed that we are dealing with some very special object, of unknown nature. The author has, on the contrary, not concurred in this view, and has regarded Cassiopeia A as the ejecta of a supernova. The point is not merely that ejecta might differ markedly from one case to the next: there are reasons to suppose that the Crab Nebula is by no means a typical object of its class. And it should be remembered that ejecta of supernovae are under observation at various stages in their evolution.

That the filaments in Cassiopeia A are confined within an almost circular periphery is an independent argument for a central origin, most probably an outburst of some type. As further evidence for an explosive origin, we have the immense internal velocities within the northern arc of the nebulosity. Finally, the discovery by Baade and Minkowski of an expansion effect compels us to recognize Cassiopeia A as the remnants of a supernova outburst.

But if this powerful discrete source corresponds to the ejecta of a supernova, then the star must have flared up no more than 2000 yr ago. Ancient observers could not have failed to notice so extraordinary a celestial phenomenon as the appearance of a very bright new star. In this way, the records of the ancients may well prove useful to the physical science of today. From 1952 to 1954 the author carried out an investigation in this area—an out-of-the-ordinary branch of astronomy, to be sure—and secured the following preliminary results.

Despite several reports which may be found in ancient Chinese, Japanese, Korean, Arabic, and European chronicles of the sudden appearance of unusually bright stars, it is strange that the astronomers of those times did not give due attention to this phenomenon. We have tended to believe that galactic supernovae have been observed only in the second millennium A.D. (in 1054, 1572, and 1604). But it may now be considered established that supernova explosions have also been observed in the first millennium A.D.

Chinese chronicles constitute the best source of information on supernova outbursts.

In ancient China, systematic observations of unusual celestial phenomena were carried out as a government service. There was incorporated into the annals of each dynasty a section devoted to astronomy, containing brief accounts of all notable celestial phenomena that had occurred during the reign of that dynasty. The authenticity of these data is incontestable; for example, the apparitions of Halley's comet have been recorded *without exception* in the Chinese chronicles, for the past 2000 yr.

The Chinese attempted to distinguish strange stars, which they called "guest stars," from the comets, "sweeping stars." There is always the danger that a nova may have been confused with a comet. Novae can sometimes be verified from the descriptions of the circumstances of observation, such as a fixed position in the sky over a considerable period of time, or proximity to the galactic equator.

Until recently, the greater part of the Chinese astronomical chronicles has remained unknown to the West. The pioneer student in this field was Biot (1846), who translated a part of the 294-volume encyclopedia by Ma Tuan-lin, the *Wên-hsien t'ung-k'ao (General Commentary on the Documents)*. Here were collected the ancient Chinese observations of comets and strange stars, from the second century B.C. to A.D. 1203. From Biot's investigations, and from various Arabic and European sources, Humboldt (1850) compiled the first catalogue of novae observed over the past 2000 years.

John Williams (1871) examined several Chinese sources, among them Ma Tuan-lin's encyclopedia; he catalogued the ancient Chinese observations of comets, introducing some observations of novae as well. His chronological tables are most useful, as are his detailed charts of the constellations which had been introduced by the astronomers of ancient China.

Zinner (1919) contributed a more complete catalogue of novae cited in various chronicles. Lundmark (1921) has compiled an extensive and interesting catalogue of these objects. Up to the time of his publication, no distinctions were made between ordinary novae and supernovae.

From the 1920's until 1942, the research of several authors gradually yielded an identification of the Crab Nebula with a supernova of A.D. 1054 (for instance, Mayall and Oort 1942). Baade (1943, 1945) and Minkowski (1943) searched for nebulae which might be identified with the supernovae of 1572 and 1604. But no attempts were made to analyze the data on earlier novae, and to segregate the supernovae from them.

We shall now consider the available data on these old supernovae, together with the characteristics of radio sources with which it might be possible to secure an identification.

1. *The supernova of* A.D. *185 December 7*. This object is indisputably an authentic nova. Williams's translation of the relevant Chinese chronicles runs as follows: "In the second year of the epoch Chung-p'ing, the 10th moon, on the day Kuei Hai, a strange star appeared in the middle of Nan-mên [α and β Centauri, and neighboring stars]. It was like a large bamboo mat. It displayed the five colours, both pleasing and otherwise. It gradually lessened. In the 6th moon of the succeeding year [A.D. 186 July] it disappeared." This remarkably bright star (Lundmark estimates -6^m at maximum) appeared very close to the horizon, since its declination was $-60°$. Optical effects due to the atmosphere may therefore account for some of the picturesque details (the comparision with a bamboo mat, the reference to changing color); Tycho Brahé had described color effects in the 1572 outburst. According to Lundmark, the approximate position of the 185 supernova was $\alpha = 14^h.2$, $\delta = -60°$ ($l = 282°$, $b = 0°$).

In the winter of 1954, the author appealed to the Academia Sinica in Peking, suggesting that a search of the ancient chronicles be made in the hope of bringing to light new information on the exploding stars of the past. Our suggestion was attentively followed up, under the auspices of the Commission on the History of Chinese Natural Sciences. Upon examining the chronicles and other sources, Aspirant Hsi Tsê-tsung found references to the outbursts of 90 novae, some of which had not been known to the West. In particular, he was able to demonstrate that the 1572 and 1604 supernovae had been systematically observed in China. Much effort was devoted to the establishment of precise dating, and to the derivation of approximate positions for the novae. Hsi's catalogue (1955) is now available in English translation (Hsi 1958), but we are very grateful to Mr. Hsi for having communicated the preliminary results of his study to us.

Reference to the A.D. 185 supernova was found to have been made in the annals *Hou Han-shu (History of the Later Han Dynasty)* as well as in the *Wên-hsien t'ung-k'ao*. The relevant excerpt from the former document reads as follows: "60th cyclical day, 10th month, 2d year of Chung-p'ing of later Han, guest star appearing at Nan-mên, big as half of a mat, five colors and different tempers, later a little diminishing, disappearing in the 6th month, next year." Although this passage has much in common with Ma Tuan-lin's description, it is important to realize that Hsi Tsê-tsung located it in a different source, a more ancient one than the *Wên-hsien t'ung-k'ao*. Indeed, the *Hou Han-shu* passage may have been used by Ma Tuan-lin in compiling his encyclopedia.

Not far from the position of the 185 supernova is located one of the most intense galactic radio sources. The coordinates of the source are (Mills 1952a) $\alpha = 13^h35^m$, $\delta = -60°$; the radio flux $F_\nu = 7 \times 10^{-24}$ w m^{-2} (c/s)$^{-1}$ at $\nu = 100$ Mc/s, fully one-third of that from the Crab Nebula. A more recent position of the source is (Mills, Little, and Sheridan 1956b) $\alpha = 13^h43^m \pm 0^m.2$, $\delta = -60°12' \pm 3'$ (1950). Unfortunately, this intense source has not yet received adequate attention; there are no available data on its angular dimensions and spectrum. Optical observations will also be necessary, in order to detect any remnants of the 185 supernova. We regard the identification as reliably established: the coordinates are in good agreement, the record of the occurrence of the outburst is trustworthy, and the high value of the radio flux would be expected on the basis of the great brightness of the supernova.

The very low galactic latitude of the object, together with its "slow" light curve, indicate that it was a true supernova, and not merely a nearby "slow nova."

2. *The supernova of* A.D. *1006*. This object is again unquestionably an authentic one. We have, for example, the testimony of the annals prepared by the St. Gallen monk Hepidannus (d. 1088); these annals extend from 709 to 1044. The Latin passage, in Humboldt's emendation (1850, p. 156), informs us that "in 1012, in the southernmost sky, in the sign of the Ram, a new star was seen for three months, beginning in the month of May. It was of exceptional magnitude, and of a brilliance which blinded the eyes. It behaved in a wondrous way, now larger, now smaller; at times it could not be seen at all." Another European manuscript gives a different date, 1006; it is this date which is now accepted for the outburst.

From an analysis of the Syriac chronicle of Barhebraeus, Chladni concluded that the star appeared in Scorpius. Here is a fragment from this chronicle, as quoted by Schönfeld (1891): "In the year 396 [of the Hegira] there appeared a star resembling Venus in size and brightness ...its rays moved, and it shone forth like the moon. It remained four months, and then vanished." Gildemeister had pointed out an even more ancient reference to this nova, in the Arabic chronicle of Ibn al-athîr (d. 1233). This is again quoted by Schönfeld: "In this year [396 of the Hegira], on the first of Sha'bân [1006 May 3], there appeared a large star, resembling Venus, to the left of the *qïbla* from 'Irāq [that is, the direction from Babylonia toward Mecca]. It shed rays on the Earth like the rays of the moon, and remained until the 15th of Dsul-ka'da [1006 August 13], when it vanished." Schönfeld concluded that the 1006 nova appeared at the southern boundary of Scorpius, and believed that, in any event, its brightness could not have fallen far short of that of the 1572 nova.

Japanese chronicles are also available; these contain a useful estimate of the brightness of the 1006 outburst (Iba 1934). Samurai Sadaiye Fujiwara had observed a nova in 1230; the experience incited him to compile a list of novae which had appeared in the past—stars whose memory had been preserved in the popular tradition. The list, incorporated into his diary *Meigetsu-ki,* refers to outbursts of novae in the years A.D. 877, 891, 930, 1006, 1054, 1166, and 1181. The novae of 1006 and 1054 were especially bright. The 1006 nova, appearing at the end of April, "was so luminous that the people could see it without any difficulty. It looked like the moon at its quarter phase." It is interesting that in the same Japanese chronicle we find the brightness of the 1054 nova compared to that of Jupiter. Shigeru Kanda (Iba 1934) calls attention to Chinese annals, which indicate that after the outburst the 1006 nova "was observed for 10 years in China, witnessed in the East at dawn in November, and dipped under horizon in S. W., in August every year." Taking into consideration the precession over almost 1000 yr, it follows that the 1006 nova appeared in Lupus.

Biot's translation (1846) of Ma Tuan-lin confirms the independent Japanese records. The Academia Sinica investigation has brought to light further Chinese sources, which also indicate that the 1006 nova appeared in Lupus; they too emphasize the exceptional brilliance and long duration of the phenomenon. According to the *Yü-hu ch'ing-hua* (Hsi 1958): "3d year of Ching-tê of southern Sung, a big star seen at the west of T'ien-ti, shining like a golden circle, no one knows this star." From the combined evidence of all these sources, we may conclude with great assurance that the star observed in 1006 was indeed a supernova. Its position, however, has not yet been established with an accuracy sufficient to warrant serious attempts at securing an identification of the supernova with a discrete radio source.

3. *The supernova of* A.D. *369.* In Lundmark's catalogue (1921), this object is allotted the highest confidence-designation (very great probability that the nova is genuine).

Ma Tuan-lin's 294-volume encyclopedia provides the following report (after Biot 1846): "In the T'ai-ho period, in the 4th year and the 2d moon [that is, A.D. 369 March], a strange star was seen along the western wall of the Blue Palace [Tzŭ-kung]." In the astronomical nomenclature of ancient China, "Blue Palace" refers to the zone of declination within which lie all stars that never set (zone of perpetual apparition).

Williams (1871) and Hsi (1958) cite other chronicles, such as the *T'ung-chih.* Hsi quotes this excerpt: "2d month, spring, 4th year of T'ai-ho of Chin, guest star seen along the west edge of Tzŭ-kung, disappearing in 7th month [that is, 369 August]."

There is no doubt that both excerpts deal with the same star. We can specify the position of this nova more accurately by noting that it had been observed near Peking, in latitude 34° N. It follows that the declination of the star was $\approx +56°$. Further, the star was first observed in March, when the sun is near zero declination. The evening hours would have been the natural time for making the observation, or, at any rate, the hours before midnight. It follows that the right ascension must be close to 23^h. Lundmark estimates the maximum brightness as -3^m, but this value is quite arbitrary.

We have proposed to identify the 369 supernova with the discrete source Cassiopeia A (Shklovsky and Parenago 1952), whose position is: $\alpha = 23^h21^m$, $\delta = +58°32'$ (1950). Cassiopeia A is thus within 5° of the site of the outburst, the position of the latter being but roughly established. Hsi, however, now questions this identification.

4. *The nova of* A.D. *1181.* Up to five novae may have appeared in Cassiopeia over the past 2000 years: in A.D. 369, 945, 1181, 1264, and 1572. The objects reported in 945 and 1264 have been assigned positions very close to that of the 1572 supernova; but Lundmark (1921) gives the observations low weight, so that these two objects probably must be discarded from our list. On the other hand, the star observed in 1181 offers considerable interest. Hsi Tsê-tsung (1958) provides the following translation of excerpts drawn from the *Sung-shih (History of the Sung Dynasty),* and from the *Wên-hsien t'ung-k'ao:* "6th cyclical day, 6th month, 8th year of Ch'un-hsi of Sung, guest star appearing at K'uei [Andromeda, Pisces], invading the Ch'uan-shê-hsing [Cassiopeia region], till the 10th cyclical day, 1st month, next year, lasting altogether 185 days [1181 August 6 to 1182 February 6]." In consequence of an error in Biot's translation (1846) of the *Wên-hsien t'ung-k'ao,* this object, indisputably a nova, has heretofore been taken for a comet.

The text of the annals permits the site of the outburst to be fixed at $\alpha \approx 1^h40^m$, $\delta \approx +70°$; this is quite far from Cassiopeia A. The 1181 star had been recorded in many contemporary chronicles, Japanese as well as Chinese, according to the Academia Sinica. In brilliance the nova had been compared with Jupiter. We have here a fully reliable object. If for some reason the source Cassiopeia A cannot be identified with the 369 nova (if, for example, it should be found that the expansion velocity of the nebula is too great), then the nova of 1181–82 would remain the sole outburst known which could afford such an identification. In this event a revision in its position would be required.

In summary, then, we may assert that over the past two millennia galactic supernovae have definitely been observed in A.D. 185, 369, 1006, 1054, 1572, and 1604. Of these, all but the third are known to be in good correspondence with sources of radio waves. It has not yet been possible to secure a unique identification with the entirely reliable supernova of 1006. There is always the possibility that still other supernovae, of which no record has been handed down to us, may have flared up during the same period.

The maximum brightness of all the supernovae we have considered (except for the 1604 star) exceeded -3^m. This suggests that the explosions have taken place relatively close to the sun, within 3 or 4 kpc; and that on the average supernovae have appeared once every 300 yr. Figure 158 gives, diagrammatically, the positions of these supernovae in the galactic plane, with respect to the position of the sun. The distance to the supernovae of 185, 1006, 1572, and 1604 has been derived on the assumption of an absolute magnitude of $-16^M.5$. These distances have undoubtedly been exaggerated: statistics of supernovae in external galaxies lead to a mean absolute magnitude near -15^M. Kulikovsky, setting aside a few bright supernovae, found the value $M = -13^M.3$; this is now subject to a correction of $1^m.5$ because of the new distance scale. A fairly reliable distance of 1.1 kpc is available for

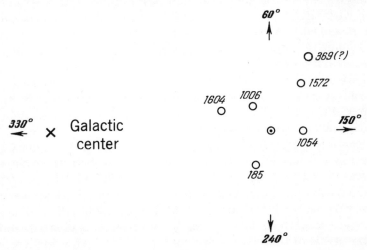

Fig. 158. Distribution of ancient supernovae in the galactic plane.

the 1054 supernova; for the 369 supernova, radio data yield a distance of 3 kpc (see Sec. 17). The absorption of light has been taken as 1^m/kpc. Figure 158 does not exhibit a preferential distribution of supernovae in the direction of the galactic center.

It is hardly likely that the sun happens to lie at the center of a local system of exploding supernovae. A far more natural assumption would be that supernovae appear with much the same frequency throughout the galactic plane. If this assumption is correct, then the frequency of supernova outbursts in our Galaxy must be considerably greater than the currently accepted value.

Let us suppose that the supernovae are not concentrated to the galactic center. Then the frequency of supernova outbursts over the entire Galaxy must be increased over the observed frequency by a factor of at least $(R_1/R_2)^2$, where $R_1 \approx 9$ kpc represents the greatest distance from the galactic center at which supernovae may occur (the 1054 supernova was 9.6 kpc from the center), and R_2 denotes the radius of the domain within which the historical supernovae have been observed. Application of this factor gives us a new mean frequency for supernova outbursts: one outburst every 30 to 60 yr! This frequency can hardly be cut down by as much as two or three times, even if all possible circumstances were pressed in that direction. We arrive at the inescapable conclusion that the supernova phenomenon in our Galaxy is at least an order of magnitude more common than had been believed: a supernova explodes every 30 to 60 yr, on the average.

Baade and Minkowski (1954b) have drawn attention to the fact that the three familiar galactic supernovae, those of 1054, 1572, and 1604, have all appeared within the solar neighborhood; but they regarded this as a coincidence, a chance fluctuation. We are unable to agree with this view. The probability that three successive galactic supernovae will appear within 1.5 kpc from the sun is $< 10^{-4}$; if we also include the supernovae that appeared in the first millennium A.D., the probability drops to 10^{-8}.

There exist galaxies which are anomalous in the frequency of their supernovae. For example, supernovae have been observed in NGC 3184 in 1921, 1937, and 1939; in NGC 6946, in 1917, 1939, and 1948; and in NGC 4321, on two occasions.

It cannot be merely a chance fluctuation that in three galaxies the frequency of supernova outbursts is so high. On the basis of the mean interval usually quoted—one outburst every 300 yr—the probability for such a fluctuation is clearly negligible. Thus there appear to be spiral galaxies with a relatively high frequency of supernovae—one outburst every 10 to 50 yr—and the impression is created that our Galaxy is one of these, a member of a species which is evidently not at all uncommon.

All the above galaxies—those with an anomalously great frequency of supernovae—are of type Sc. The great frequency of supernovae in our own Galaxy therefore argues in favor of an Sc classification for our system. It is of interest that other arguments can be adduced which gainsay the usual classification of our Galaxy as type Sb. For example, our system is considerably richer in stellar associations and H II zones than is M 31; these are characteristics of Sc spirals. However, in a very important characteristic—the relative abundance of interstellar gas—the Galaxy is very similar to M 31, and is quite unlike such typical Sc spirals as M 33 and M 51.

The frequency of supernova outbursts in the Galaxy is of vital significance for radio astronomy. We have already emphasized that the ejecta of supernovae persist as radio sources for a relatively brief period. As the nebulosity expands and dissipates, the flux and, especially, the surface brightness of the corresponding radio source will gradually diminish. The number of discrete sources of age $< t$ will evidently follow a relation of the form

$$N(< t) = \nu t,$$

where ν is the frequency of supernovae in the Galaxy. In particular, the number of "young" sources such as the Crab Nebula—sources for which $t < 3000$ yr—can hardly exceed 100 over the entire Galaxy; the number is probably even smaller. The radio observations are in accord with this conclusion, as is evident from the statistics of Class I sources (Sec. 8).

"Older" objects, such as IC 443 + S 40, must be considerably more common. From the data to be presented in the next section, we may estimate their galactic population as roughly 200 to 300. Very old objects must be still more abundant; an example is the extended low-intensity source in the Cygnus Loop region, the nebulosity NGC 6960, 6992, 6995. There must be two or three thousand such objects in our stellar system.

The frequency of galactic supernovae is of exceptional importance in questions of cosmic-ray origins, and in the closely related problem of the general field of nonequilibrium galactic radio waves (see Sec. 23).

20. THE NATURE OF THE RADIO EMISSION BY THE EJECTA OF SUPERNOVAE

In this section and the next we shall investigate the emission mechanism in the radio nebulae formed by the remnants of supernova outbursts. We shall deal with both the radio and the optical emission by these nebulae: the two are intimately connected, as in the case of the Crab Nebula.

A distinctive property of the spectrum of the Crab Nebula, the most intense and hence the most thoroughly investigated source of this type, has been considered to be the independence of spectral flux density of frequency, at meter wavelengths. (We recall that most investigators have heretofore tended to reject a supernova origin for the source Cassiopeia A; see Sec. 19.) But thermal emission from an optically thin layer of gas possesses a similar spectrum (Sec. 10). Hence, for several years after the discovery of the Taurus A source, it had been widely believed that its emission is of thermal character. Since a continuous optical spectrum is very prominent in the Crab Nebula, the radio emission, believed to be thermal, had been regarded as an extension of the thermal optical continuum. But it soon became clear that the state of affairs is quite different.

Were we to assume that the radio emission by the Crab is thermal, we would find upon analyzing the observations that the kinetic temperature of the amorphous nebular mass would be $> 4 \times 10^6$ °K, with an electron concentration $N_\varepsilon \approx 5 \times 10^4$ cm^{-3}; this contradicts an essential interpretation of the optical observations, according to which $T_\varepsilon \leq 150\,000$°K, and $N_\varepsilon \approx 2 \times 10^3$ cm^{-3}.

But apart from this evident contradiction there is another circumstance, one which is decisive for the point at issue here. For thermal emission, the intensity per unit frequency interval should be of the same order of magnitude in the optical range as in the radio range. The only difference possible arises from the inconsequential logarithmic factor $g = \ln(b_2/b_1)$, since by Sec. 9 the intensity is proportional to g. From the expression for b_2/b_1 [see Eq. (9-32)] we find that g will be only three times as great at $\nu \approx 2 \times 10^8$ c/s as at $\nu \approx 7 \times 10^{14}$ c/s (the photographic region of the spectrum). Thus if the radiation from the Crab were of thermal origin, the continuum intensity per unit frequency interval would be no more than three times as small at $\lambda = 4300$ A as at radio wavelengths. The radio flux F_ν from the Crab is $\approx 1.9 \times 10^{-23}$ w m^{-2} (c/s)$^{-1}$ = 1.9×10^{-20} erg sec^{-1} cm^{-2} (c/s)$^{-1}$; hence the optical intensity would be fully 6×10^{-21} erg sec^{-1} cm^{-2} (c/s)$^{-1}$. In fact, the optical intensity would attain an even higher value, for free–bound transitions must be considered in addition to free–free transitions.

The true situation is not at all like this. The color temperature of the Crab is near 8400°K for λ between 4000 and 5000 A (Minkowski 1942); allowance has been made for interstellar absorption. The bolometric correction is $0^m.9$ at $T = 8000$°K. Taking for the photographic magnitude of the nebula $m_{\text{pg}} = 9^m$, we have a theoretical apparent bolometric magnitude of $m_{\text{bol}} = 8^m.1$ for the Crab. Now for the sun, $m_{\text{bol}} = -26^m.95$; hence the theoretical bolometric flux of optical radiation from the Crab is 10^{-14} times that from the sun, and is equal to

$$1.32 \times 10^6 \cdot 10^{-14} = 1.32 \times 10^{-8} \text{ erg sec}^{-1} \text{ cm}^{-2}.$$

At $T = 8000°$K, 30 percent of the total radiation from a black body falls within the interval 3500 A $< \lambda <$ 5000 A. The flux from the Crab will therefore be 4×10^{-9} erg sec^{-1} cm^{-2} in this interval; and this is an *observed* value for the flux, not a theoretical one, for the color temperature does equal $8000°$K in this portion of the spectrum. We have introduced a theoretical bolometric correction only in order to pass to an absolute value for the energy flux.

The frequency range over the interval 3500 A $< \lambda <$ 5000 A is 2.6×10^{14} c/s. It follows that the observed flux of photographic radiation from the Crab is $F_\nu \approx 1.5 \times 10^{-23}$ erg sec^{-1} cm^{-2} (c/s)$^{-1}$, a value *400 times as small* as is obtained from an assumption that the radio emission is of thermal character. This great discrepancy cannot be removed merely by introducing an accepted value for interstellar absorption of light. We conclude that the emission of radio waves by the Crab Nebula cannot be a thermal process.

Recent observations in the wavelength interval 3.2 cm $< \lambda <$ 10 cm have shown (Sec. 6) that the radio flux from the Crab is three times as small in this range as at $\lambda = 3$ m. Clearly this is further evidence that the radio emission is nonthermal.

Two nonthermal mechanisms of radio emission are known: (*a*) plasma oscillations; (*b*) radiation by relativistic electrons in magnetic fields.

As early as 1946 we had proposed plasma oscillations as a mechanism which could account for the radio waves emitted by the disturbed sun (Shklovsky 1946). Since that time a number of efforts have been made to develop this idea; in fact, an overwhelming majority of investigators have invoked this mechanism to explain much of the radio emission by the disturbed sun. Reliable spectral analyses have been made which confirm the mechanism as the one responsible for chromospheric flare emission (Wild, Murray, and Rowe 1953, for example). But may we also regard plasma oscillations as the mechanism operative in the ejecta of supernovae?

It is very tempting at first to apply this mechanism to the remnants of supernova outbursts. Nebulae of this type are composed of gaseous masses moving at immense velocities. Powerful plasma oscillations can arise from interpenetrating clumps of charged particles. However, if we lay aside the complicated matter of the possible emission of transverse electromagnetic waves under the processes and physical conditions obtaining in the nebulae, we find that the frequencies of these oscillations must be many times as low as the frequency of the radio waves observed. The oscillation frequencies must in fact be of the same order as the *plasma frequency* ω_L, whatever the prevailing conditions; here

$$\omega_L = (4\pi N_\varepsilon \varepsilon^2/m)^{1/2}. \tag{20-1}$$

Even for a free-electron concentration $N_\varepsilon = 10^4$ cm^{-3}, we would have $\omega_L = 5.6 \times 10^6$ c/s, which is tens or hundreds of times below the observed values.

There is also a direct observational demonstration that the plasma mechanism is essentially inoperative in radio nebulae: there is no correlation between the brightness distributions at radio and optical wavelengths for these nebulae. Such a correlation would be a necessary consequence of a plasma mechanism, since the optical brightness is proportional to the emission measure $\int N_\varepsilon^2 ds$.

We conclude, then, that the plasma-oscillation mechanism must be discarded. On the other hand, in deceleration radiation by relativistic electrons in magnetic fields we have a mechanism which can fully explain all the peculiarities of the radiation from the ejecta of supernovae. In their pioneering paper, Alfvén and Herlofson (1950) had appealed to this mechanism in order to account for the radio emission by discrete sources (see Sec. 11); at that time hardly anything was known of the nature of discrete sources. Ginzburg (1951) also applied the mechanism to discrete sources; but here as well it was not possible to study specific astronomical properties of the sources, for their nature was still not understood.

By 1952 the picture had brightened considerably: the most intense sources had been identified optically, and their angular dimensions had been measured for the first time. In the following year, we turned to a special type of nebula—the remnants of supernova outbursts—and applied the mechanism of radiation by relativistic electrons (Shklovsky 1953*d*, 1953*e*). The intensity of the radiation, the character of the internal motions, the density of matter in the nebulae, and other observational data were all taken into account. Our calculations were refined by Ginzburg (1953*a*), who studied the energy spectrum of the relativistic electrons; this determines the spectral composition of the radio emission.

The basic formulae of this "synchrotron-radiation" theory have been introduced in Sec. 13. To apply them to the radio nebulae formed in supernova outbursts, we must first evaluate the exponent γ appearing in the expression for the differential energy spectrum of the relativistic electrons,

$$N(E)dE = KE^{-\gamma}dE$$

[see Eq. (13–6)]. According to Eq. (13–7), we have for the spectral density of the radio emission from the relativistic electrons

$$F(\nu) \propto \lambda^{(\gamma-1)/2} = \lambda^n.$$

Observations of Cassiopeia A yield $n = 1.2$ at meter wavelengths and $n = 0.6$ at decimeter wavelengths; hence $\gamma = 3.4$ and 2.2, respectively. The radio spectrum of the Crab Nebula may be expressed in the form

$$F(\nu) \propto \nu^{0.2}$$

throughout the observed frequency range; thus here $\gamma \approx 1.5$.

We must now turn to the magnetic-field intensity prevailing in the ejecta of supernovae. It follows from general considerations of magnetohydrodynamics that a magnetic field is to be expected within an ionized gas in turbulent motion. The velocity v of the internal motions in the Crab is of the order of 300 km/s. From the formula

$$\tfrac{1}{2}\rho v^2 = H^2/8\pi$$

we find, for $\rho = 5 \times 10^{-22}$ gm/cm³ and $v = 3 \times 10^7$ cm/s, that $H \doteq 2 \times 10^{-3}$ gauss. This is evidently an upper bound on the intensity of chaotic magnetic fields in the Crab. It is important to note that the source of the radio emission is localized in the interior of the nebula where the density of matter is low, possibly of the order of 10^{-24} gm/cm³ (see Sec. 21). Furthermore, the magnetic field may be determined by the turbulent velocities of small eddies, velocities less than the observed velocities of internal motion. A more probable value is thus $H \approx 3 \times 10^{-4}$ gauss. For Cassiopeia A we shall take $H \approx 5 \times 10^{-4}$ gauss. The distance R to the Crab is ≈ 1.1 kpc, and to Cassiopeia A, ≈ 3 kpc (Sec. 17). For the effective volumes V of these nebulae we shall take 5×10^{55} cm³ and 6×10^{56} cm³, respectively.

It is now possible to compute, from Eq. (13–7), that $K \approx 5 \times 10^{-9}$ for the Crab, and $\approx 1.8 \times 10^{-11}$ for Cassiopeia A. The observed radio emission will essentially be generated by relativistic electrons with energy E in the interval $E_1 < E < E_2$; the concentration of these electrons will be

$$N = \int_{E_1}^{E_2} N(E)dE = K \int_{E_1}^{E_2} E^{-\gamma}dE.$$

For the Crab, $E_1 \approx 2 \times 10^7$ ev and $E_2 \approx 2 \times 10^{10}$ ev, whence $N \approx 2 \times 10^{-6}$ cm⁻³. The total population of relativistic electrons with 2×10^7 ev $< E < 3 \times 10^9$ ev in the Crab is therefore $5 \times 10^{55} \times 2 \times 10^{-6} = 10^{50}$. For Cassiopeia A, we again take $E_1 \approx 2 \times 10^7$ ev; then $N(E > E_1) \approx 2 \times 10^{-4}$ cm⁻³. The number of relativistic electrons in the entire nebula having $E > 2 \times 10^7$ ev will then be $\approx 10^{53}$.

In this way a comparatively small number of relativistic electrons can ensure powerful emission of radio waves from the nebular ejecta of supernovae.

A substantial portion of the energy released in a supernova outburst is thereby converted into energy of relativistic particles. We may in fact estimate the total energy of the relativistic electrons in the Crab Nebula as

$$E = V \int_{E_1}^{E_2} N(E)E \, dE \approx 8 \times 10^{46} \text{ erg}$$

(see also Sec. 21), and for Cassiopeia A we may similarly find that $E \approx 5 \times 10^{48}$ erg. The energy radiated during a supernova outburst may amount to 10^{49} to 10^{50} erg; the kinetic energy of the gas ejected during the outburst may be of the same order. The corresponding energy $E_H = VH^2/8\pi$ localized in the magnetic field of the nebula will then be 10^{47} to 5×10^{48} erg.

Our estimate of the quantity K, which determines the population of relativistic particles (and their total energy) in ejecta of supernovae, may at first seem quite arbitrary, for we are unable to quote a value for H with any confidence. Actually, however, this is not so: the quantities K and H cannot be regarded as independent. A relation of the following form must subsist between them:

$$P_{\cos} = \tfrac{1}{3} \int_{E_0}^{E} KE^{1-\gamma} dE = \tfrac{1}{3}K(\gamma - 2)^{-1} E_0^{1-\gamma} \leq H^2/8\pi;$$

this means that the relativistic-electron pressure must be less than the magnetic-field pressure (see Sec. 23). Otherwise the relativistic particles would not be constrained by the magnetic field. But, on the other hand, the total magnetic energy over the entire nebula will become very great if $H^2/8\pi > P_{\cos}$. For a given intensity of radio emission, the total energy of the magnetic field *plus* that of the relativistic particles will be at a minimum under the condition $P_{\cos} \approx H^2/8\pi$. The approximate equality of the energies of the magnetic field and the relativistic particles in sources of radio waves has a deep physical meaning; it is evidently to be accounted for by the conditions under which relativistic particles are generated in peculiar nebulae. In our calculations we shall proceed from the condition $P_{\cos} \approx H^2/8\pi$.

In view of the presence of relativistic electrons in the Crab Nebula, as well as in the remnants of other supernova outbursts, the following question arises: since the velocities of the electrons approach the velocity of light, why are they not dissipated into the surrounding space? There can be only one answer: a *magnetic field* confines them within the nebula. Since a regular field could not retain charged relativistic particles within a bounded region of space, it follows that the magnetic field in radio nebulae possesses a highly irregular, patchy structure.

We shall estimate the dimensions of the domains in the Crab

within which the magnetic field may be regarded as quasi-homogeneous. The advance of a relativistic particle diffusing along some direction through irregular magnetic fields will be equal to

$$S = \langle l \rangle (n/3)^{1/2} = \langle l \rangle (tc/3)^{1/2} \, ; \qquad (20\text{-}2)$$

here $\langle l \rangle$ is the mean path-length for a particle—the extent of the cells of homogeneity of the magnetic field, in our case; $n = tc$ is the number of paths traversed; and t is the diffusion time. We shall take $t = 2 \times 10^{10}$ sec, a span of time over which the Crab has suffered no essential change. Then if $\langle l \rangle = 10^{14}$ cm, $S = 1.4 \times 10^{17}$ cm; if $\langle l \rangle = 10^{16}$ cm, $S = 1.4 \times 10^{18}$ cm; if $\langle l \rangle = 10^{17}$ cm, $S = 4.4 \times 10^{18}$ cm. The radius of the domain of the Crab now occupied by relativistic electrons is $\approx 3 \times 10^{18}$ cm, corresponding to $\langle l \rangle \approx 5 \times 10^{16}$ cm. This value of $\langle l \rangle$ exceeds the mean free path of the particles in the Crab by only a small factor (see below). It is to be noted that the dimensions of even the smallest cells of turbulence must exceed the mean free path. Our estimate of $\langle l \rangle$ must therefore be quite reliable.

In Sec. 22 we shall consider the reasons for the abundance of relativistic particles in the ejecta of supernovae.

Looking briefly ahead, we may formulate the following basic conclusion: at an early stage in the evolution of the ejected nebulosity, conditions were for some reason highly favorable for the acceleration of a portion of the existing particles to relativistic energies. These particles are virtually unaccelerated throughout the subsequent expansion of the nebula. They are confined within it by the more or less chaotically oriented magnetic fields which develop as the nebula expands. It may, however, be shown that the particles experience no significant loss of energy, even over thousands of years (provided that their energies are not excessive; see Sec. 21). As the envelope expands, the particles "rove" over a continually increasing volume. A continual decrease in the magnetic-field intensity in the expanding envelope sets in simultaneously, simply as a consequence of the diminishing mean density.

The surface brightness of a radio nebula must be proportional to the inverse square of the radius, simply because of the expansion of the radiating volume. But the diminishing magnetic-field intensity in the expanding envelope results in a further decrease in the radio flux. If the intensity H of the field decreases by an order of magnitude, then at meter wavelengths electrons will radiate whose energy is three times as great as for the electrons radiating prior to the decrease in H. But such electrons will be far less abundant in the radio nebula, and hence the flux of radiation will drop sharply. According to Eq. (13-7),

$$I_\nu \propto H^{(\gamma+1)/2},$$

and if, for example, $\gamma = 3$, then, as H decreases by an order of magnitude, I_ν will drop by a factor of 100. Thus the "old" radio nebulae, the ejecta of supernovae which flared up in the relatively distant past, must be objects of low surface brightness, and of linear dimensions of the order of 10 pc. The radio nebula IC 443 + S 40 in Aquarius is evidently such an object, as is the filamentary nebulosity in Cygnus, whose radio emission we have discussed in Sec. 8.

If H remained constant, the surface brightness I_ν of a radio nebula would clearly fall off with increasing radius as r^{-2}, until the interstellar medium begins to decelerate the nebula; but since H decreases, I_ν will fall off considerably more rapidly. As deceleration of the nebula sets in, the expansion will proceed more and more slowly, the decrease in H will similarly be retarded, and the surface brightness will fall off slowly. The nebula will gradually dissipate into the interstellar medium, and the magnetic field will eventually coincide with the ambient fields of interstellar space. The relativistic electrons, finding themselves in interstellar space, will wander for a long time in weak, chaotically oriented magnetic fields. This conclusion, which we have obtained in a natural way, is of the greatest importance in connection with the origin of cosmic rays, and of the general field of nonequilibrium galactic radio waves.

We pass now to the problem of the deceleration by the interstellar medium of the expanding envelopes of supernovae. Beginning with the law of conservation of momentum, it is elementary to derive the formulae expressing the deceleration of an expanding spherical envelope. Let v_0 denote the initial velocity of the envelope, v the velocity at time t after the outburst, r the radius of the envelope, ρ the density of the interstellar medium, and $\mathfrak{M}$ the mass of the envelope; we then have

$$\frac{v}{v_0} = \frac{\mathfrak{M}}{\mathfrak{M} + 4\pi r^3 \rho/3}, \tag{20-3a}$$

and

$$\mathfrak{M}r + \pi r^4 \rho/3 = \mathfrak{M}v_0 t. \tag{20-3b}$$

For a strongly decelerated envelope, when $\pi r^3 \rho/3 \gg \mathfrak{M}$, we may write approximately:

$$\frac{v}{v_0} \approx \frac{\mathfrak{M}}{\mathfrak{M} + 4\mathfrak{M}v_0 t/r} \approx \frac{r}{4v_0 t}, \tag{20-4}$$

whence

$$r \approx 4vt. \tag{20-5}$$

This formula permits us to determine the radius of an expanding, but strongly decelerated, envelope, if we merely know the instantaneous velocity of expansion and the time elapsed since the outburst. From Eq. (20-3a) we find

$$r = \left[\frac{3(v_0/v - 1)\mathfrak{M}}{4\pi\rho} \right]^{1/3}. \tag{20-6}$$

We shall now apply these formulae to actual ejecta of supernovae, taking three objects, all at different stages in their evolution: the Crab Nebula, IC 443 + S 40, and the group of filamentary nebulae forming the Cygnus Loop. The basic data for these radio nebulae are listed in Table 23.

TABLE 23. Radio nebulae.

Nebula	Angular size	Flux, $\lambda = 3$ m $[10^{-24}$ w m^{-2} (c/s)$^{-1}$ sterad$^{-1}]$	T_b, $\lambda = 3$ m (°K)
Crab	5′	18	4 000 000
IC 443 + S 40	48′	4	10 000
Cygnus Loop	3°	2	300

The brightest and, we may suppose, the youngest of these—the Crab—is some 400 times as bright as IC 443 + S 40, which in turn is some 30 times as bright as the very old Cygnus Loop. In fact the background cosmic radio emission at $\lambda = 3$ m in the part of the sky where the Cygnus nebulosity is found exceeds the brightness of the extended source localized in the Loop by a factor of 5 to 10. Thus the source is almost at the limit of detection.

The distance to the filamentary nebulae in Cygnus has been estimated by various authors as from 300 pc to 1000 pc; we shall adopt 500 pc. The linear dimension r of the Cygnus system of nebulae is then ≈ 25 pc. The Crab extends over ≈ 2 pc. Finally, if we take the distance to IC 443 + S 40 as ≈ 1000 pc (the value is poorly determined), we obtain a linear dimension of ≈ 14 pc. It will be noted that the dimension of the IC 443 + S 40 nebula indeed falls between those of the Crab and of the Cygnus Loop; the nebula consequently occupies an intermediate position with respect to radio surface brightness and radio luminosity. If $\gamma \approx 3$, then the theoretical ratios of the surface brightness of the Crab to that of IC 443, and of the surface brightness of IC 443 to that of the Loop, must be ≈ 100 [see Eqs. (22-25) and (22-26)]; this is in qualitative accord with the observational results.

What may we say of the age of the IC 443 and Loop radio nebulae? From a radial-velocity analysis of the filamentary nebulosity in Cygnus, we may conclude that the entire system is expanding with a velocity ≈ 45 km/s. This value of v may now be entered into Eq. (20-5) to estimate the interval of time which has elapsed since the outburst: we obtain $\approx 65\,000$ yr. If a value for the rate of expansion of IC 443 + S 40 can be secured (and this is entirely possible), we shall be able to estimate the age of this radio nebula as well. As a rough conjecture we may take 5000 or 10 000 yr. We have employed these age estimates in the foregoing section, where we derived the approximate population of objects like IC 443 + S 40 and the Loop from the frequency of supernova outbursts in the Galaxy.

Turning to an "old" radio nebula such as the Cygnus Loop, we shall estimate the total number of relativistic particles. Let us take $\gamma = 2.6$, $H = 1.3 \times 10^{-5}$ gauss, and the distance to the filamentary nebulosity as ≈ 500 pc; we then find from Eq. (13-7) that $K \approx 5 \times 10^{-15}$ cgs, or $\approx 4 \times 10^4$ ev$^{1.6}$/cm^3. This leads to a mean concentration of relativistic electrons having $E > 2 \times 10^7$ ev of $\approx 5 \times 10^{-8}$ cm^{-3}, and an energy density for the relativistic electrons of $\approx 5 \times 10^{-12}$ erg/cm^3. On the other hand, the energy density $H^2/8\pi$ of the magnetic field is 7×10^{-12} erg/cm^3. The volume V of the Cygnus system of filamentary nebulae is $\approx 2 \times 10^{59}$ cm^3. Hence the relativistic electrons of this radio nebula command a total energy of $\approx 10^{48}$ erg, greater than for the Crab Nebula.

To conclude this section, we call attention to the following matter. Since relativistic electrons are generated in abundance in a supernova outburst, it is natural to suppose that a large number of heavy high-energy nuclei, such as protons, will be formed simultaneously. We shall return to this point in Secs. 22 and 23. Because of this effect the expression P_{cos}, introduced above, must subsume the pressure of *all* high-energy particles; this should be kept in mind when calculating models of radio sources. The electron component of P_{cos} apparently constitutes ≈ 10 percent of the total value of P_{cos} (see Sec. 23). Upon taking this circumstance into account it is found that H is increased, but not substantially (by a factor of about two), and that K is decreased by a factor of about three.

21. THE OPTICAL CONTINUUM AND POLARIZATION OF THE CRAB NEBULA

The Crab Nebula is the only envelope of a supernova which has been adequately investigated by the techniques of optical astronomy. Important efforts in this direction are reported in papers by Minkowski (1942), Baade (1942), Greenstein and Minkowski (1953), and Barbier (1945).

Until quite recently, the opinions which have been advanced as to the nature of the Crab Nebula have all been based essentially upon these papers. Generalization may be made to other supernova envelopes as well, insofar as the Crab may be taken as a typical remnant of a supernova explosion. We shall briefly consider these proposals.

Baade and Minkowski have suggested that the Crab comprises two interpenetrating systems: a system of relatively slender *filaments,* forming an envelope at the periphery of the nebula, and expanding with a velocity of 1000 to 1300 km/s; and an *amorphous mass,* wholly filling the interior of the nebula. Whereas the expanding network of filaments radiates a line spectrum, the spectrum of the amorphous mass is strictly continuous. Barbier has pointed out that the luminosity of the Crab in the emission lines constitutes only a few percent of the luminosity of the amorphous mass in the continuous spectrum. It follows from the values of the color temperature (corrected for interstellar absorption) given by Minkowski (1942) that the intensity of the radiation in the continuum falls off with frequency approximately as ν^{-1}.

Figure 159 presents a photograph of the Crab Nebula, obtained with the 200-in. reflector through a filter admitting wavelengths from 6400 to 6700 A. The filamentary structure of the nebula is outlined particularly well, since in this spectral range the intense [N II] lines are found. In Fig. 160 we have a photograph of the Crab taken through a filter which does not admit the emission lines from the nebular filaments. Here only the amorphous mass, which has a continuous spectrum, is visible.

Fig. 159. Photograph of the Crab Nebula, taken with a filter admitting the radiation from the filaments (Baade 1956*a*).

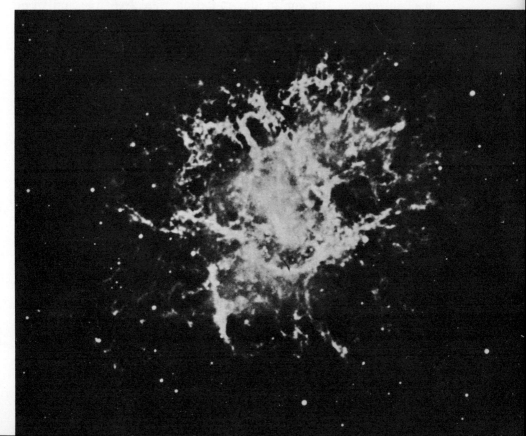

Fig. 160. Photograph of the Crab Nebula, taken with a filter admitting the radiation from the amorphous mass of the nebula (Baade 1956*a*).

All proposals regarding the physical processes in the Crab have until recently been founded on interpretations of its continuous spectrum. The only possibilities that Baade and Minkowski admitted as mechanisms for the continuum radiation were free–free and free–bound transitions in the highly ionized gas of which the nebula is composed. They believed that this emission was excited by an exceptionally hot star, namely the former supernova. At first sight it appears quite natural to make such a proposal. Of course scattering processes cannot account for the observed luminosity of the Crab, since the brightness of the nebula is about 9^m while the brightness of the two little stars at the center of the nebula is only 16^m. The radiation we observe from this nebula must be exclusively its own. It is quite understandable that Minkowski (1942) interpreted the radiation from the Crab and that from the planetary nebulae in the same way: at that time no other mechanism was known which might lead to continuum radiation by a gaseous medium.

Later on, A. Ya. Kipper introduced a very successful mechanism to explain the faint continuum in the planetaries—the "splitting" of Lyman-α quanta. But if this mechanism were responsible for the radiation from the Crab, then intense hydrogen lines would also be radiated, and this is not observed. The energy distribution in the continuum would furthermore differ from the distribution observed. For a similar reason, a modification of Kipper's mechanism, the splitting of the quanta of helium resonance lines, cannot explain the continuum of the Crab.

Were we to assume, then, that the continuum of the Crab arises from free–free transitions, we would nevertheless meet with considerable difficulties. Attempts were not made to analyze these difficulties critically, even though they were recognized at an early stage; for full confidence was placed in the Baade–Minkowski proposal for the continuum radiation. We shall now outline these difficulties.

The following conclusions would be inescapable if the above mechanism were accepted: (a) the kinetic temperature of the Crab would be very high, of the order of 100 000°K or even higher (the observations do not reveal a jump in intensity at the Balmer limit; see Barbier 1945); (b) the electron concentration in the nebula would be of the order of 10^3 cm^{-3}; (c) the mass of the nebula would be $\approx 20 \mathfrak{M}_\odot$. If we were to regard the luminosity of the nebula as being excited by the hot central star, the former supernova, then the surface temperature of the star would be excessive ($>500\,000°K$), while the radius would be very small ($< 0.02\,R_\odot$).

On the basis of these quite fantastic characteristics of a hypothetical central star, Oort (1951), adhering to the mechanism of free–free and free–bound transitions for the radiation from the amorphous mass of the Crab, proposed that the hot amorphous mass was not excited by any source. According to Oort, the hot nebula may have retained the temperature of the deep interior of the exploded central star. He pointed out that the cooling process for such a very hot extended mass of gas would be a sufficiently slow one. It is, however, difficult to agree with a hypothesis that a supernova disintegrates completely and dissipates upon explosion. This would contradict the spectral and photometric observations of supernova outbursts in external galaxies.

W. H. Ramsey has advanced the hypothesis that the high temperature of the amorphous mass of the Crab may be supported by radioactive decay of some unstable isotopes formed during the outburst of the supernova. But this hypothesis has also been unable to withstand criticism (Cowling 1954).

We therefore see that there can be no serious ground for supposing that the amorphous mass represents the fully exploded stellar mass, and that no external agent is required to maintain its exceptionally high kinetic temperature. But there is even less reason to suppose that the central star, the former supernova, possesses the characteristics which must follow from the Baade–Minkowski interpretation of the continuous spectrum of the Crab.

The morphology of the Crab would remain entirely unexplained under such an interpretation. The kinetic temperature of the filaments is quite low ($\approx 10\,000°K$); the density cannot be very great. The most intense lines in the spectra of the filaments are $\lambda\,3727$ [O II] and λλ 6548 and 6584 [N ɪɪ]. But the transition probabilities for these lines are very small, so that they are strengthened in nebulae of low density. The concentration of particles in the filaments can hardly exceed 300 or 400 cm^{-3}. In this event, however, it would remain for us to explain the protracted coexistence of the relatively cold, quite rarefied filaments on the one hand, and the very hot, but no less dense diffuse mass on the other. How could these filaments have moved through the amorphous mass for fully 900 yr?

Furthermore, it would not be possible to account for the rather low state of ionization and excitation in the filaments. We would expect fast, energetic electrons from the amorphous mass to find themselves in the filaments, ionizing the atoms there. In just the same way it would remain unexplained why intense ultraviolet radiation from the central star or from the hot diffuse mass does not strongly ionize the atoms in the filaments. To be sure, in the solar atmosphere we observe very hot coronal material along with relatively cool prominences; but here the situation is entirely different, for the density of the "cool" prominences exceeds that of the "hot" corona by two or three orders of magnitude. In the Crab, on the other hand, the density of the filaments is always of the same order as the density of the amorphous mass.

Finally, it would be difficult to understand why the nebular ejecta of the 1572 and 1604 supernovae are so faint in comparison with the Crab Nebula. If a complete disintegration and dissipation of the star takes place during the explosion of a supernova, then why do we fail to observe quite luminous nebulae with a continuous spectrum at the sites of these supernovae?

We emphasize that all these difficulties follow from the Baade–Minkowski interpretation of the continuous spectrum of the Crab. Since the obstacles appeared insurmountable, we set about to derive a completely new explanation (Shklovsky 1953*f*).

Some important new data, throwing light on the nature of the Crab, have appeared since the publication of Baade and Minkowski's investigations. We refer to the detection of radio emission from the nebula. More than once we have pointed out that the spectrum of this emission is far smoother than for other sources. Over the immense spectral range from 7.5 m to 3.2 cm, almost eight octaves, the radio flux per unit frequency interval diminishes by only two and one-half or three times. Of course it would be absurd to regard the radio emission from the Crab as terminating at λ = 3.2 cm, with $F_\nu = 0$ for λ < 3 cm. There can be no doubt that a radio flux does exist for λ < 3 cm, but observing techniques have not yet been developed to the point where it may be detected.

We naturally inquire whether the optical continuum of the Crab represents an extension of the radio continuum. Or, alternatively, are the radio and the optical emissions a consequence of the same mechanism? In the foregoing section we have shown that the radio emission from the Crab cannot be thought of as a "long-wavelength tail" of the optical continuum, provided that the latter is taken as thermal (determined by free–free transitions). But now the problem is a different one: can the radio emission from the Crab be explained as an extension of its nonequilibrium optical emission, or the optical emission as an extension of the nonequilibrium radio emission? Under such an interpretation, the mechanism responsible for the optical continuum of the Crab must be a peculiar one, differing fundamentally from all (equilibrium) mechanisms of optical emission known to astrophysics.

In the preceding section we have indicated that the radiation flux in the optical continuum of the Crab is some 400 times as small (per unit frequency interval) as in the meter wavelength range. Now it is quite reasonable to suppose that if the radiation flux diminishes by a factor of three over the eight octaves which have been investigated by radio, it might diminish by another factor of 100 over the 15 octaves remaining until the optical region is reached (at, say, 8000 A). In the optical range, the dependence of intensity on frequency may be represented approximately by the relation $F_\nu \propto \nu^{-1}$; this is considerably steeper than at decimeter and centimeter wavelengths.

Inasmuch as the radio emission from the Crab arises through radiation from relativistic electrons in magnetic fields, we may now regard the emission in the optical continuum as arising in the same way. But if electrons of energy $< 10^9$ ev are responsible for the radio emission, then the optical emission will primarily be determined by electrons of energy $\approx 10^{11}$ to 10^{12} ev. We shall now attempt to estimate the concentration of these radiating relativistic electrons.

We draw attention to the fact that about 40 percent of the luminosity of the amorphous mass of the Crab is concentrated within a domain about its center of $1'$ radius. The results of the computations reported below therefore refer to the central portion of the Crab. At $\lambda = 4250$ A ($\nu = 7 \times 10^{14}$ c/s), $F_\nu = 2.25 \times 10^{-23}$ erg sec^{-1} cm^{-2} (c/s)$^{-1}$. For the exponent in the energy spectrum of the fast electrons, we have $\gamma \approx 3$, whence $u(\gamma) = 0.087$ (Sec. 13). By Eq. (13-7), $K = 1.2 \times 10^{-9}$ when $H = 3 \times 10^{-4}$. For the concentration of electrons of energy $E > E_0 = 3 \times 10^{11}$ ev we find

$$N(E > E_0) = K \int_{E_0}^{\infty} E^{-3}\, dE \approx 2.4 \times 10^{-9}\ \text{cm}^{-3}.$$

The energy density of electrons with $E > 3 \times 10^{11}$ ev will be

$$K/E \approx 3.2 \times 10^{-9}\ \text{erg/cm}^3.$$

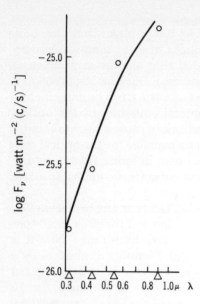

Fig. 161. Continuous spectrum of the Crab Nebula from 3300 A to 9000 A (Shcheglov 1957).

The energy density of the "softer" relativistic electrons, those responsible for the radio emission, will be of the same order. Taking a volume of $\approx 5 \times 10^{55}$ cm³ for the Crab, we find a total energy for all the relativistic electrons in the nebula of $\approx 2.4 \times 10^{47}$ erg.

P. V. Shcheglov (1957), at the Sternberg Astronomical Institute in Moscow, has recently employed an electronic image converter to measure the radiation flux from the Crab in the near infrared (8000 A $< \lambda <$ 10 000 A). The infrared stellar magnitude of the Crab was found to be $6^m.5$ (in the system used, the infrared magnitudes of A stars were taken to be the same as their photographic magnitudes). Shcheglov also determined the photographic magnitude of the nebula. From all the available determinations he established the spectrum of the Crab over the range 9000 A $> \lambda >$ 3300 A, as shown in Fig. 161. It appeared that this portion of the spectrum is considerably steeper than had previously been believed. Here the absorption of light was taken as $1^m.2$ in the photographic region (following Baade and Minkowski). After allowing for the absorption, the relation $F \propto \nu^{-2}$, or possibly even $F \propto \nu^{-2.5}$, was derived for the above frequency range. This implies that the exponent γ in the differential energy spectrum of the relativistic electrons is ≈ 5.

We have computed that, for $\gamma = 4.5$ and $H = 3 \times 10^{-4}$, the concentration of "radiating" relativistic electrons will be 2.5 times as small as for $\gamma = 3$; and the energy density will be 6.4 times as small. The total energy of all relativistic electrons in the Crab having $E > 3 \times 10^{-11}$ ev will be $\approx 2 \times 10^{46}$ erg if $\gamma = 4.5$, a comparatively small value.

For a definitive reconstruction of the spectrum of the Crab, it would be very desirable to expand the spectral range of the observations. It should be possible to derive the radiation flux from the Crab in the range $1.9\,\mu < \lambda < 2.3\,\mu$, where a window of transparency again appears in the Earth's atmosphere. Such observations may be carried out with photoresistors. The flux in the millimeter range might also be accessible. Figure 162 presents the spectrum of the Crab Nebula throughout the frequency range which has thus far been investigated.

We have seen that a negligible number of relativistic electrons appears to be responsible for the rather intense optical emission from the Crab. There follows the important conclusion that the mass of the amorphous portion of the Crab cannot be very great: if the interior of the nebula were occupied by relativistic particles alone, its mass would be of the order of $10^{-6}\mathfrak{M}_\odot$.

But one cannot expect a vacuum to admit of magnetic fields which are tangled to any extent. Within the nebula there must be a gas so rarefied that it does not reveal itself optically. This gas may owe its origin to the ejection of matter from the supernova after maximum light. In Sec. 20 we obtained an estimate for the dimension $\langle l \rangle$ of the cells of turbulence in the Crab: $\langle l \rangle \approx 3 \times 10^{16}$ cm. The quantity $\langle l \rangle$ must still be at least several times the mean free path. Consequently we may set a lower limit of 10^{32} gm, or $0.05\mathfrak{M}_\odot$, to the mass of the amorphous portion of the Crab; this must be close to the true value.

Similarly, the mass of the filamentary system can hardly exceed several hundredths of a solar mass. This follows from the estimated

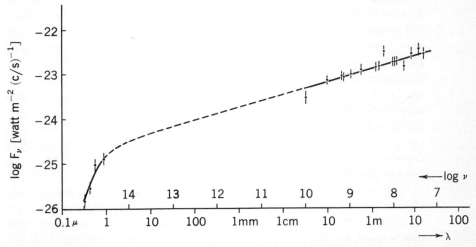

Fig. 162. Continuous spectrum of the Crab Nebula from $\lambda = 3300$ A to $\lambda = 15$ m.

volume occupied by the filaments, and from the density of the filaments, $\rho \approx 7 \times 10^{-22}$ gm/cm³.

We conclude that the mass of the gas ejected in the supernova explosion of 1054 is evidently of the order of $0.1\mathfrak{M}_\odot$, a value a hundred times as small as that previously accepted. The following considerations, developed recently by Pikelner (1956), provide independent support for this estimate. Pikelner regards the magnetic field and relativistic particles as exerting a pressure upon the nebula, thus accelerating its motion. Baade (1942) has called attention to the remarkable acceleration in the nebular expansion. If the nebula had continually expanded along the major axis with the presently observed velocity, then the date of the supernova explosion would have been fixed at 1184 rather than 1054. We obtain as an estimate of the acceleration

$$g = 0.0011 \text{ cm/sec}^2.$$

Until new explanations were offered for the radiation from the Crab, it was not possible to account for this acceleration. Since the energy densities of the relativistic particles and of the magnetic field in the nebula are of the same order, the difference in pressure within and without the nebula will be of the order of

$$2H^2/8\pi \approx 10^{-8} \text{ dy/cm}^2.$$

This pressure difference is equal to the force exerted on a column of unit cross-sectional area traversing the nebula, a force imparting an acceleration to the mass of gas contained within the column. In this way the mass of gas in such a column may be derived (it is $\approx 10^{-5}$ gm), and hence the mass of the entire nebula. This is found to be $0.1\mathfrak{M}_\odot$.

Osterbrock (1957) has derived the electron concentration in the filaments of the Crab by means of the intensity ratio of the lines at 3729 and 3726 A. In the filaments investigated, N_ε lies between 630 and 3700 cm⁻³. Osterbrock estimates the total mass of the bright filaments as $0.02\mathfrak{M}_\odot$, while the mass of the entire system of filaments, including the faintest, is between 0.05 and $0.1\mathfrak{M}_\odot$. This result is in good accord with our own estimate for the mass of the Crab, as well as with the later estimate by Pikelner.

The apparent stellar magnitude of the filamentary system of the Crab is 12 or 13, only 3^m brighter than the nebular ejecta of the 1604 supernova. But the latter star was 4^m fainter at maximum than the 1054 supernova.

It is highly significant that an envelope of relatively small mass is ejected during the supernova outburst. As a result, the explosion by no means represents a disintegration and dissipation of the star: in this respect a supernova outburst, in essence, does not differ from the ordinary nova phenomenon. The distinction is simply one of degree.

Of course one must expect peculiarities in supernova outbursts which differentiate them qualitatively from the novae. For example, the brightest novae (in absolute magnitude at maximum) are the fast novae (Shklovsky 1955*a*), whereas the extreme luminosity of the supernovae at maximum is accompanied by a rather slow decline.

As the primary features of the nova and supernova phenomenon have now become known, it might be suggested that the nebulae generated by the outburst of nearby novae may also be sources of radio waves (Shklovsky 1953*d*, 1953*e;* Gordon 1954). The observations appear to confirm this prediction. Bolton, Stanley, and Slee (1954) have detected quite an intense source near Nova Aquilae 1918. In the Cambridge catalogue, Source No. 1595 has the coordinates $\alpha = 18^h47^m10^s$ $\pm5^s$, $\delta = -0°2' \pm26'$ (1950), and the flux at $\lambda = 3.7$ m is 7.2×10^{-25} w m^{-2} sterad^{-1}. The coordinates of Nova Aquilae are: $\alpha = 18^h46^m20^s$, $\delta = 0°30'$ (1950). The authors of the catalogue do not regard this identification as decisive. Nova Aquilae attained $-1^m.1$ at maximum. Attempts might well be made to detect radio emission from Nova Persei 1902 and Nova Puppis 1942, both of which were sufficiently bright at maximum. The remaining novae which have appeared in recent decades have been too faint. Mills, Little, and Sheridan (1956*b*) have been unable to detect radio sources of measurable intensity $[> 10^{-25}$ w m^{-2} (c/s)$^{-1}]$ at the sites of former novae.

In the light of our interpretation of the optical emission from the Crab, we can now understand why bright nebulae with continuous spectra, like the amorphous portion of the Crab, have not been detected at the positions of other galactic supernovae. Sufficiently many relativistic particles with $E > 5 \times 10^{11}$ ev are certainly not created in the envelopes of all supernovae; thus a strong optical continuum can arise only for a small number of radio nebulae. Special conditions are also necessary in order that energetic relativistic electrons, formed at a certain stage in the evolution of the nebula, may not lose a significant fraction of their energy and become invisible.

I. M. Gordon (1954) and, independently, V. L. Ginzburg (1954) have shown that if our proposal is correct, the optical continuum from the Crab being induced by relativistic electrons, we should expect this radiation to be *polarized*. Special observations were carried out to verify this effect. The first to detect polarization in the Crab was V. A. Dombrovsky (1954), who employed photoelectric techniques. The low surface brightness of the Crab compelled him to measure the mean polarization over an extensive portion of the nebula—the dimension was 3', and about 50 percent of the area of the brightest part was included. Considerable polarization, about 13 percent, was detected through these observations; the polarization was directed (as determined by the electric vector) at the relatively small angle of

20° to the galactic equator. Moreover, it became clear that over the entire nebula the polarization retained approximately the same character. M. A. Vashakidze (1954) soon detected polarization in the Crab photographically.

The detailed distribution of brightness and direction of polarization in the Crab is of great interest. Analysis of this distribution would yield valuable information on the physical conditions in the nebula. The first efforts in this direction were carried out photographically by Shajn, Pikelner, and Ikhsanov (1955). Observations were made with the 40-cm astrograph (160-cm focal length) of the Crimean Astrophysical Observatory. Polaroids were placed in front of the plate, and photographs were taken for three positions of the polaroids. The diaphragm of the microphotometer was 0.1 mm, or 13″. It was found that the mean polarization is 8 percent, and the angle between the electric vector and the galactic equator is $\approx 10°$.

It was shown statistically that the polarization differs at various points in the nebula, so that it cannot be ascribed to an interstellar source. At certain positions the degree of polarization was so great that the form of the nebula seemed to depend on the orientation of the polaroids. Quantitative determinations of the polarization at various points were regarded as unreliable, since the transparency of the two objectives of the double astrograph did not appear to be the same. The polarization was therefore derived for only four of the measured points, at which it was especially high. The arrows in Fig. 163 designate the direction of polarization at these points.

We wish to emphasize that the polarization observed in the Crab cannot be attributed to the interstellar medium. In general, the interstellar absorption (and hence the correlated interstellar polarization) is small in this region of the sky. The nebula itself is practically transparent, for the lines radiated by the "front" and "back" portions

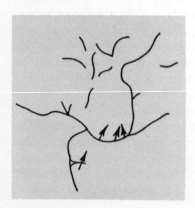

Fig. 163. Distribution of filaments in the Crab Nebula; arrows indicate the directions of polarization (Shajn, Pikelner, and Ikhsanov 1955).

of the extended filamentary system have approximately the same intensity.

Khachikyan (1955) has also investigated the polarization in the Crab, using the 8-in. Schmidt camera of the Byurakan Observatory in Armenia. Photographs were taken for three orientations of the polaroids, but at a smaller scale than that employed by Shajn, Pikelner, and Ikhsanov. The diaphragm of the microphotometer isolated a $20'' \times 20''$ area. The degree of polarization was found to vary over the nebula from the threshold of photographic sensitivity to 50 or 60 percent; it was observed to be most intense where the continuous emission attained maximum brightness. The direction of vibration of the electric vector was similar at different points in the nebula. Over the entire nebula, the mean polarization was about 20 percent. The correlation of the degree of polarization with individual details in the amorphous mass of the nebula demonstrates that the polarization arises from internal causes, and not from the interstellar medium.

One rather unexpected result of the observations just described is the quite regular character of the direction of vibration of the electric vector. A polarization of this character might arise if an additional, more or less homogeneous, field were superimposed on the chaotic field. The presence of such a supplementary weak field would have no essential effect on the diffusion velocity of relativistic particles, but would contribute to a spreading of the particles along the regular field. This may perhaps account for the drawn-out form of the amorphous mass. In this connection, we note that Gaze and Shajn (1954) have pointed out preferential directions in the extension of the filaments of the radio nebula IC 443 + S 40, as well as for the Crab Nebula.

A quasi-regular field of this type may be the consequence of certain internal regularities peculiar to the nebula and the supernova which generated it. Mustel (1956) showed that magnetic fields can explain the kinematics and morphology of expanding nova shells. In a number of cases, well-defined directions of material ejection have been observed at the outburst, directions depending on the character of the magnetic field of the star. It would be natural to expect similar regularities to occur in supernova outbursts as well.

In 1954, we informed Professor J. H. Oort, then visiting the U.S.S.R., about the above considerations regarding the nature of the optical continuum in the Crab. Upon his return to The Netherlands, Oort, together with Walraven, embarked on an investigation of the polarization of the optical emission from the nebula. The results of these observations were published in an extensive paper (Oort and Walraven 1956), which also contains some highly important and interesting facts concerning the Crab, unpublished until that time.

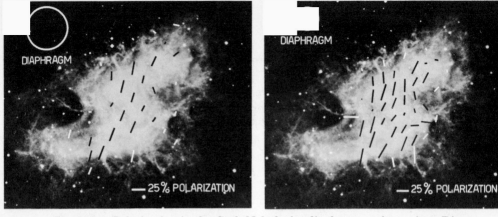

Fig. 164a. Polarization in the Crab Nebula for diaphragms of two sizes. Direction of electric vector and degree of polarization are indicated (Oort and Walraven 1956).

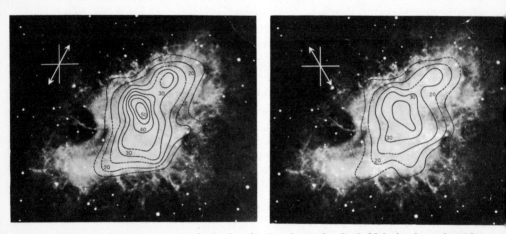

Fig. 164b. Isophotes of polarized radiation from the Crab Nebula, for polaroids oriented as indicated by the arrows (Oort and Walraven 1956).

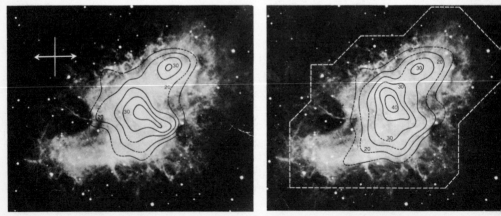

Fig. 164c. Isophotes of polarized radiation (left), for a polaroid oriented as indicated by the arrows, and of the unpolarized radiation (right) (Oort and Walraven 1956).

The observations of the Crab polarization were carried out photo-electrically. Two large diaphragms were employed, cutting out portions of $\frac{1}{2}'$ and $1'$ diameter from the nebula. According to these observations the mean polarization over the entire nebula is 9.3 ±0.3 percent.

The degree of polarization varies from several percent to 34 percent in individual regions of dimension $< \frac{1}{2}'$. Possibly still smaller diaphragms might permit the detection of total polarization of the radiation from some sections. One direction of vibration prevails, on the whole; but all directions are observed. Within a region of $\approx 1'$ radius at the center of the nebula, no correlation between distance from the center and degree of polarization is detected; here the degree of polarization is independent of the surface brightness. In the external portions of the nebula, $> 1'$ or $2'$ from the center, the polarization (as smoothed over the diaphragm of the photometer) drops rapidly.

The results of Oort and Walraven's photoelectric measurements of the polarization have been plotted on a photograph of the Crab, obtained by Baade (Figs. 164a, b, c). Circles in Fig. 164a designate the sizes of the diaphragms employed. The degree of polarization and the direction of vibration of the electric vector are indicated on the same two photographs. The right-hand photograph of Fig. 164c presents the isophotes of the optical radiation from the Crab, as obtained without polaroids. Isophotes obtained with polaroids, in the three orientations marked by arrows, are given in Fig. 164b and Fig. 164c (left).

The strongest polarization is observed near the center of the Crab. Figure 165 shows the intensity distribution along a certain direction, for two mutually perpendicular orientations of the polaroids. The peak in the solid curve graphically illustrates the very high degree of polarization.

Comparison of the distributions of optical and radio emission over the nebula is especially interesting. Figure 166 reproduces Baldwin's photoelectric observations of the optical and the radio brightness (see Sec. 7). Evidently the optical intensity diminishes considerably more rapidly with distance from the center of the nebula than does the radio intensity. This circumstance is highly significant for any attempt to construct a quantitative theory for the Crab.

Of the new facts relating to the Crab that appeared for the first time in Oort and Walraven's paper (1956), the most startling was the announcement that, in 1942–43, Baade had discovered some formations in the Crab which are highly variable in appearance and brightness, and which move quite rapidly.

Figure 167 is a rough diagram of the central portion of the Crab. The circles denote stars in this portion of the nebula, a distance of 4''.9 apart. Baade and Minkowski have expressed the opinion that the

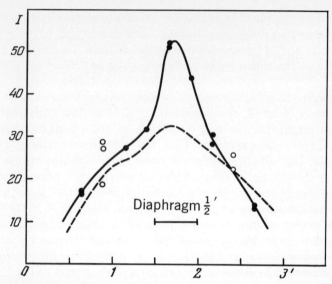

Fig. 165. Intensity distribution across the Crab Nebula for two perpendicular orientations of the polaroid (Oort and Walraven 1956).

more southerly of these stars is the former supernova. This need not be the case, however: it is possible that the supernova is now too faint to be observable.

The bright region b, outlined in Fig. 167, represents a more or less stable configuration in the central portion of the amorphous mass of the nebula. On the other hand, the region a is highly unstable, and is manifestly moving about within the central portion. At times the formation a is absent altogether. Then it will appear quite suddenly, either as an extended bright wisp (as illustrated in Fig. 167), or as several bright round knots. The formations always appear at least 7″ from the line joining the central stars. They then move toward the region b, meanwhile diminishing in brightness, usually to the point where the wisps a have ceased to be visible by the time they have reached b. A 1″ displacement in a was observed after a 68-day interval. It follows that its velocity component in the plane of the diagram was 26 000 km/s, or 20 to 25 times the expansion velocity of the Crab Nebula. Baade estimates that the luminosity of wisp a constitutes roughly 1/1500 of the total luminosity of the Crab. This remarkable phenomenon is in complete contradiction to earlier ideas concerning the nebula's radiation. It is typical that over a period of 13 years Baade had nowhere published his notable discovery, which appeared to him extremely puzzling and, perhaps, irrational. In the years that followed he observed the phenomenon on more than one occasion. The new interpretation of the optical radiation from the Crab permits us to

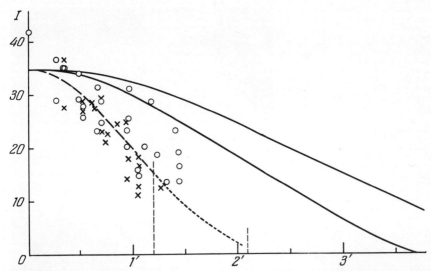

Fig. 166. Distribution of optical (broken curve) and radio brightness (solid curves) in the Crab Nebula (Oort and Walraven 1956). The two solid curves indicate the uncertainty in Baldwin's observations (1954a).

understand the phenomenon, although a number of details still remain unclear.

Lampland (1921) was the first to detect changes in the nebulosity. He found definite differences between photographs of the Crab taken with the 40-in. Lowell reflector between 1913 and 1921. Oort and Walraven, comparing an 1899 Lick photograph of the Crab with Baade's 1942 photograph, noted major changes in the amorphous mass of the nebula; these consist of variations in the form and brightness of individual details, whereas the large-scale structure of the amorphous mass remained almost unchanged. It is almost as though the Crab Nebula is "breathing"—as if "ripples" are running through it. Superficially, all this recalls the variability in the fine structure of the

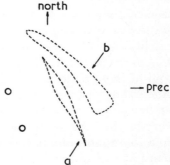

Fig. 167. Sketch of the central part of the Crab Nebula, indicating the light-ripples (Oort and Walraven 1956).

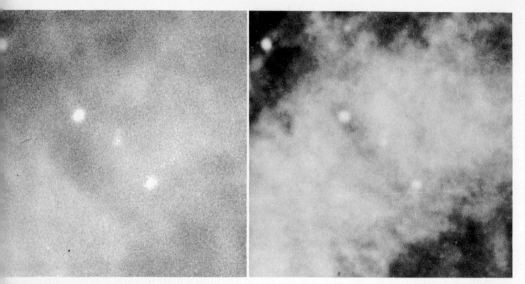

Fig. 168. Two photographs of a region in the northwestern part of the Crab
Nebula, by Hubble in 1924 (left) and Baade in 1938 (right) (Oort and Walraven
1956).

solar chromosphere and corona, but now on an incomparably greater
scale.

Figure 168 presents two photographs of a section of the Crab to the
northwest, obtained with the 100-in. reflector in 1924 and 1937. Upon
comparing these photographs, we may confirm the variability of the
radiation from the amorphous mass beyond any question.

We emphasize once again that the general form of the nebula is constant despite the strong variation in individual details of the amorphous mass. A number of characteristic structural features are
unchanged, such as the prominent dark bay to the west, and the dark
band in the middle.

In September 1955, the proprietors of the greatest telescope in the
world finally expressed active interest in the problem of the polarization of the Crab. At the 200-in. Palomar reflector, Baade (1956a)
secured several magnificent photographs of the nebula, under different
orientations of the polaroids. He employed a filter admitting radiation in the interval 5400 to 6400 A. In Figs. 169a, b, c, d we reproduce
four of these photographs, obtained with four different orientations.
A startling picture emerges. Individual features will vanish entirely
on one or two of the photographs, while on the others they will be
very bright. The negatives of these photographs have been subjected
to a thorough examination at Leiden; but even a preliminary treatment of this unique material readily permits us to reconstruct the

distribution of magnetic fields over the Crab, as Oort and Walraven have done (Fig. 170). From this chart we see that the relation between the direction of the magnetic field and the principal filaments in the Crab closely resembles that derived by Shajn *et al.* (compare Fig. 163). As Baade has pointed out, the observed structure of the amorphous mass of the Crab is fully determined by the structure of its magnetic field.

The Leiden measurements of the polarization in the Crab, as derived from Baade's photographs, have recently been published (Woltjer 1957), and are given in Fig. 171.

We are only now beginning to understand the singular state of affairs presented by the Crab Nebula radiation. Much is still unexplained; for example, the role of the central star of this nebula, the former supernova, is not entirely clear. Is it continuing to eject relativistic particles at the present time? Are the rapidly moving and changing formations which we have described the result of such ejections?

Recently we considered this question and arrived at a negative result (Shklovsky 1957). For one thing, it follows from Fig. 170 that the magnetic field in a wisp is normal to the direction of its motion. If we were to regard a wisp as a cloud of relativistic electrons ejected by the former supernova, then the magnetic field would necessarily be parallel to the motion of the wisp. Secondly, a wisp emerges quite rapidly, after a few weeks. But the linear dimension of such a feature is 2×10^{17} cm, and, as may readily be computed, relativistic particles would require no less than 2.5 months to traverse it.

Aside from the wisps, there are other variable formations in the amorphous mass, located sufficiently far from the former supernova. The velocity of these formations, although rather less than that of the wisps, is very considerable, and may be as much as 10 000 km/s.

We have therefore to deal with fluctuations in the luminosity per unit volume in the amorphous mass of the Crab, and to seek a suitable explanation for this by means of the synchrotron mechanism.

These fluctuations in the Crab Nebula reach an immense value: in a wisp, for example, the luminosity per unit volume is some 300 times the mean value for the central part of the nebula. Yet there is a region in the central part whose luminosity per unit volume is several times as small as the mean.

There are two possible causes for the fluctuations: variations in the intensity of the magnetic field, and variations in the concentration of relativistic electrons within a given element of the nebula. The two mechanisms are interrelated: an increase in the magnetic-field intensity within some volume element would entail an increase in the concentration of relativistic particles. One may show

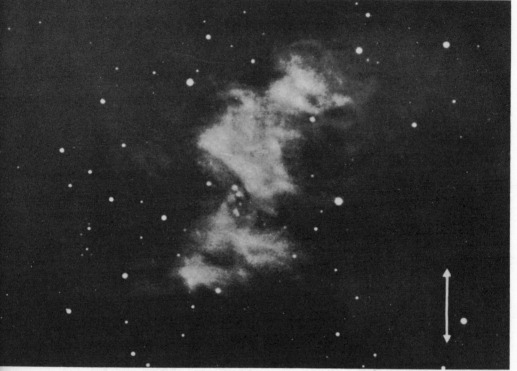

Fig. 169a. Photograph of the Crab Nebula, taken through a polaroid filter; the arrows indicate the direction of the electric vector (Baade 1956a).

Fig. 169b. The Crab Nebula in polarized light (Baade 1956a).

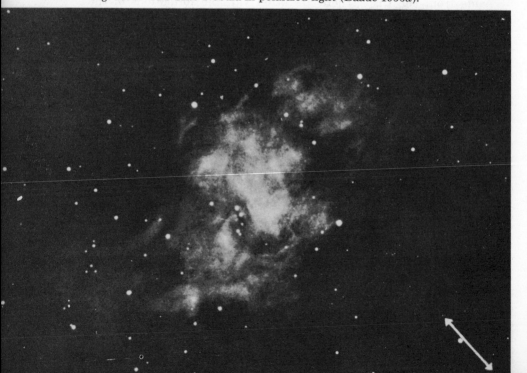

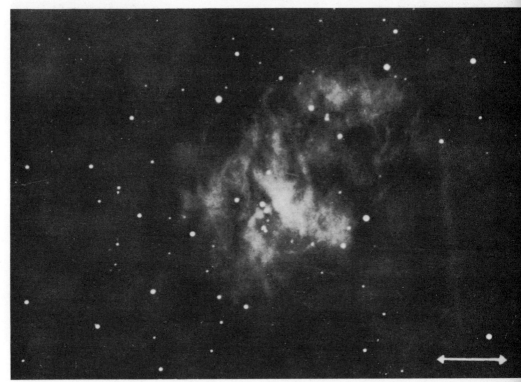

Fig. 169c. The Crab Nebula in polarized light (Baade 1956a).

Fig. 169d. The Crab Nebula in polarized light (Baade 1956a).

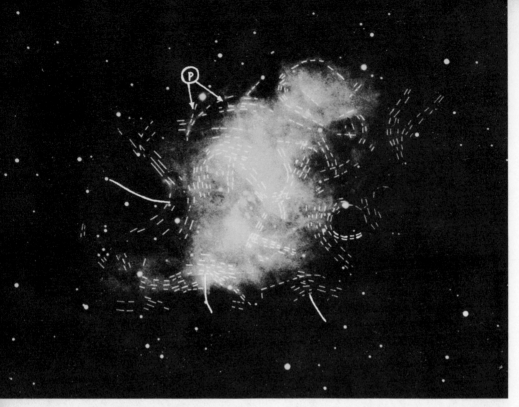

Fig. 170. Distribution of magnetic fields in the Crab Nebula (Oort and Walraven 1956).

that if the field intensity H suffers, for some reason, a local amplification by a factor m, then the luminosity per unit volume increases by the factor $n = m^{\gamma+1}$, where γ is the exponent in the differential energy spectrum for the relativistic electrons. According to Shcheglov, $\gamma > 4$. Hence if H were to be augmented by, say, a factor of only 3, the luminosity per unit volume would rise by more than 200 times.

We see, then, that the initial cause of separate fluctuations in the brightness of the amorphous mass (in particular, the wisps) must be corresponding fluctuations in the magnetic field in the Crab. This explanation accounts for the survival of high-energy electrons in the Crab up to the present time—the electrons responsible for the optical emission. A local increase in H leads to an increase in the relativistic-electron energy in the same region, by virtue of the invariant expression

$$P_\perp{}^2 \cdot H^{-1} = \text{const};$$

here $P_\perp$ is the component of momentum normal to H. Electrons which formerly emitted effectively no optical radiation now begin to radiate. Then, as the fluctuations abate, the electron energy automatically diminishes, and the electrons cease to be luminous. In this way the relativistic electrons lose their identity, so to speak; they remain

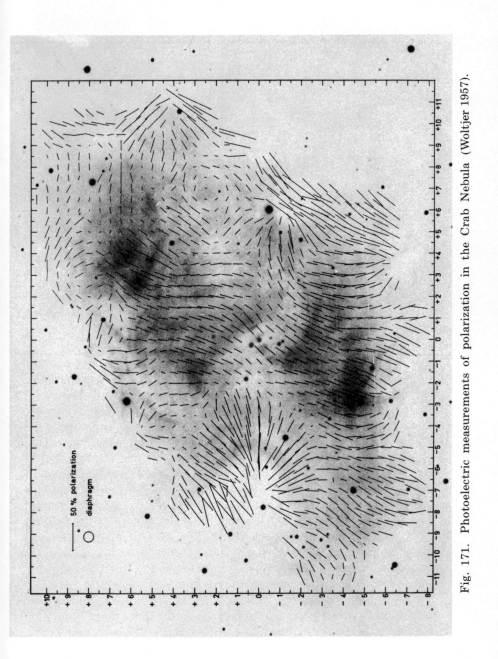

Fig. 171. Photoelectric measurements of polarization in the Crab Nebula (Woltjer 1957).

Fig. 172. Sketch of the filaments and the lines of force in the Crab Nebula (Shklovsky 1957).

luminous only for the limited time that they find themselves in a fluctuating field.

What is the reason for the fluctuations in the fields of the Crab? We have pointed out that the magnetic field becomes highly tangled at the periphery of the nebula. This is naturally to be explained by the presence of gaseous filaments there. The magnetic field of the Crab cannot be dissociated from the expanding system of dense gaseous filaments.

A good illustration of this is afforded by an analysis of the relative position of the magnetic lines of force and the gaseous filaments in the Crab, as shown schematically in Fig. 172. The lines of force clearly seem to avoid, or bypass, the filaments. The bending of the lines of force into gigantic arcs in the space between the filaments is highly characteristic. We evidently observe, in this situation, a tendency of the magnetic field and of the relativistic particles moving in this field to "pierce" the nebula wherever it may easily be done, namely in the space between the filaments. Since the field is interlaced and entangled with gaseous filaments at the base of the arcs, it cannot cut its way through at all. The field is simply deformed strongly, is bent, and attempts to "draw out" the filaments. But the energy of the field is inadequate to sweep the gaseous filamentary system about, although it may impart some acceleration to the system at certain points. It was on the basis of just this phenomenon that Pikelner was able to estimate the mass of the entire Crab Nebula (see above).

Different fluctuations can arise in the magnetic field under these conditions. In particular, magnetohydrodynamic waves may be propagated along the system of magnetic lines of force. We may even envision the formation of standing magnetohydrodynamic waves upon reflection from the boundary of the nebula. The radiation density will attain its maximum at the junction of such waves.

We therefore conclude that the magnetic field in the Crab may change its character quite rapidly from time to time. It always has a tendency to break through the network of gaseous filaments. If in some region favorable conditions are for any reason created for such a breakthrough, the field will rapidly be deformed at this position of weakness in the filamentary structure.

Although some of the properties of the magnetic field in the Crab have now become more or less clear, we are still far from an explanation of the origin of the field. It is to be hoped that before long this difficult and fundamental problem will be fully solved.

The detection of the polarization of the optical radiation from the Crab, as well as the remarkable variability within its amorphous mass, is of exceptional significance for both radio astronomy and astrophysics. We can scarcely overestimate the value which this discovery of polarization has for the problem of the origin of cosmic rays. The discovery convincingly demonstrates that there are relativistic particles in the nebular ejecta of supernovae, and it is upon this fact that the whole development of the nature of cosmic rays as outlined in this book is founded.

Inasmuch as the same mechanism—radiation by decelerated electrons in a magnetic field—is responsible for both the optical continuum and the radio emission from the Crab, one might expect the radio waves to be polarized as well. However, as we have pointed out in Sec. 13, Getmantsev and Razin (1956) have stressed that a magnetized plasma constitutes a magnetically active medium for radio waves. Because of the Faraday effect, the plane of polarization of a beam traversing a homogeneous magnetic field of dimension $\langle l \rangle$ will rotate through the angle ψ given by Eq. (13-13). For an assumed dimension $\langle l \rangle = 5 \times 10^{16}$ cm of the regions wherein a chaotic field may be regarded as quasi-homogeneous (see Sec. 20), we find, taking $N_\varepsilon \approx 1$ cm^{-3}, $\nu = 10^8$ sec^{-1} ($\lambda = 3$ m), and $H = 10^{-4}$ gauss, that $\psi \approx 10$ radians! Thus no polarization will be detected at all. But for optical frequencies $\psi \approx 0$ (no rotation of the plane of polarization), so that the polarization is not "smeared out."

Vitkevich attempted to detect polarization in the Crab at meter wavelengths (Ginzburg 1954) but was unsuccessful, in accord with the Getmantsev–Razin prediction. Westerhout (1956) also failed to detect polarization, at 22 cm.

Nevertheless we should not exclude the possibility that polarization of the radio emission from the Crab may yet be observed. It would be necessary to work at considerably higher frequencies than used by Vitkevich and Westerhout. By Eq. (13-13) the angle ψ becomes negligible at centimeter or even decimeter wavelengths. The approximate constancy of the direction of vibration of the electric vector in the Crab indicates that polarization will be present even in measurements smoothed over the entire nebula, as required for radio observations. We will recall that Dombrovsky, in his photoelectric investigation of the optical polarization, had introduced just such a smoothing.

It would in fact be interesting to determine the frequency ν_1 at which the polarization of the nebula's radio emission vanishes. The angle corresponding to ν_1 would be $\psi_1 \approx 1$. Entering Eq. (13-13), we could now evaluate the parameter $HN_\varepsilon \langle l \rangle$, which manifestly is of importance for the theory.

Mayer, McCullough, and Sloanaker (1957) have detected, for the first time, a polarization in high-frequency radio waves from the Crab, as predicted by the foregoing theory. The observations were carried out at 3.15 cm with the 50-ft reflector of the Naval Research Laboratory. The degree of polarization was about 7 percent; the position angle of 149° for the electric vector differs by 11° from that of the mean optical polarization of the Crab. The Faraday effect may well account for this departure in the position angle.

Kuzmin and Udaltsov have employed the 30-m fixed reflector of the Institute of Physics of the Academy of Sciences of the U.S.S.R. to observe the polarization in the Crab at a wavelength of about 10 cm. They detected a 3.5 percent polarization at a position angle for the electric vector of 142° ±9° (compare Udaltsov and Vitkevich 1958).

Since Westerhout observed no polarization at $\lambda = 22$ cm, we may assume that the frequency at which $\psi = 1$ is of the order of 2×10^9 c/s ($\lambda = 15$ cm). We find from this value that $HN_\varepsilon \langle l \rangle \approx 3 \times 10^{14}$ in the Crab. If we take $H = 3 \times 10^{-4}$ as an appropriate mean for the nebula, we have $N_\varepsilon \langle l \rangle \approx 10^{18}$. If now the Faraday rotation takes place in the filamentary system at the periphery of the nebula, where $N_\varepsilon \approx 300$ cm^{-3}, then the dimension of the quasi-homogeneous region becomes $\approx 3 \times 10^{15}$ cm, an entirely reasonable value. And if the Faraday rotation takes place in the amorphous mass of the nebula, where $N_\varepsilon \approx 1$ to 10 cm^{-3}, then $\langle l \rangle$ becomes quite large, namely $\approx 3 \times 10^{17}$ cm. This result also agrees with a high homogeneity of the field in the Crab Nebula.

22. ACCELERATION OF PARTICLES IN RADIO NEBULAE

We have shown in Sec. 13 that the most probable mechanism for the nonequilibrium radio emission from the Galaxy and from discrete sources is the radiation from relativistic electrons in cosmic magnetic fields. The detection of polarization in the optical radiation from the Crab Nebula, as predicted by theory, constitutes a direct proof that this hypothesis is correct.

There remains the basic question: why are relativistic particles (electrons) formed in certain cosmic objects—in particular, in the ejecta of supernovae? It is natural to tie the presence of relativistic particles to special physical conditions obtaining in discrete sources. But what are these conditions?

One peculiarity of discrete cosmic radio-wave sources is the immense macroscopic velocity of the gaseous masses. This is especially evident in sources related to supernova outbursts. The expansion velocity of the Crab Nebula exceeds 1000 km/s, while the velocities of internal motions in the nebula are of the order of 300 km/s. An even more striking example is the most powerful source, Cassiopeia A, evidently corresponding to the supernova of A.D. 369. Here the random velocities in individual filaments of the nebula attain 4000 km/s; the expansion velocity, on the other hand, is relatively small, about 500 km/s. In such "old" supernova radio nebulae as IC 443 + S 40, the internal motions have velocities of the order of 100 km/s, and thus are still considerable. Yet the macroscopic velocities of the gases must have been much greater when these radio nebulae were "young," shortly after the explosions of the supernovae which generated them.

On an entirely different scale there are the sources Cygnus A, NGC 1275, NGC 4486, and the like, which have been identified with certain peculiar galaxies: these also exhibit macroscopic motions of high velocity. For example, the relative velocity of the colliding galaxies NGC 1275 exceeds 3000 km/s, while the random internal motions of the interstellar gas in these galaxies approach a velocity of 1000 km/s. Therefore it is natural to suppose that the relativistic particles in radio sources are in some way related to the high macroscopic gas velocities.

How, then, do the relativistic particles found in discrete sources acquire such an enormous energy? The answer is that the primary cosmic-ray particles and relativistic electrons which are found in supernova ejecta have not acquired their high energies all at once, but gradually, continually accelerated by some mechanism active in the nebula or in the star itself.

The example of the colliding galaxies in Cygnus clearly shows that relativistic particles can be generated during the collision of cosmic masses of gas at a sufficiently high relative velocity. Such processes as nuclear energy release are not effective in this event. Each of the Cygnus galaxies had evidently been an ordinary large late-type spiral prior to the collision, a system resembling our own Galaxy. We arrive at the important conclusion that the radio energy of the powerful source Cygnus A, which arises from the relativistic electrons confined within this object, is in the final analysis drawn from the kinetic energy of collision—the collision of the gaseous components of the galaxies, for the stars of the two systems interpenetrate with virtually no interaction. Thus in radio sources such as Cygnus A, a considerable part of the kinetic energy of macroscopic gas motion is converted into the energy of a comparatively small number of relativistic particles. The formation of these particles is accounted for by some accelerating mechanism which continually effects a transfer to the particles of the energy in the macroscopic motions.

We must now identify the acceleration mechanism. It might, for example, be supposed that it is associated with the rise of extended electromagnetic fields, in which particles are accelerated to relativistic energies. Various authors, such as Terletsky (1946, 1948, 1949), have worked out models for "cosmic generators," based on the phenomenon of electromagnetic induction. Indeed, during a supernova explosion individual clouds of gases, moving at great speed, must become magnetized, and vortical electric fields must arise in the surrounding space. We shall show in the following section, however, that this induction mechanism cannot ensure the acceleration of particles to energies as high as those observed. This is not to say that electromagnetic fields of inductive origin play *no* role in the acceleration of the particles: at certain stages of the acceleration, the induction mechanism may be of considerable importance, and yet remain subsidiary.

Fermi (1949) proposed another mechanism for the acceleration of separate charged particles to relativistic energies, a *statistical mechanism*. He suggested this mechanism to explain the origin of the high energies in primary cosmic rays. We shall first discuss the mechanism in the context of the problem of accelerated particles in the interstellar medium; then we shall consider the possibility of applying the statistical mechanism to the nebular remnants of supernova explosions.

A cloud of interstellar gas is intimately affected by interstellar magnetic fields (Sec. 12). Within such a cloud the magnetic field may be taken as quasi-homogeneous. When an interstellar cloud attains a peculiar motion of the order of 10 km/s, the associated magnetic field will be carried along with it. We now turn to the motion of charged particles in the interstellar medium. In clouds where the magnetic field is quasi-homogeneous, the particles will move in helices, winding about the magnetic lines of force. The radius of curvature of the trajectory of a charged relativistic particle is given by the formula

$$r_H = \frac{E}{300H}, \tag{22-1}$$

where E is the energy of the particle in electron volts. For $H = 10^{-5}$ gauss and $E = 10^9$ ev, $r_H = 3 \times 10^{11}$ cm, which is negligible compared with the dimension of a cloud of interstellar matter over which the field may be taken as quasi-homogeneous. We may therefore assume that the particle moves along the lines of force—that is, along a straight line—within such a cloud. At the boundaries of regions of homogeneous field, the particle will change its direction of motion. It is as though the particle had collided with the region of homogeneous magnetic field. The motion of a charged particle in interstellar magnetic fields therefore resembles Brownian motion, to a certain extent.

If the magnetic fields associated with interstellar gas clouds were stationary, these "collisions" of charged particles with regions of homogeneous field would lead to a diffusion of particles, in which the particle energy would remain unchanged (elastic "collisions"). Here the dimension of the homogeneous-field region plays the part of the mean free path. According to diffusion theory, the root-mean-square distance traversed by particles in time t is given by the expression

$$\langle L \rangle = (6Dt)^{1/2} = (2lvt)^{1/2}, \tag{22-2}$$

where the diffusion coefficient $D \approx lv/3$. For relativistic particles $v \approx c$.

The fact that an interstellar gas cloud and the associated magnetic field are in a state of random motion does not affect the character of the diffusion, to a first approximation. However, as Fermi had pointed out, collisions of charged particles with moving cosmic magnetic fields can induce a gradual *acceleration* of the particles. In each such collision a particle can either acquire or lose energy, according to whether it is overtaken by, or overtakes, the magnetic field. But the collisions in which the energy of the particle is increased will predominate. This may readily be seen from the following simple analogy. The region of quasi-homogeneous magnetic field may be likened to a single gigantic particle moving with the cloud velocity u. Although u is small, the kinetic energy of the gigantic "particle" is extremely great because of its large mass. Let us regard the moving cosmic magnetic field and the charged particles colliding with it as a mixture of two "gases," in which the kinetic energy of a "molecule" of one "gas" (the magnetized clouds) greatly exceeds that of a "molecule" of the other "gas" (the charged particles). In collisions between these "molecules," there will clearly be a tendency toward a smoothing out of the energies of the two types of "molecules," such that energy will continually be transferred from the moving magnetic field to the charged particles. It is this effect which admits of the continual acceleration of the particles.

The acceleration mechanism just described might well be termed statistical. It differs fundamentally from the induction mechanism we referred to above. With the statistical mechanism of acceleration, the magnetic field may remain constant, on the average. The source of the energy of the accelerated charged particles is the kinetic energy of moving gaseous masses having associated magnetic fields. The magnetic field itself plays the role of a "driving belt" in the realization of this energy exchange.

Thus the statistical acceleration mechanism in principle enables kinetic energy of randomly moving gaseous masses to be converted into the energy of a comparatively small number of relativistic particles. At the beginning of this section we examined the physical properties of discrete radio sources and concluded that such a mechanism must necessarily exist. In any explanation of the nature of discrete sources of nonequilibrium cosmic radio emission, the theory of statistical acceleration is therefore especially important.

One may appeal to the laws of conservation of energy and momentum in connection with the "collision" of a charged particle of energy $E = Mc^2(1 - v^2/c^2)^{-1/2}$ with a particle of infinite mass moving with velocity u (the magnetized cloud of interstellar matter), and show that the energy acquired by the charged particle in one collision is, on the average,

$$\Delta E \approx (u^2/c^2)E. \qquad (22\text{-}3)$$

Equation (22-3) may be derived if we assume the velocity v of the charged particle to vanish, that is, that the particle is stationary until the collision. In this event, the velocity of the particle after collision will be approximately equal to the velocity u of the magnetized cloud, and its increase in energy will be

$$\Delta E = Mu^2 = (u^2/c^2)Mc^2 = (u^2/c^2)E,$$

where M is the mass of the charged particle.

As a result of these "collisions" the energy of the particle will continually increase. The mean increase in energy per unit time will be

$$\frac{dE}{dt} = \alpha E \approx \frac{\Delta E}{\tau} \approx \frac{u^2}{c^2\tau} E \approx \frac{u^2 v}{c^2 l} E, \qquad (22\text{-}4)$$

where $\tau = l/v$ is the time required for the charged particle to traverse a mean free path, and $v/l = 1/\tau$ is the number of "collisions" per unit time. In the simplest case, when the Fermi parameter α is independent of time, and when $E = Mc^2$ at $t = 0$ (that is, the acceleration sets in in the nonrelativistic region),

$$E = Mc^2 \exp{(\alpha t)} \approx Mc^2 \exp{\left(\frac{u^2 v}{c^2 l} t\right)}. \qquad (22\text{-}5)$$

We shall assume that each accelerated particle is accelerated by the statistical mechanism over the time T, and, furthermore, that a particle may with equal probability begin to accelerate at any arbitrary instant of time. As a result, particles of various energies will be present among the accelerated particles; the greater the time over which a particle has experienced acceleration, the greater will be its energy. The probability of finding a particle which has experienced a nonvanishing acceleration over a time interval between t and $t + dt$ is

$$dW = (e^{-t/T}/T)dt.$$

By Eq. (22-4),

$$dt = dE/\alpha E, \qquad t = \alpha^{-1} \ln(E/Mc^2);$$

we therefore have:

$$dW = \text{const} \cdot N(E)dE = (\alpha T)^{-1}(Mc^2)^{1/\alpha}E^{-1-1/\alpha T} dE. \qquad (22\text{-}6)$$

Here $N(E)$ is the differential energy spectrum of the accelerated particles. Thus charged particles accelerated by the statistical mechanism possess a differential energy spectrum obeying the power law

$$N(E) = KE^{-\gamma},$$

where

$$\gamma = 1 + (\alpha T)^{-1}.$$

On the other hand, if the nonequilibrium cosmic radio emission from discrete sources is interpreted as radiation by relativistic electrons in a magnetic field, then it follows directly from the spectrum of this radiation that the differential energy spectrum of relativistic electrons follows a power law. Thus once again we find good agreement between the theory of radio emission by relativistic electrons and the statistical acceleration mechanism.

It must be emphasized that charged particles can be accelerated by moving magnetic fields only if the energy lost by the particles during the time between two "collisions" with magnetized clouds is less than the increment of energy acquired, on the average, in a single "collision." For electrons, the energy loss consists in the following:

(*a*) losses due to the ionization of particles with which the accelerated electron collides; in the nonrelativistic case, these losses are given by the formula

$$-(dE/dt)_{\text{ion}} = 7.62 \times 10^{-9}n(2mc^2/E_{\text{kin}})^{1/2}[11.8 - \ln(mc^2/E_{\text{kin}})] \text{ ev/s}, \tag{22-8}$$

and in the relativistic case by

$$-(dE/dt)_{\text{ion}} = 7.62 \times 10^{-9}n[20.1 - 3\ln(mc^2/E)] \text{ ev/s}, \quad (22\text{-}9)$$

where n is the number of particles per unit volume with which the electron collides;

(*b*) losses by deceleration radiation in the collision of electrons with protons (radiation losses):

$$-(dE/dt)_{\text{rad}} = 8 \times 10^{-16}nE \text{ ev/s}; \tag{22-10}$$

(c) losses due to the inverse Compton effect:

$$-(dE/dt)_{\text{Comp}} = 5.6 \times 10^{-16}(E/mc^2)^2 \text{ ev/s}, \qquad (22\text{-}11)$$

where the numerical coefficient refers to the conditions of the interstellar medium, taking the color temperature of the radiation as $6000°K$, the dilution coefficient $W = 2 \times 10^{-13}$, and the photon concentration $\langle n \rangle = 2 \times 10^{-2} \text{ cm}^{-3}$;

(d) losses from radiation by relativistic electrons in a magnetic field:

$$-(dE/dt)_{\text{mag}} = 3.8 \times 10^{-15}E^2H^2 \text{ ev/s}. \qquad (22\text{-}12)$$

For small values of the energy of the accelerated electron (that is, early in the acceleration process), the ionization losses predominate. Equation (22-8) shows that these losses are almost independent of the energy of the electron, whereas the growth in energy per unit time as determined by the statistical mechanism for acceleration is proportional to the energy of the electron, by Eq. (22-4). It follows that particles will be accelerated by the statistical mechanism only if their energy exceeds a critical "energy of injection" E_{inj}, given by the condition

$$\alpha E_{\text{inj}} = (dE/dt)_{E=E_{\text{inj}}}. \qquad (22\text{-}13)$$

The following expression holds for the kinetic energy of injection (Getmantsev and Razin 1956):

$$E_{\text{inj, kin}} \approx 4 \times 10^{-19} n^2/\alpha^2 \text{ ev}. \qquad (22\text{-}14)$$

The energy of injection depends essentially on n and α, which in turn are governed by the physical conditions in magnetized gaseous masses which are pervaded by random macroscopic motions. For a sufficiently large value of the Fermi parameter, αE_{inj} may be held to a rather small level, and the acceleration of the more rapid charged particles may commence.

We have mentioned that Fermi had initially introduced the statistical mechanism of acceleration in 1949 in connection with the problem of the acceleration of relativistic particles in the interstellar medium. Through this mechanism he sought to explain the exceptionally high energies in cosmic rays. V. L. Ginzburg (1953b) soon employed the statistical mechanism to account for the acceleration of electrons in sources of cosmic radio waves, namely the envelopes of supernovae with their associated turbulent motions. Although in the interstellar medium α appears to be very small, of the order of 10^{-17} (corresponding to an acceleration time of hundreds of millions of years), in the nebular ejecta of supernovae α may become as large as 10^{-10}, or even more, leading to a duration of the acceleration process of hundreds of years or less.

Since the Crab Nebula was formed 900 yr ago, the acceleration process cannot have gone on for a longer period than this.

Following Ginzburg's work, the statistical mechanism explanation was advanced for supernova envelopes by other authors as well. However, it should be noted that neither in Ginzburg's paper (1953*b*) nor in the later discussions had the statistical mechanism received further development: Fermi's idea, which had been developed for application to the interstellar medium, has been carried over without essential change to the envelopes of supernovae.

Yet there is a difference in principle between the circumstances of statistical acceleration in the interstellar medium and in such an envelope. The point is that the supernova envelopes are *expanding,* and it is essential in many cases to take the expansion into account. Ginzburg, Pikelner, and Shklovsky (1955) have considered this problem, together with a number of other questions related to the mechanism of particle acceleration in the ejecta of supernovae.

It was found that, for an envelope expanding with velocity V, the mean energy received by nonrelativistic particles in collisions with wandering magnetic fields is given by

$$\Delta E' \approx Mv(u + V), \qquad (22\text{-}15)$$

where v is the velocity of a particle, and u is the random velocity of the wandering magnetic fields. If $V = 0$, then upon smoothing we find the mean energy acquired by a particle in one collision to be

$$\langle \Delta E \rangle = \Delta E = Mu^2 = (u^2/c^2)Mc^2 = (u^2/c^2)E. \qquad (22\text{-}16)$$

If $u = 0$ as well, then

$$\Delta E \approx -Mv\,\Delta V \approx -Mv(Vl/r), \qquad (22\text{-}17)$$

where $\Delta V \approx Vl/r$ represents the velocity with which adjacent elements of the nebula, a distance l apart, retreat from each other, and r denotes the radius of the expanding envelope. The negative sign in Eq. (22-17) indicates that in each collision with the magnetic field the energy of the particle is diminished; that is, the particle will be decelerated. This process is closely analogous to the cooling of a gas in adiabatic expansion. In the general case, where both u and V are nonvanishing,

$$\Delta E \approx Mc^2 \left(\frac{u^2}{c^2} - a\frac{vVl}{c^2r} \right), \qquad (22\text{-}18)$$

where $a \approx 1$. The change in the particle's energy per unit time is given by the expression

$$dE/dt = \Delta E/\tau = \beta E \approx (u^2 - avVl/r)(v/lc^2)E. \qquad (22\text{-}19)$$

The Crab Nebula illustrates how important it is to take the expansion of supernova envelopes into account. We shall set $l = 5 \times 10^{15}$ cm (several times the mean free path for $n = 1$ cm^{-3}), $u = 10^7$ cm/s (the velocity of small pulsations), $V = 10^8$ cm/s (the observed expansion velocity for the nebula), and $r = 3 \times 10^{18}$ cm; it follows that

$$\beta \approx - V/r \approx - 5 \times 10^{-11} \, \text{sec}^{-1}.$$

This means that, at the presently observed stage of evolution in the Crab, acceleration of charged particles by the statistical mechanism has not yet taken place. On the contrary, the relativistic particles have *decelerated;* over the 900 yr since the formation of the Nebula, the energy of these particles has diminished by more than an order of magnitude, quite apart from any energy due to nebular expansion that they may have possessed when the statistical deceleration began.

At any rate, we must conclude that favorable conditions for a statistical acceleration could have occurred only at an *early* phase in the development of the Crab. Indeed, l must be proportional to the free path λ, that is, to n^{-1}. In turn, $n \propto r^{-3}$, whence $l/r \propto r^2$. On the other hand, if the maximum dimension L of the turbulence is of the order of r, and if the velocity u of the large-scale turbulence elements remains constant throughout the expansion process, then we may apply the Kolmogoroff law, because of the locally isotropic character of the turbulence:

$$u^3/l = U^3/L \propto r^{-1}, \tag{22-20}$$

whence $u \propto r^{2/3}$. The quantity $aVul/r$ [Eq. (22-19)] will vary with r as $r^{2/3}$, but $u^2 \propto r^{4/3}$. Thus at some epoch during the expansion, β may vanish, so that conditions favorable for acceleration may have obtained at earlier epochs. For the Crab Nebula we find that $\beta = 0$ for $r \approx 10^{17}$ cm, a stage which occurred about 30 yr after the explosion in 1054. Yet we must stress that these considerations are valid only on the assumption of locally isotropic turbulence in the Crab; this can hardly be the actual state of affairs.

If, then, statistical acceleration of charged particles has in fact taken place in the Crab Nebula, it can only have arisen very early in the development of the envelope.

One peculiarity of the statistical mechanism is the presence of a threshold of energy injection. A charged particle can be accelerated by wandering magnetic fields only under conditions in which its energy exceeds the injection energy. This also just accounts for the fact that only a small fraction of particles accelerated in a turbulent nebula may acquire relativistic energy: only the most rapid, and hence the most energetic, particles will begin to accelerate. What then are the conditions favoring the appearance of these rapid particles?

The author of this volume has shown that such conditions can arise at a shock front in the collision of two gas masses having a high relative velocity (Shklovsky 1954). At the shock front the pressure suffers a jump (on one side of the front it is four times as great as on the other). The temperature in the region adjoining the shock front becomes extremely high. It is given by the relation

$$T \approx (m_H v^2/2k)(\gamma - 1)/4, \tag{22-21}$$

where γ is the ratio of the specific heats and v the collision velocity. At $v = 1000$ km/s, $T = 10^7$ °K; at $v = 3000$ km/s, $T = 10^8$ °K, corresponding to a particle energy of $\approx 10^4$ ev. A few particles in the "tail" of the Maxwellian velocity distribution will have an energy of 10^5 ev, already in excess of the injection energy. For example, Eq. (22-14) yields $E_{inj, kin} \approx 10^5$ ev if $\alpha = 10^{-6}$ sec^{-1} and $n \approx 10^6$ cm^{-3}. We may obtain the same value for $E_{inj, kin}$ if $\alpha = 10^{-1}$ sec^{-1} and $n \approx 10^{11}$ cm^{-3}; these are the conditions in stellar atmospheres.

Greenstein (1955) has shown that at these high kinetic temperatures, of the order of 10^7 or 10^8 °K, thermonuclear reactions can proceed behind a shock front. Fast electrons can be formed in this way, and they may subsequently be accelerated by the statistical mechanism.

For example, in every pair of deuterium–deuterium (d–d) reactions, one electron of 20-kev energy is formed in tritium decay, one electron of 800-kev energy is formed in neutron decay, and one proton of 1.5-Mev energy is formed. Every proton–deuterium (p–d) reaction yields a γ-quantum of 5.5-Mev energy. In Thomson scattering these quanta will generate electron recoils of ≈ 3-Mev energy.

The (d–d) reaction rate per deuterium nucleus will be

$$P_{dd} = 4 \times 10^{10} \, \rho x_D T^{-2/3} \exp(-42.6 \, T^{-1/3}), \tag{22-22}$$

where ρ is the density of the medium and x_D is the relative abundance of deuterium. At $T = 2.5 \times 10^8$ °K, corresponding to $v \approx 5000$ km/s, we have

$$P_{dd} \approx 10^6 \, \rho \, x_D.$$

The (p–d) reaction rate per deuterium nucleus is

$$P_{pd} = 7 \times 10^4 \rho x_H T^{-2/3} \exp(-37.2 \, T^{-1/3}); \tag{22-23}$$

at $T = 2.5 \times 10^8$ °K, $P_{pd} = 5\rho x_H$. The total number of reactions over the entire nebula will be

$$L = p x_D \mathfrak{M}/m_H, \tag{22-24}$$

where $\mathfrak{M}$ is the mass of the gas within which the reactions take place. Taking the deuterium abundance x_D relative to hydrogen as 10^{-3}, Greenstein finds that, for $\rho = 3 \times 10^{-22}$ gm/cm^3 and $\mathfrak{M} = \mathfrak{M}_\odot$, the frequency of all the above thermonuclear reactions, for the entire nebula, may be as much as 10^{34} sec^{-1}. And if we were to consider an earlier stage of evolution in the nebula, or indeed the atmosphere of the supernova itself, where the density would have been many orders of magnitude greater, then the number of these reactions would be considerably in excess of the above value.

It should be kept in mind that the question of deuterium content in supernova envelopes is completely open at the present time.

In summary, Greenstein's proposal seems a highly promising one, and deserves serious attention.

In our interpretation of the radio and optical emission from the Crab Nebula there arises a difficulty of a fundamental character, first pointed out by Gordon (1954). The total energy of the relativistic electrons in this nebula is $\approx 10^{47}$ to 10^{48} erg, according to Eq. (20-1) and the results given in Sec. 21; this value approaches the kinetic energy of the gaseous matter ejected by the supernova. If the Fermi statistical mechanism is to account for the acceleration, then by Eq. (22-5) the energy of the particle is proportional to the mass. We might at first suppose that in a given magnetized turbulent medium electrons and protons would be accelerated equally, so that to each relativistic electron of energy E there would correspond a proton of energy some 1800 times as great. But the energy of the relativistic particles in the Crab Nebula would then amount to 10^{51} rather than 10^{48} erg, and this is unacceptable.

The electrons, however, have a number of advantages over the protons in the statistical acceleration process. Thus the energy of the electrons tends to catch up with that of the protons. A detailed discussion has been given by Ginzburg, Pikelner, and Shklovsky (1955). We shall mention here just one of the interesting effects responsible for this behavior. High-energy protons collide with the nuclei of the atoms in the nebula, forming mesons, which decay, yielding electrons and positrons. The energy of these secondary particles may approach 10 percent of the energy of the primary protons. If the density of the medium early in the expansion is relatively high in the regions where acceleration takes place, almost all the high-energy protons will form relativistic electrons and positrons. Under these circumstances the positrons will have a long lifetime and will radiate in a magnetic field just as the relativistic electrons do. Light relativistic particles subsequently formed in nuclear collisions will be accelerated by the statistical mechanism; as a result, the total energy of the relativistic electrons and positrons will increase to a value considerably closer to that of the heavy particles.

Let us now turn briefly to the evolution of radio nebulae (see Sec. 20). The intensity of the magnetic field in such a nebula will continually decrease with time. In the presence of locally isotropic turbulence, $H \propto \rho^{1/2}$; since $u \propto r^{2/3}$, we have $H \propto r^{-5/6}$. By Eq. (13-7), the flux of radiation from the source follows the relation

$$F_\nu \propto H^{(\gamma+1)/2},$$

where for the exponent of the differential energy spectrum of the relativistic electrons we have $3 > \gamma > 2$.

If the formation of new relativistic electrons in a radio nebula should cease at some particular time, then the number of relativistic electrons in the nebula may be taken as constant for all succeeding time. In this event,

$$F_\nu \propto r^{-5(\gamma+1)/12},$$

for the energy of the relativistic electrons emitting radio waves remains practically the same over the evolution of the nebula.

The surface brightness will be given by

$$I_\nu \propto F_\nu/r^2 \propto r^{-(5\gamma+29)/12}.$$

For $\gamma = 3$, $\qquad F_\nu \propto r^{-5/3}, \quad I_\nu \propto r^{-3.7};$

for $\gamma = 2$, $\qquad F_\nu \propto r^{-5/4}, \quad I_\nu \propto r^{-3.3}.$ $\qquad$ (22-25)

In "younger" radio nebulae we must take into account the continual formation of new relativistic particles. If we assume that this process operates at a constant velocity, independent of the development occurring in the nebula, then we have for the number N of relativistic electrons in the expanding radio nebula,

$$N \propto t \propto r \propto H^{-6/5}.$$

For the radio flux,

$$F_\nu \propto N H^{(\gamma+1)/2} \propto r^{(-5\gamma+7)/12} ;$$

and for the brightness,

$$I_\nu \propto r^{(-5\gamma-17)/12}.$$

In this event:

for $\gamma = 3$, $\qquad F_\nu \propto r^{-2/3}, \quad I_\nu \propto r^{-8/3} \approx r^{-2.7};$

for $\gamma = 2$, $\qquad F_\nu \propto r^{-1/4}, \quad I_\nu \propto r^{-2.25}.$ $\qquad$ (22-26)

At the present time it is still not possible to define the evolutionary stage of a radio nebula at which formation of new relativistic particles ceases, with one law of decrease of F_ν and I_ν replacing the other. Moreover, our assumption of locally isotropic turbulence is in fact a very crude approximation to the true situation. The magnetic field in a radio nebula must nevertheless diminish with time, whatever the assumptions; this will entail a gradual decrease in F_ν and, in particular, in I_ν. Even though our formulae are very rough, they appear to be in satisfactory agreement with the observational results for radio nebulae (see Sec. 20).

In summary, we can explain a variety of phenomena by means of the mechanisms we have considered for the acceleration of charged particles as a result of supernova explosions. But we nevertheless remain far from a full understanding of the unusually complex behavior which breaks out with a supernova.

23. THE ORIGIN OF COSMIC RAYS

The problem of the origin of primary cosmic rays has unquestionably been one of the outstanding enigmas of natural science. Although cosmic rays have been known for more than 40 years, this problem has not received a satisfactory solution until quite recently. The composition of the primary component of cosmic rays had long been uncertain. Little more than 15 years ago it was firmly established that protons are the most abundant particles in the primary component. By 1948 helium nuclei had been detected in primary cosmic rays, together with oxygen and other elements as far as iron, and even beyond. In this way the "chemical composition" of cosmic rays was clarified. Until this matter had been decided, no theory of the origin of cosmic rays could possibly be accepted. It is now established that primary cosmic rays comprise 80 to 85 percent protons, 15 to 20 percent α-particles, and ≈ 1 percent nuclei of heavy elements. One remarkable property of this composition is that the nuclei of lithium, beryllium, and boron are well represented; these elements are cosmically very rare, but the relative abundance in cosmic rays is a million times the mean cosmic abundance.

An important experimental result was the establishment of the energy spectrum of the primary component, which was found to be of the form

$$I(E) = KE^{-\gamma},$$

$$I(E > E_0) = \int_{E_0}^{\infty} I(E)dE = (\gamma - 1)^{-1}KE_0^{-(\gamma-1)}, \qquad (23\text{-}1)$$

where I is the intensity of a given kind of particle and E is the total energy of a proton or nucleon. For E between 5×10^8 and 3×10^{10} ev we have $\gamma \approx 1.9$ to 2.2 both for protons and for nuclei. At greater energies only the integrated spectrum of the primary component is known. For $5 \times 10^{10} < E < 10^{12}$ ev, $\gamma \approx 2.5$; for $10^{13} < E < 10^{18}$ ev, $\gamma \approx 2.7$ to 3.0.

The intensity of cosmic radiation depends on the geomagnetic latitude. For example, at geomagnetic latitude 58° the total intensity of the primary component is 0.3 particles $\text{cm}^{-2}\,\text{sec}^{-1}\,\text{sterad}^{-1}$. Since it has been established observationally that primary cosmic radiation is *isotropic,* the concentration of cosmic-ray particles at the Earth is given by

$$N(E > 1.5 \times 10^9 \text{ ev}) = (4\pi/c) I(E > 1.5 \times 10^9 \text{ ev})$$

$$\approx 10^{-10} \text{ cm}^{-3}. \tag{23-2}$$

The energy density of primary cosmic rays is given by the expression

$$W = \int_{E_1}^{\infty} N(E) E \, dE \approx 1.2 \times 10^{-12} \text{ erg/cm}^3 \approx 1 \text{ ev/cm}^3, \tag{23-3}$$

where $E_1 = 1.5 \times 10^9$ ev. An important circumstance is that the cosmic-ray intensity hardly increases at all poleward of geomagnetic latitude 58°. We must conclude that for some reason protons of energy $< 1.2 \times 10^9$ ev fail to reach the Earth. Investigations have shown that this high-latitude cutoff is magnetic in character; low-energy particles are not incident upon the Earth near the magnetic poles because of the influence of a magnetic field external to the Earth. Ginzburg (1956) has proposed that this field is localized in the solar system.

As soon as the "chemical composition" of the primary component had been derived, the first serious theoretical efforts were directed toward an interpretation of the origin of cosmic rays. Since a metagalactic localization of primary cosmic rays had been rejected from energy considerations (Sec. 12), two possibilities remained: (*a*) cosmic rays are localized in a relatively small region surrounding the solar system; (*b*) they are localized in the Galaxy. A persuasive argument for the former hypothesis is the undoubted fact that the sun generates primary cosmic rays during periods of high activity (such as the time of powerful flares). Thus, for example, at the great flare of 1956 February 23 the intensity of the primary component increased by several times. The intensity level of cosmic radiation normally remains constant, however. Variations in intensity, as determined both by radiation from relativistic particles in active regions on the sun and by variations in the Earth's magnetic field, are usually held to within 1 or 2 percent.

In 1949 several important theoretical papers appeared in connection with the origin of cosmic rays. Fermi's investigation (1949) is of special significance, for it provided the first clear demonstration that, under the conditions prevailing in the interstellar medium, cosmic rays survive in the Galaxy for a limited time ($\approx 10^8$ yr) because of collisions with nuclei of interstellar atoms. This time is in any event less than the evolutionary time scale for the Galaxy. This paper furthermore destroyed the speculative ideas being circulated at that time as to a "relic" origin for primary cosmic rays. According to these hypotheses, cosmic rays were formed at the beginning of the expansion of the observable universe, when conditions in the Galaxy were entirely different from those at present.

Fermi also developed the theory of statistical acceleration: cosmic particles are accelerated in interstellar space by moving magnetic fields (Sec. 22). The longer a particle wanders in interstellar magnetic fields, the greater will be its energy. Proceeding from elementary considerations Fermi obtained the theoretical energy spectrum for statistically accelerated particles:

$$N(E) \propto E^{-1-1/\alpha T},$$

where the Fermi parameter $\alpha \approx u^2 v / c^2 l$, and T is the lifetime of cosmic rays in the interstellar medium. Thus the theoretical spectrum corresponds to an observed spectrum $N(E) \propto E^{-\gamma}$ with $\gamma = 1 + 1/\alpha T$.

We have shown in Sec. 22 that the statistical mechanism can accelerate only those particles whose energy exceeds a minimal "injection energy" E_{inj}, which depends on the kind of particle. For protons in the interstellar medium $E_{\mathrm{inj, kin}} \approx 10^8$ ev, while for heavy particles it is considerably greater. For oxygen nuclei $E_{\mathrm{inj, kin}} = 10^9$ ev per nucleon; and for iron nuclei, 5×10^9 ev. The lifetime T of relativistic particles, as defined by their effective cross section for nuclear collisions, also varies with the particle, being considerably shorter for heavy nuclei than for protons. This circumstance leads to certain difficulties [see, for example, Ginzburg (1953a)]. If the statistical mechanism actually obtains in interstellar space, then we may derive α from an observed value of γ and a value of T estimated from the characteristics of the interstellar medium. We shall take the mean concentration of particles in the interstellar gas as ≈ 0.01 cm^{-3} (this value will be justified at the end of the present section), and the effective cross section for nuclear collisions as $\approx 2.5 \times 10^{-26}$ cm^2; we then obtain $T = 4 \times 10^9$ yr $= 1.25 \times 10^{17}$ sec. Setting $\gamma = 2.7$ (for $E \approx 10^{13}$ ev), we find that $\alpha = 5 \times 10^{-18}$ sec^{-1}. Since T is considerably smaller for heavy nuclei, with the same α the value of γ will be considerably greater. For oxygen $\gamma = 18$, so that at sufficiently high energies heavy nuclei will be practically absent from the primary component. Observations indicate that, for $E < 3 \times 10^{10}$ ev per nucleon, the energy spectrum is the same for protons as for nuclei. Consequently the statistical mechanism cannot be valid for cosmic particles of energy less than 3×10^{10} ev. On the other hand, because of the injection-energy threshold, nuclei of energy less than a few billion electron volts cannot, in general, be accelerated by the statistical mechanism at all.

We see, then, that the study of questions of the origin of cosmic rays demonstrates that the Galaxy must contain sources which somehow generate particles of energy exceeding the injection energy for the interstellar medium. Carrying this great energy, the particles abandon their sources and are released into interstellar space, where they wander in cosmic magnetic fields for hundreds of millions of years. It is possible that they are subject to further acceleration by the statistical mechanism in the interstellar medium, yet such acceleration need not take place. A choice between these two possibilities can only be made upon further analysis of observational data and development of theory.

To a considerable degree the problem of the origin of cosmic rays therefore reduces to the identification of the sources in which they arise, and of the acceleration mechanism in these sources. The latter problem has a long history. Even when hardly anything was known of the nature of the primary component, there was much speculation as to the reason for the high energy of the primary particles; in particular, a number of variants of the "annihilation" hypothesis were advanced. But the discovery of heavy nuclei in primary cosmic radiation annihilated all hypotheses of this type. Gradually it became clear that the high energies of primary particles are acquired through acceleration in certain cosmic electromagnetic fields. Various authors, such as Swann (1933), Alfvén (1939), and especially Terletsky (1946, 1948, 1949), treated several definite models for cosmic generators of ultra-high–energy particles. However, all these hypotheses were merely qualitative. Authors would frequently investigate one mechanism or another for the generation of high-energy particles and, as soon as such a mechanism was discovered, they would regard their task as completed. Terletsky, for example, developed in some detail a mechanism for the acceleration of charged particles by electric fields arising in rotating magnetic stars and in stars with a variable magnetic moment; but he failed to consider the quantitative side of the question—whether these cosmic accelerators actually possess sufficient power.

We must also mention one further circumstance. In all earlier models of cosmic accelerators it had been assumed that the charged particles move in a vacuum under the influence of a vortical electric field. Such a scheme involves an error in principle. However paradoxical it may seem, it is a fact that under cosmic conditions there is hardly any place at all where a vacuum exists, for in all cases of interest the mean free path of the particles is considerably less than the characteristic dimensions of the cosmic objects concerned. Stellar interiors, atmospheres, and envelopes; nebulae; interstellar gas clouds; and even the highly rarefied medium between the clouds are *plasmas*. We may say that under cosmic conditions matter exists primarily in the form of a plasma, a *continuous* medium. Solid bodies of the universe (planets, cosmic dust) are exceptions, but for the problems of interest here these objects are of no significance.

If for any reason an electric field is created in a plasma, this field will affect *all* charged particles in the region where it is localized. In this event a quantity of energy will be transmitted to *all* particles, so that only a very small share of energy will be allotted to an individual particle. The field will accelerate particles only over a mean free path. As a result, each joule of heat will only be apportioned over the plasma, heating the plasma to some extent, but there will by no means be formed a relatively small number of corpuscles of immense energy. In short, the case for all the proposed cosmic induction accelerators appears to be closed.

Until recently the problem of the origin of cosmic rays has remained in a very unsatisfactory state, for the following basic reason. The problem is essentially an *astronomical* one, for primary cosmic rays are not formed on the Earth, but in certain cosmic sources. Thus we cannot reach a solution without taking the properties of the cosmic objects carefully into account. And it is at just this point that we are confronted with a fundamental difficulty. Primary cosmic rays could, up to a very short time ago, be observed only in the immediate neighborhood of the Earth; and the very possibility of *observing* primary cosmic rays in the depths of the universe had seemed so fantastic that no one had even thought of it. Yet it was evident that until methods for such observations were found it would hardly be possible to secure a decisive result not only about the origin, but even about the localization of primary cosmic rays. However convenient it may have been to construct purely theoretical arguments, and to analyze assorted mechanisms for particle acceleration, all the clever theory remained only hypothesis, for no observational data on primary cosmic rays were available at the sites of their formation.

Radio astronomy achieved almost its most signal success in making possible the observation of primary cosmic rays far from the environs of the Earth. Since deceleration radiation from relativistic electrons in interstellar magnetic fields is responsible for nonequilibrium galactic radio emission, the existence in interstellar space of particles having energies from 10^8 to 10^9 ev is an established fact. But this is none other than the electron component of primary cosmic rays! This conclusion is corroborated by the fact that the concentration of relativistic electrons of energy $> 5 \times 10^7$ ev is found to be of the same order (a) from the observed intensity of cosmic radio waves, using a value for the intensity of the magnetic field derived from analysis of observed interstellar polarization of light, and (b) from the concentration of primary cosmic rays (protons and nuclei) as obtained experimentally in the neighborhood of the Earth. The spherical system formed by the sources of the general field of nonequilibrium galactic radio emission outlines the domain of space within which the primary cosmic rays are localized. The same observations demonstrate that primary cosmic rays are localized essentially in the Galaxy, rather than in the neighborhood of the sun.

We have pointed out in Sec. 13 that relativistic electrons have not been detected in the primary component near the Earth. But the reason for their absence has already been made clear in Sec. 13. In contrast to heavy particles, cosmic electrons wandering in interstellar magnetic fields continually lose energy by radiation, so that their energy is in the mean considerably less than that of heavy particles. The presence of the high-latitude cutoff hinders the arrival at the Earth of the abundant relativistic electrons of energy $< 10^9$ ev, while electrons of energy $> 10^9$ ev are rare because of the circumstance just mentioned. From the study of galactic radio waves we may, however, conclude that if the experimental accuracy were raised by an order of magnitude, electrons, and perhaps positrons, might be detected in the primary component. This has been emphasized by Ginzburg (1953c).

Radio astronomy has not merely settled the matter of the localization of primary cosmic rays in the spherical component of the Galaxy. It has identified the *sources* in which the primary component is generated, and from which it is disseminated throughout interstellar space. The author called attention to this several years ago (Shklovsky 1953d, 1953e).

Inasmuch as relativistic electrons moving in the magnetic fields associated with supernova ejecta are responsible for the radio emission from these peculiar nebulae, it is evident that a number of the relativistic particles in these objects attain energies of 10^7 to 10^{12} ev. Estimating the magnitude of the magnetic field in such ejecta, together with their radiation flux and the distance from the sun, we may employ established formulae for synchrotron radiation to find the population of relativistic electrons in the nebula. In the Crab Nebula the number of relativistic electrons of energy $>10^8$ ev is of the order of 10^{50}.

The discussion of Sec. 22 led us to conclude that the radio-emitting phase in supernova ejecta is a comparatively brief one. The expanding nebulosity will gradually be decelerated by interstellar matter and will eventually be dissipated. It can be shown that over this period of time the energy of relativistic electrons wandering in the nebula, and spreading out with it, remains essentially the same (Sec. 22). Consequently, relativistic particles in expanding supernova envelopes will enter the interstellar medium after some tens of thousands of years, come under the influence of interstellar magnetic fields, and gradually begin to fill the entire domain in which these fields are localized. This process will continue for several hundred million years, until through repeated collisions with interstellar atomic nuclei the relativistic particles have lost a significant part of their energy, and have "dropped out of play."

If the foregoing picture of the origin of primary cosmic rays is correct, then dynamic equilibrium of relativistic particles must prevail in the Galaxy: the mean frequency with which the particles are released to the Galaxy in supernova outbursts must equal the frequency with which they are "quenched" in nuclear collisions. We have made calculations (Shklovsky 1953d, 1953e) which indicate that to a first approximation this equality actually holds.

Let us assume that on the average supernovae appear in the Galaxy once every 50 yr, and that 10^{50} relativistic particles are released in every such explosion. Then the supernovae supply the Galaxy with $\approx 6 \times 10^{40}$ relativistic particles per second.

We shall now estimate the mean number of relativistic particles which are lost to the Galaxy each second because of nuclear collisions. The number of such processes per unit volume is equal to

$$Z_1 = c n_1 n_2 \sigma. \tag{23-4}$$

Here n_1 and n_2 denote the mean concentrations of relativistic particles and of interstellar gas atoms, σ the effective collisional cross section for protons, and c the velocity of light. Taking $n_1 \approx 10^{-10}$ cm^{-3}, $n_2 \approx 0.01$ cm^{-3}, and $\sigma = 2.5 \times 10^{-26}$ cm^2, we have $Z_1 = 7.5 \times 10^{-28}$ cm^{-3} sec^{-1}. The domain occupied by the spherical component of the Galaxy, in which the magnetic fields and the cosmic particles wandering within them are localized, has a radius of ≈ 10 kpc, and hence a volume of $\approx 10^{68}$ cm^3. Hence the mean frequency with which relativistic particles are lost to the Galaxy as a whole is $\approx 7.5 \times 10^{40}$ sec^{-1}; within the accuracy of the parameters involved, this agrees with our estimate of $\approx 6 \times 10^{40}$ sec^{-1} for the frequency with which they are supplied to the Galaxy by supernovae. We note that energy losses in the comparatively dense interstellar clouds near the galactic plane have little effect on our estimate, for the thickness of the cloud layer is not great (≈ 100 pc).

Thus it may now be regarded as established that supernova outbursts maintain dynamic equilibrium of the cosmic rays in our Galaxy. By taking into account the energy spectrum of primary cosmic radiation, we find that, as a result of nuclear collisions with the interstellar gas, about 10^{39} erg/s are lost in the Galaxy.

The following more careful analysis enables us to differentiate the balance of cosmic electrons from that of heavy particles. For electrons of energy $E > 5 \times 10^8$ ev, basic energy losses in interstellar space lead to deceleration radiation in magnetic fields. In this case $(dE/dt)_{rad} > 4 \times 10^{-7}$ ev/s, whereas ionization losses, taking $E = 5 \times 10^8$ ev and $n = 0.1$ cm^{-3}, amount to 3.2×10^{-8} ev/s, and radiation losses through collisions amount to $\approx 4 \times 10^{-8}$ ev/s. As the energy increases, losses through deceleration radiation in magnetic fields soon overwhelm all other losses.

We shall assume that the energy of relativistic electrons in their sources (supernova envelopes) is essentially distributed over electrons of energy $>10^8$ ev. Then the nonequilibrium radio power of the Galaxy immediately gives the share of energy in the sources of the electron component of cosmic rays.

Let us now evaluate the radio power of the Galaxy. At $\lambda = 3$ m, the brightness temperature T_b of the spherical component at the galactic poles may be taken as $450°$K. Therefore we may write

$$2kT_b/\lambda^2 = \epsilon_\nu l/4\pi, \tag{23-5}$$

where ϵ_ν is the emission per unit volume in the solar neighborhood and $l \approx 10$ kpc is the effective extent of the Galaxy in the direction of the galactic pole (see Sec. 11). From Eq. (23-5) we find that $\epsilon_\nu = 6 \times 10^{-40}$ erg cm^{-3} sec^{-1} (c/s)$^{-1}$. Introducing a spectral distribution $F_\nu \propto \lambda^{0.8}$ (Sec. 4), we obtain

$$\int_{\nu_1}^{\nu_2} \epsilon_\nu d\nu = 10^9\epsilon_{100} = 6 \times 10^{-31} \text{ erg cm}^{-3} \text{ sec}^{-1},$$

where $\nu_1 = 9$ Mc/s, $\nu_2 = 3000$ Mc/s. Since the volume of the Galaxy is $\approx 10^{68}$ cm^3, we find that the power of galactic radio emission is 6×10^{37} erg/s. This is less by about an order of magnitude than the energy losses of heavy cosmic particles (see above). According to Baldwin's models (Sec. 5), the radio power of the Galaxy, integrated over the spectrum, is 0.8 to 1.7×10^{38} erg/s.

This may signify that in supernova ejecta the energy of the relativistic electrons is an order of magnitude less than that of the "heavy" component of primary cosmic rays. A conclusion of this nature agrees with a peculiarity of the statistical mechanism, namely that accelerated heavy particles will acquire greater energy than light particles. However, as we have remarked in Sec. 22, the energy of the electrons then would not fall short of the energy of the heavy particles in the ratio of their masses (about 10^3), but only by a factor of a score or two.

A more probable explanation is the following. As indicated above, the power of the energy losses of the heavy component of cosmic rays through nuclear collisions is $\approx 10^{39}$ erg/s. Mesons are formed in the collisions, and these in turn decay into relativistic electrons and positrons. According to available data, 10 to 15 percent of the energy of the primary heavy particles is transformed into the energy of the relativistic electrons and positrons in this process. This amounts to $\approx 10^{38}$ erg/s. But this is of just the order of the energy losses by the relativistic electrons in the Galaxy, which is equal to the power of its nonthermal radio emission (see above). We conclude that the majority of the relativistic particles in the Galaxy may be of secondary origin, being formed continually in the interstellar medium upon nuclear collision of the primary heavy cosmic particles.

Ginzburg (1954) has suggested an interesting possibility for experimental confirmation of this hypothesis of a secondary origin for the "light" component of primary cosmic rays. If it is correct, the primary component, as observed at the Earth, should contain along with the relativistic electrons approximately the same number of relativistic positrons.

We turn now to the following important circumstance. The energy density $H^2/8\pi$ of the interstellar magnetic field is $\approx 10^{-12}$ erg/cm^3. The kinetic energy density is also close to 10^{-12} erg/cm^3. Finally—and this is the essential point—the energy density of the heavy component of primary cosmic rays is ≈ 1 ev/cm^3, which is once again of the same order! It is difficult to attribute this remarkable coincidence to chance. If the aggregate energy of cosmic particles in the Galaxy did not happen to equal the kinetic and the magnetic energies of the interstellar clouds, then an effective exchange of energy would necessarily take place between the cosmic rays and the magnetized moving clouds of interstellar matter. Fermi considered the problem of the acceleration of *one* relativistic particle by *all* the moving clouds. Such a formulation of the problem is overidealized, in our opinion. In fact, let us assume that the number of particles accelerated is quite large. Then under Fermi's proposal these particles, accelerated independently, may acquire arbitrarily great energy. But it is clear that the interstellar medium cannot transfer to the accelerated cosmic particles more energy greater than the medium itself possesses.

It is vital to consider the effect of the *pressure* of primary cosmic rays on the magnetized interstellar medium. Assuming that primary cosmic radiation is isotropic, we have

$$P_{\cos} \approx \tfrac{1}{3} E_{\cos} \approx 10^{-12} \text{ dy/cm}^2.$$

If cosmic rays were found to be sufficiently abundant in the interstellar medium, so that $P_{\cos} > H^2/8\pi$, then they could not be retained by the medium. Cosmic rays would tend to "squeeze out" the interstellar matter, distending it and elevating it to great distances from the galactic plane; the cosmic rays would then spread over a wider volume.

Because of fluctuations in the cosmic-ray pressure, such as those induced by the activity of individual supernovae, the interstellar medium will approach a state of chaotic motion which ensures an adequate value for H. This affords us an opportunity for explaining the vast spherical distribution of rarefied intercloud diffuse matter. It is apparent that the formation of a gaseous *corona* of this nature around our Galaxy (similar to those around other galaxies) is in the final analysis conditioned by the sources of cosmic rays.

Quite apart from the question of the origin of this halo in our own and in other galaxies, it is of interest to consider the conditions for equilibrium in such a highly rarefied gaseous corona. This matter has been treated by Pikelner and Shklovsky (1957). In Sec. 5 we have emphasized that the division of the sources of nonthermal galactic radio emission into a halo component and an Oort–Westerhout disk is an artificial one. Actually we have a single system of sources, the relativistic electrons in interstellar fields; and this system is concentrated to the plane of the Galaxy and to its center.

In the first approximation we may assume that the halo of our Galaxy is in a state of equilibrium. The condition for mechanical equilibrium in the external layers of the halo, at a distance $z = 10$ kpc from the galactic plane, may be written in the form

$$\int_{z=10} g\rho \, dz = (H^2/8\pi)_{z=10} + P_{\cos} + \tfrac{1}{2}\rho v^2 + P_g, \qquad (23\text{-}6)$$

where g denotes the gravitational acceleration due to the Galaxy, v the velocity of macroscopic motions, and P_g the gas pressure. It is readily shown that the quantity P_g may be neglected. Furthermore, we must have the relation $\tfrac{1}{2}\rho v^2 = H^2/8\pi$; at $z = 10$ kpc, H may be determined from the following considerations.

For deceleration radiation from relativistic electrons in magnetic fields the radiation per unit volume is given by $\epsilon_\nu \propto KH^{(\gamma+1)/2}$ (see Sec. 13). On the other hand, we would expect that the distribution of the concentration of relativistic particles (electrons, in particular) moving along the lines of force will be proportional to H. It follows that $\epsilon_\nu \propto H^{(\gamma+3)/2}$. If $\gamma = 2.6$ (Sec. 4), then $\epsilon_\nu \propto H^{2.8}$, whence

$$H \propto K \propto \epsilon_\nu^{1/2.8}. \qquad (23\text{-}7)$$

If the relation $H^2/8\pi = P_{\cos}$ holds in the "lower" strata of the halo or in the neighborhood of the disk, then for $z \geq 10$ kpc we would have $P_{\cos} > H^2/8\pi$, and equilibrium would be violated. Cosmic rays would break through the magnetic field and escape from the Galaxy. To preserve the equilibrium concept we shall assume that the condition $P_{\cos} = H^2/8\pi$ is fulfilled in the outer layers of the halo. Then in the lower strata $H^2/8\pi > P_{\cos}$, although not by a significant margin. For the energy density of cosmic radiation near the galactic plane we shall adopt $E_{\cos} \approx 1$ ev/cm^3 $= 1.6 \times 10^{-12}$ erg/cm^3, whence $P_{\cos} = E_{\cos}/3 = 0.5 \times 10^{-12}$ erg/cm^3. Since at $z = 10$ kpc ϵ_ν is five or ten times as small as near the galactic plane (excluding the region of the galactic center), then by Eq. (23-7) the energy density of cosmic radiation at $z = 10$ kpc will be $\approx 0.8 \times 10^{-12}$ erg/cm³. This leads to a value for H of $\approx 3 \times 10^{-6}$ gauss, while near the galactic plane $H \approx 6 \times 10^{-6}$ gauss.

The quantity g entering into Eq. (23-6) is known from stellar dynamics. This equation then yields the gas concentration in the halo at $z = 10$ kpc:

$$n_{z=10} = 0.6 \times 10^{-2} \text{ cm}^{-3}.$$

From the condition that in layers below $z = 10$ kpc the halo must also be in equilibrium, we can estimate that the mean value of n over the interval 10 kpc $> z > 1$ kpc is $\approx 10^{-2}$ cm⁻³. This corresponds to a velocity v of turbulent motions in the halo of ≈ 100 km/s.

Spitzer (1956) has advanced an objection to the presumption of a comparatively dense gaseous corona surrounding the Galaxy. He has pointed out that in the presence of velocities greater than the velocity of sound the macroscopic gas motions would go over into shock waves whose energy would be transformed quite rapidly into heat. Spitzer therefore regarded the halo as a statistical phenomenon, whose great extent is a consequence of a high kinetic temperature ($T \approx 10^6$ °K). In this respect the Spitzer galactic corona is very like the solar corona. From the condition that the pressure in dense H I regions be equal to that in the rarefied coronal gas, the particle concentration in the halo would become 5×10^{-4} cm⁻³, a value smaller by a factor of 20 than our estimate given above.

However, this objection does not take into account the important circumstance that the velocity of sound is increased by the presence of a magnetic field. If the magnetic and kinetic energy densities are equal, the velocity of motion is equal to the sound velocity. In this event it can be shown that the energy dissipation decreases by a factor of several tens, although it by no means becomes negligible. The ionization of hydrogen in the halo is significant, but not total.

The presence of a halo with a density and temperature corresponding to our model follows from 21-cm radio observations of external galaxies (Sec. 27). Nevertheless we are very far from a full understanding of the halo phenomenon. It is not quite clear what agency is responsible for the maintenance of the energy in the macroscopic gas motions in the halo. We estimate the time t_0 for the dissipation of the kinetic energy in the halo as $\approx 30 \times 10^6$ yr; this leads to an estimate for the power of the turbulent mechanism. The power is given by

$$t_0^{-1} \int (H^2/8\pi + \tfrac{1}{2}\rho v^2) dV \approx 3 \times 10^{41} \text{ erg/s},$$

where the integral is extended over the entire volume of the halo. The turbulent mechanism is most likely active in the neighborhood of the galactic nucleus, where a number of peculiarities have been observed (see end of Sec. 17).

We have given some qualitative arguments in support of the proposal that the halo has been formed as a result of cosmic-ray pressure. Quantitative treatment of this problem is extremely difficult. It is possible that the assumption of an equilibrium mechanism for the outermost layers of the halo must be given up. In this connection, there arise a large number of interesting problems with regard to a possible nonstationary condition in the halo, but we shall not attempt to discuss them here.

Much is still unexplained in the radio-astronomical theory of the origin of primary cosmic rays; various difficult questions await a full investigation experimentally as well as theoretically. In particular there is the question of the spectrum of relativistic electrons, and its relation to the spectrum of heavy cosmic particles. It has been shown (Ginzburg and Fradkin 1953) that if sources of cosmic rays generate electrons and heavy nuclei, both of which are delivered to the Galaxy with the same spectrum $N(E) \propto E^{-\gamma}$, then because of synchrotron radiation the spectrum of the electrons in the interstellar medium will reduce to $N(E) \propto E^{-(\gamma+1)}$.

Proceeding from these considerations, Pikelner (1956) has developed an elegant method for deriving the intensity of the magnetic field in the Crab Nebula. If relativistic electrons are continually discharged from former supernovae, then as a result of their radiation in a magnetic field the spectrum of the Crab should exhibit a break at some frequency ν_m. The reason for this is that, because of the brief lifetime of the nebula, not all of the electrons will have suffered a substantial loss of energy through synchrotron radiation. The spectrum must become steeper for $\nu > \nu_m$. Analysis of spectra of the Crab indicates that $\nu_m \approx 3 \times 10^{13}$ c/s. It follows that in the central part of the nebula $H = 5 \times 10^{-4}$ gauss, with a mean value of $H = 3 \times 10^{-4}$ gauss for the entire nebula. This value of H was employed by Pikelner in a determination of the mass of the Crab Nebula (see Sec. 21).

Since $\gamma \approx 2$ for heavy cosmic particles with $E < 10^{10}$ ev, we would expect that for relativistic electrons $\gamma_{el} = \gamma + 1 = 3$. The quantity γ_{el} is actually smaller, between 2.0 and 2.6 (Sec. 13). This difficulty may perhaps be removed by allowing for ionization and radiation losses by the electrons.

An interesting problem arises here. If relativistic electrons are generated at supernova outbursts, then in principle we might well find the energy spectrum to depend upon distance from the galactic plane, for the following reason. Supernovae exhibit a rather flat distribution, but the relativistic electrons travel to extreme distances from the galactic plane. Therefore in the neighborhood of the disk the relativistic electrons may be comparatively young, and their energy spectrum, which has not yet felt the full influence of the interstellar magnetic fields, may show a more gentle slope than in the halo region. Hence it will be of particular interest to carry out a thorough search for differences between the disk and the halo spectra of nonequilibrium radio emission. This effect may have been observed by Adgie and Smith (Sec. 5).

One important fact merits careful attention, namely that in the mean the spectrum of the spherical component of galactic radio waves coincides with that of individual discrete sources (the ejecta of supernovae); in Cassiopeia A, for example, $F_\nu \propto \lambda^{0.8}$. This may perhaps signify that the transformation of the relativistic-electron spectrum through synchrotron radiation takes place not in the interstellar medium, as proposed by Ginzburg and Fradkin (1953), but in the source itself. This agrees with our conclusions at the end of Sec. 22.

To conclude this section, we shall consider the following rather curious problem (Krasovsky and Shklovsky 1957). Let us inquire what would happen if one of the stars nearest to the sun were to explode as a supernova. The optical effect of such an outburst would be very striking, although not of a catastrophic character. For several days there would blaze forth in our sky an object of stellar magnitude -15 or even -20. After some hundreds or thousands of years a shock wave would pass through the solar system—the gaseous envelope of the supernova. We may readily satisfy ourselves that during this event our solar system would experience no palpable physical or dynamical changes. Far more important is the fact that in the course of a few thousand years the solar system would have found itself in a region where the density of cosmic rays was tens, possibly even hundreds, of times as great as it is now. This could have most serious genetic consequences for a wide variety of species. For some species, these consequences might well be catastrophic; but for others the irradiation might be of little effect, or even favorable.

Given the frequency with which supernovae appear in the Galaxy and the mean number of relativistic particles formed in such an outburst, we may estimate the probability that the solar system will enter a region of excess cosmic-ray density. It turns out that every few hundred million years the sun must find itself, for some tens of thousands of years, in a region where the cosmic-ray density exceeds the present density by a factor of a score or two. In other words, over the span of its evolution our solar system has repeatedly passed through radio nebulae, the remnants of supernova explosions.

Who knows whether this might not be the explanation for such paleontological phenomena as the wholesale extinction of the reptiles at the close of the Cretaceous period, or the exceptionally luxuriant growth of vegetation in the Carboniferous period! Indeed, the rise of life itself on the Earth may have been stimulated by a prolonged, powerful irradiation of primeval organic molecules, not yet living.

At any rate, paleontology today must take into account, along with other basic factors which have governed the evolution of life on the Earth, such an extremely vital genetic factor as the general level of hard radiation. Radio astronomy can provide an estimate, although a crude one, of the magnitude of the variation in this factor over the geological history of the Earth.

We may say, in closing the chapter, that despite many perplexing questions which are being and will be investigated, radio astronomy has beyond doubt opened a new epoch in our understanding of the nature and origin of primary cosmic rays.

VI

Extragalactic Radio Waves

24. EXTRAGALACTIC SOURCES

By 1949 it had become clear that the observed cosmic radio waves had an extragalactic component. Among the first discrete sources detected, two proved to be peculiar galaxies: NGC 5128, identified with Centaurus A, and NGC 4486, identified with Virgo A. It is quite obvious that if, for example, the radio flux from NGC 4486, whose apparent visual magnitude $m_{vis} = 10$, is only one-twentieth of the radio flux from the quiet sun ($m_{vis} = -27$), then the expected integrated extragalactic radio flux may be very considerable.

After the publication of Westerhout and Oort's work (1951) it was almost universally accepted that a large fraction of cosmic radio waves was of extragalactic origin (the so-called isotropic component, Sec. 4). Subsequently, however, it became clear that this is not so, although about 20 or 30 percent of the intensity measured in the vicinity of the galactic poles *is* due to extragalactic sources.

The identification of Cygnus A, one of the strongest discrete sources, with two faint colliding galaxies more than 100 Mpc from us marked the beginning of a new epoch in our understanding of the nature of cosmic radio waves. We can assume that among the weak discrete sources there may be a considerable number which are of the same character as Cygnus A, but are at even greater distances from us.

Statistical investigations of catalogues of discrete sources made possible a division into two classes. Class II sources, which comprise the overwhelming majority of weak objects, are distributed isotropically—that is, they do not show any tendency to concentrate toward the galactic equator. Such a spatial distribution may be explained in two ways: (*a*) the sources are nearby objects, at distances not greater than those of the nearest stars; (*b*) the sources are extragalactic. Possibility (*a*) is essentially one of the various "radio-star" hypotheses about the nature of radio emission, which at present have little support. Furthermore, in Sec. 26 we shall present observational evidence which makes such an interpretation of the distribution of Class II sources untenable and shows that they must be extragalactic.

Which extragalactic objects are sources of radio waves? What is the ratio of their optical to their radio flux? In order to answer the latter question we need to investigate the extragalactic objects of high apparent luminosity, such as the nebula in Andromeda (M 31) and the Magellanic Clouds.

In the course of his pioneering work Reber made an unsuccessful attempt to detect radio waves from M 31. The first positive results in this direction were obtained by Hanbury Brown and Hazard (1951), working with the main lobe of the 218-ft fixed paraboloid of the Jodrell Bank Station at $\nu = 158.5$ Mc/s. By setting the antenna at different declinations they made continuous records of the level of radio emission as a function of right ascension for the region $+43° > \delta > +38°47'$.

Of seventy reliable records, fifty showed the presence of a source at $\alpha = 0^h40^m$. The maximum flux was observed at $\delta = +40°50' \pm 20'$; the angular extent of the source (between the half-intensity points) was $3°$ in declination and $3°.5$ in right ascension. The total flux at the frequency in question, integrated over the angular extent of the source, was 10^{-24} w m^{-2} (c/s)$^{-1}$ (to an accuracy of 25 percent). The coordinates of the center of gravity of the radio source agreed with the coordinates of the optical center of M 31.

During the following year the same authors made an attempt to detect radio emission from the brightest galaxies within reach of their instrument (Hanbury Brown and Hazard 1952b) by the same technique as used on M 31. They detected two sources whose coordinates —within observational error—agreed with those of NGC 5194–5 (M 51) and NGC 4258. Both these galaxies are of apparent visual magnitude $m_{vis} \approx 9$, and their radio fluxes were about 5×10^{-26} w m^{-2} (c/s)$^{-1}$ (to an accuracy of ± 30 to 50 percent). In a subsequent investigation they also detected NGC 3031 (M 81), NGC 2841, and NGC 891, and in all cases the fluxes were again about 5×10^{-26} w m^{-2} (c/s)$^{-1}$ (Hanbury Brown and Hazard 1953c). All these galaxies are late-type spirals—Sb and Sc. On the basis of their rather limited data, Hanbury Brown and Hazard concluded that the radio fluxes from normal late spirals were proportional to the optical fluxes, and that a spiral of $m_{vis} = 10$ gives a radio flux $F_\nu = 4.2 \times 10^{-26}$ w m^{-2} (c/s)$^{-1}$ at $\nu = 158$ Mc/s.

Notice, however, that the peculiar galaxy NGC 4486, identified with the strong source Virgo A, although of magnitude $m_{vis} = 10$, has a flux $F_\nu \approx 1.2 \times 10^{-23}$ w m^{-2} (c/s)$^{-1}$, or 250 times as great as the flux from a "normal" galaxy of that visual magnitude. Thus, the ratio of radio to optical intensities for different galaxies may vary between wide limits: there exist peculiar galaxies for which this ratio is one hundred times as large as for ordinary ones. In the future we shall call these peculiar systems *radio galaxies*.

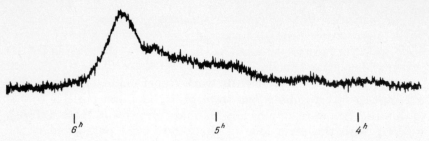

Fig. 173. A record of radio emission received from the Large Magellanic Cloud (Mills 1955).

The galaxies closest to us seem to be the Magellanic Clouds. Since their angular extent is larger, they cannot be observed by the usual interferometric methods, particularly because their surface brightness in the radio range is not large compared with the fluctuations of the background radiation. Thus, our nearest neighbors are difficult to observe by radio-astronomical methods and only an antenna of the Mills-cross type, having a relatively high resolving power, makes possible a systematic study of the Magellanic Clouds.

Mills (1955) carried out such an investigation using an antenna with a main lobe 50' wide, at $\lambda = 3.5$ m. The fixed antenna scanned a region of the sky from declination $-62°.5$ to $-75°.5$. The interval $\Delta\delta$ between scans was one-half the width of the main lobe for the central regions of the Magellanic Clouds, and considerably larger for the periphery. Figure 173 shows a typical record of the Large Magellanic Cloud obtained at declination $-69°26'$. After reducing the original records Mills derived a system of isophotes for the region of the sky in question (Fig. 174); the interval between the isophotes is 125 K deg above a certain level T_0, which was estimated as $700°$K.

The optical centers of the Large and Small Clouds are: $\alpha = 5^h24^m$, $\delta = -69°.8$, and $\alpha = 0^h51^m$, $\delta = -73°.1$, so that there can be no doubt from the isophotes in Fig. 174 that the radio emission comes from the clouds. Figure 175 shows the outermost isophotes of monochromatic radiation (in Sec. 27 we shall describe in detail observations of these galaxies in the 21-cm line of neutral hydrogen), and the intensity distribution of the continuous radiation at $\lambda = 3.5$ m. The cross-hatching corresponds to the regions where the emission was detected with certainty, while the plain hatching marks a region where emission may occur. The similarity between the intensity distribution of the monochromatic and continuous radiations is immediately apparent; this may show the close connection between interstellar matter and discrete sources of radio waves.

It is well known that the Magellanic Clouds contain a large number of hot stars and a considerable mass of interstellar gas. Is it

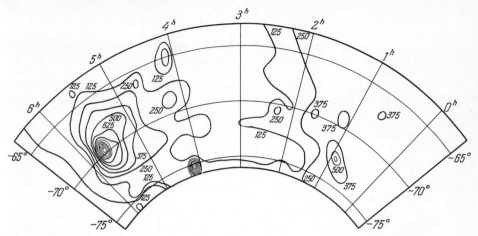

Fig. 174. Isophotes of radio emission in the vicinity of the Magellanic Clouds (Mills 1955).

possible that a significant part of the observed radio emission may be due to thermal radiation from the interstellar gas ionized by the hot stars?

Figure 174 shows a strong maximum at $\alpha = 5^{\mathrm{h}}40^{\mathrm{m}}$, $\delta = -69°.3$, which coincides with the giant gaseous nebula 30 Doradus. This nebula should be optically thick to $\lambda = 3.5$ m radiation; taking its angular diameter to be about 15′, and assuming $T_{\mathrm{b}} = T_{\mathrm{kin}} = 10\,000°$K, we can compute its flux (see Sec. 9) to be $F_{\nu} = 3.4 \times 10^{-25}\,\mathrm{w\,m^{-2}\,(c/s)^{-1}}$. The excess flux above the background of the cloud at the position of

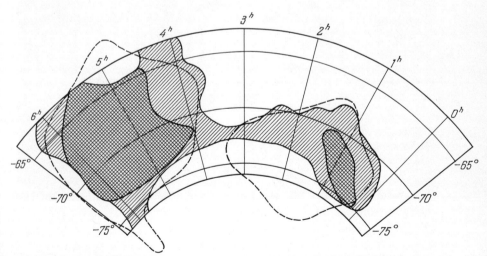

Fig. 175. Isophotes of radio emission at $\lambda = 3.5$ m (hatched) and of monochromatic emission (broken curves) from the Magellanic Clouds (Mills 1955).

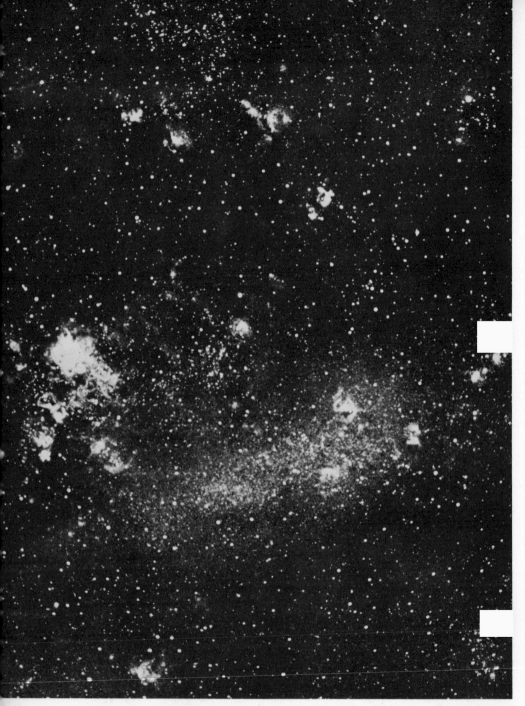

Fig. 176. The Large Magellanic Cloud in Hα light. To the left of the center is the emission nebula 30 Doradus (Mills 1955).

30 Doradus is 3×10^{-25} w m^{-2} (c/s)$^{-1}$, so we can assume that it is due to thermal emission by the nebula. The total fluxes from the Large and Small Clouds are, respectively, 2×10^{-23} and 3×10^{-24} w m^{-2} (c/s)$^{-1}$. If they were entirely due to thermal emission by ionized gas, then the total amount of ionized interstellar hydrogen in the Large Cloud should be many times as great as that contained in 30 Doradus; this, however, is not in agreement with observations. Figure 176 is a photograph of the Large Cloud in the light of Hα, showing that 30 Doradus is the dominating detail.

Piddington and Trent have recently observed the Magellanic Clouds at 50 cm. They detected a measurable flux only from the region of 30 Doradus whose value, 1.5×10^{-24} w m^{-2} (c/s)$^{-1}$, is approximately equal to that of the Orion Nebula, which is about one-hundredth as far away. These measurements first of all fully confirm the thermal character of the radio emission from 30 Doradus, so that this nebula becomes the first extragalactic source reliably identified with thermal emission. Secondly, they show that the spectrum of the bulk of the radiation from the Magellanic Clouds is very similar to that of non-equilibrium sources and in no way resembles the spectrum of a thermal source.

Shain (1958a) has made new observations of the Magellanic Clouds at $\nu = 19.7$ Mc/s, using a Mills cross. The nebula 30 Doradus was observed in absorption, thus making it possible to single out directly the extragalactic emission from the observed cosmic radio waves at high galactic latitudes. The H II region in 30 Doradus serves as a screen for extragalactic radiation; therefore by comparing the brightness of the region in 30 Doradus with the surrounding sky it is possible, in principle, to separate the cosmic radio waves into a galactic and an extragalactic component. At the present time this is difficult to achieve in practice since the optical characteristics of 30 Doradus are still badly known. All the preliminary observations give us reason to believe that the extragalactic brightness temperature lies between 7000° and 22 000°K. Thus at least two-thirds and most probably 80 or 90 percent of the sky brightness at high galactic latitudes is due to the galactic corona.

In order to compare the optical and radio fluxes of different galaxies we can assign to them a *radio magnitude* m_r. Hanbury Brown and Hazard (1952b) introduced this parameter and defined the zero-point of the scale as follows: they adopted the value $9^m.7$ for the apparent photographic magnitude of NGC 5194 + 5195 and set the radio magnitude of this galaxy at $\nu = 158$ Mc/s equal to the same value. At $\lambda = 3.5$ m the radio magnitude is defined by

$$m_r\,(\lambda = 3\ \text{m}) = -\,5.34 - 2.5 \log F_\nu\,(\lambda = 3.5\ \text{m}).$$

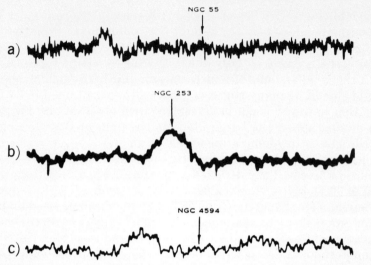

Fig. 177. Records of radio emission received from three galaxies (Mills 1955).

It follows, then, that for the Large and Small Clouds m_r is, respectively, $3^m.35 \pm 0^m.15$ and $5^m.40 \pm 0^m.35$. To refer these quantities to the standard frequency $\nu = 158$ Mc/s we need to know the dependence of F_ν on frequency. Assuming $F \propto \nu^{-0.7}$, the difference in magnitude between $\lambda = 3.5$ m and $\lambda = 1.89$ m ($\nu = 158$ Mc/s) is $0^m.47$. Therefore, at $\nu = 158$ Mc/s, $m_r = 3^m.8$ and $5^m.9$ for the Large and Small Clouds.

Using his cross-shaped antenna, Mills detected radiation from a whole series of galaxies observable from the Southern Hemisphere. Then he compared the radio and photographic magnitudes for different galaxies using de Vaucouleurs' revision of the photographic magnitudes of the well-known Shapley and Ames (1932) catalogue of bright galaxies (see de Vaucouleurs 1956). Figure 177 reproduces several records obtained with the Mills cross: a shows that NGC 55 can be detected only at the limit of sensitivity of the equipment [its flux is about 1.8×10^{-26} w m^{-2} (c/s)$^{-1}$]; b shows a record for a considerably stronger source, NGC 253 (notice that the photographic magnitudes of these two objects are almost identical: $7^m.8$ and $7^m.6$ respectively); c shows that no source of measurable flux is present in the region of the sky occupied by the bright galaxy NGC 4594. Table 24 presents data on the optical and radio emission from a group of normal galaxies observed by Mills in the Southern Hemisphere and by Hanbury Brown and Hazard in the Northern Hemisphere.

Figure 178 depicts the relation between m_{pg} and m_r; Fig. 179 gives the dependence of $m_r - m_{pg}$ on galactic type (Mills's results are indicated by circles, Hanbury Brown and Hazard's by triangles). We must keep in mind that the two hemispheres were investigated with different instruments and there may be systematic errors due to the difficulties involved in the calibration of the equipment. Nevertheless we can conclude that galaxies of intermediate type have the smallest *radio index* $m_r - m_{pg}$, that is, the largest relative radio emitting power. The mean value of the radio index for galaxies of type Sa is $+3^m.0 \pm 0^m.4$; for type Sb it is $+1^m.6 \pm 0^m.2$, and for irregulars (such as the Magellanic Clouds) it is $+3^m.6 \pm 0^m.2$.

So far it has not been possible to detect any radio waves from the largest known globular cluster — 47 Tucanae—whose radio index reaches the record value of $> +8^m.0$. Since globular clusters belong to galactic population II we can conclude that discrete sources in the Galaxy are not related to population II objects. On the other hand, nonequilibrium sources form a spherical system. This system, however, cannot be identified with radio stars because if that were the case we would expect the radio stars to form part of the population of the globular clusters. This gives us still another argument against the idea of radio stars being the sources of nonequilibrium radiation in the Galaxy, while it strengthens the concept of a "continuous" distribution of sources in the interstellar medium which, according to the available data, is absent in globular clusters.

Mills (1958) has recently reported on new identifications of radio

TABLE 24. The radio index of cosmic objects.

Galaxy	Type	m_r	m_{pg}	$m_r - m_{pg}$
Large Mag. Cloud	Ir	3.8	0.5	3.3
Small Mag. Cloud	Ir	5.9	2.0	3.9
M 31	Sb	6.0	4.0	2.0
NGC 55	Ir	11.4	7.8	3.6
NGC 253	Sc	9.4	7.6	1.8
NGC 300	Sc	10.4	8.5	1.9
NGC 5236 (M 83)	SBc	8.9	7.4	1.5
NGC 4945	SBc	9.2	7.8	1.4
NGC 6744	SBbc	10.7	9.1	1.6
NGC 1068 (M 77)	Sb	8.9	9.6	−0.7
IC 5267	Sb	11.1	10.8	0.3
NGC 5194–5 (M 51)	Sc	9.7	8.5	1.2
NGC 3031 (M 81)	Sb	8.9	7.7	1.1
NGC 4258 (M 106)	Sb	9.8	9.1	0.7
NGC 2841	Sb	10.4	10.2	0.2
NGC 891	Sb	9.0	10.7	−1.7
NGC 4594 (M 104)	Sab	>11.8	8.9	>2.9
NGC 1291	SB0	>11.6	9.5	>2.1
NGC 3115	E7	>11.9	10.15	>1.7
47 Tucanae	Glob. cluster	>11.0	3.0	>8.0
NGC 362	Glob. cluster	>10.8	6.0	>4.8

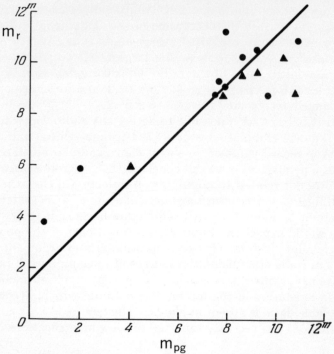

Fig. 178. Dependence of m_r on m_{pg} (Mills 1955).

sources with galaxies in the declination zone $+10° < \delta < +20°$. The criterion for identification was the agreement of the coordinates of the radio sources with those given by the Skalnate Pleso *Atlas of the Heavens* to within $0^m.7$ in α and $13'$ in δ. The number of observed agreements several times exceeded the number of expected chance coincidences; Table 25 lists the most reliable identifications.

Mills shows that a considerable number of normal galaxies may have a high radio emitting power. On the average, one out of 30 galaxies has a radio index $m_r - m_{pg}$ two to five times the mean radio index of normal galaxies. The dispersion in the radio indices can be explained by the lack of agreement between the coordinates of relatively bright galaxies and radio sources. In our opinion, the elliptical galaxies identified by Mills with radio sources are peculiar objects with large negative radio indices, -3^m or -4^m. In comparison, notice that the radio indices of such unusual spheroidal galaxies as NGC 4486 and 5128 are respectively $-5^m.5$ and $-3^m.5$. Apparently the "normal" ellipticals emit weakly in the radio range.

It would be useful to evaluate the ratio of the radio flux (integrated over all frequencies) to the optical flux. This quantity we shall call "relative radio emitting power". The apparent photographic magni-

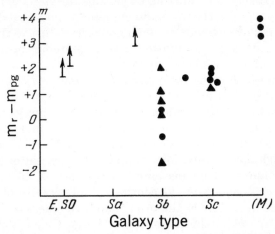

↑ Undetected southern galaxies
● Southern galaxies
▲ Northern galaxies

Fig. 179. Dependence of the radio index $m_r - m_{pg}$ on the type of galaxy (Mills 1955).

tude of NGC 5194–5 is $8^m.5$; assuming that the energy distribution over its optical spectrum is the same as for the sun (and this is fully justified), we find that the optical flux from this galaxy is about 10^{-8} erg cm^{-2} sec^{-1}. To compute the integrated radio flux we assume the spectral distribution of this galaxy to be the same as for the spherical component, that is, $F_\nu \propto \lambda^{0.8}$. Then

$$\int F_\nu d\nu \approx 10^{-13} \text{ erg cm}^{-2} \text{ sec}^{-1}.$$

TABLE 25. The radio index of some galaxies.

NGC	Type	$m_r - m_{pg}$
474	E0	−4.0
533	E3	−3.6
584	E3–4	−1.9
1068	Sb	−1.0
1417	S	−2.3
4038–39	Sc	−0.7
4234	Ir	−3.6
4261	E2–3	−3.8
5792	S	−2.5
7171	SBb	−2.3

Therefore the relative radio emitting power of NGC 5194–5 is about 10^{-5}. This value corresponds to a radio index $m_r - m_{pg} = 1^m.2$.

Knowing the radio index we can now easily compute the ratio of radio to optical flux for different objects. Notice, however, that neither the relative radio emitting power nor the radio index is sufficient to specify completely the radio emission from any galaxy, since the optical emission depends on such characteristics as the number of young stars of high luminosity, the interstellar medium from which these stars are formed, and so on.

It is well known that the ratio of mass to luminosity varies widely with galactic type. For example, for elliptical galaxies it may reach several hundred, for late spirals about twenty or thirty, and for the Magellanic Clouds it is about four. Therefore the small value of the relative radio emitting power of the Magellanic Clouds indicates an abnormally high optical flux. For example, the total integrated radio flux from the Large Cloud (assuming a distance of 46 kpc) is 5×10^{36} erg/s, which is about one-fifteenth to one-twentieth of that from the Galaxy (see Sec. 23). On the other hand, the mass of the Galaxy is $\approx 6 \times 10^{10} \mathfrak{M}_\odot$, while the mass of the Large Cloud is $3 \times 10^9 \mathfrak{M}_\odot$ (Sec. 27). Thus, the radio fluxes per unit mass from the Galaxy and the Large Cloud are about the same.

At first sight it seems more natural to relate the radio flux from a galaxy not to its optical flux but to its mass; indeed, optical and radio sources are entirely different. Furthermore, since sources of relativistic particles are identified with some types of stars (particularly supernovae), we would expect that their number would be determined by the mass of the galaxy. In reality the situation is much more complex: for example, the radio flux from the Large Cloud is six times that from the Small Cloud, although its mass is only 1.5 times as large (Sec. 27). The radio flux from M 31 is about half that of our Galaxy, although its mass is three times as large. Furthermore, the most massive normal elliptical galaxies do not emit a measurable flux.

These examples show that although the mass of a galaxy is undoubtedly an important factor in the determination of its radio flux, it is by no means the only one. Apparently another such factor is the presence of a considerable amount of interstellar gas embedded in a magnetic field. The elliptical galaxies are a good example: these galaxies often contain supernovae, that is, they have sources of relativistic particles. However, the lack of dense regions of interstellar gas and of associated magnetic fields makes it impossible for these galaxies to retain the relativistic particles. This may be the reason why their emission in the radio range is so small.

Table 24 lists the "normal" galaxies found among the weak discrete sources. The Cambridge catalogue contains 1936 sources in the zone

$+83° > \delta > -38°$; we can assume that in the entire region of the sky accessible to the Cambridge interferometer there may be about 2500 sources, of which at least 2000 would be Class II objects, that is, extragalactic. According to Table 24 only 14 among them are normal galaxies, so that the overwhelming majority are peculiar objects, which we shall call radio galaxies. Although their total number in the universe is relatively small, their enormous radio emission makes them observable at very large distances. A similar phenomenon occurs in optical astronomy: the majority of the stars observed are giants, although their spatial density is much lower than that of dwarfs.

The radio index for radio galaxies may reach large negative values. For example, the relative radio emitting powers of Cygnus A, NGC 4486 (Virgo A), and NGC 5128 are, respectively, 10, 10^{-3}, and 3×10^{-4} and their radio indices are -14^m, $-5^m.5$, and -3^m. The index of Cygnus A is particularly striking.

Together with the normal and the radio galaxies, clusters of galaxies also seem to be radio sources. Hanbury Brown and Hazard (1952b), while making a survey with the 218-ft fixed reflector, detected two clusters of galaxies. The first one was in Perseus ($\alpha = 3^h15^m15^s$ $\pm 1^m30^s$, $\delta = +41°22' \pm 30'$); its diameter is about 2° and it contains about 500 galaxies, whose mean photographic magnitude is $16^m.4$. The expected radio magnitude (assuming that the radio index for each galaxy is the same as for M 31) is $11^m.2$, while according to the observations it actually is $7^m.2$. Baldwin and Elsmore (1954) have shown that three-fourths of the total radio flux from this cluster is due to NGC 1275 (Sec. 8). Therefore the radio magnitude of the cluster in Perseus is actually $8^m.7$, and its radio index is $m_r - m_{pg} = -0^m.5$.

Radio waves were also detected from the cluster centered at NGC 911. The computed radio magnitude on the assumption that $m_r - m_{pg} = 2^m$ (as for M 31) is approximately $13^m.5$, while the observations give $9^m.3$. In this case the radio index becomes $-2^m.2$.

The Cambridge catalogue lists, under No. 886, a source whose co-ordinates are nearly the same as those of the cluster in Leo. At $\lambda = 3.7$ m its flux is 1.2×10^{-25} w m^{-2} (c/s)$^{-1}$, so that in the system of Hanbury Brown and Hazard its magnitude is $9^m.7$. This cluster is composed of about 200 galaxies, whose mean photographic magnitude is about $17^m.9$. This means that if the radio indices for these galaxies were the same as for M 31 the radio magnitude of the cluster would be $13^m.9$. Therefore, for this cluster $m_r - m_{pg} = -4^m.2$.

From the preceding examples it would seem that the radio emissions from clusters of galaxies have anomalous intensities. A possible explanation of this effect will be given in Sec. 25.

Our Galaxy, M 31, the Magellanic Clouds, and several other nearby systems form the so-called Local Group. In its turn the Local Group seems to be a member of a large cluster of galaxies, composed of several hundred such groups. This cluster forms a highly flattened system whose center is in the direction of Virgo. De Vaucouleurs (1953*b*) pointed out that two-thirds of the bright (and therefore relatively nearby) galaxies are concentrated in a fairly narrow strip along a great circle, whose area is only one-tenth the area of the sphere. This great circle is almost perpendicular to the galactic equator and seems to be the intersection of the plane of symmetry of the flattened cluster referred to above with the celestial sphere.

This large cluster of galaxies has received the name of "Supergalaxy," and the great circle toward which the bright galaxies concentrate is called the "supergalactic equator." Figure 180 shows the distribution of bright galaxies and their well-marked concentration toward the supergalactic equator.

Kraus and Ko (1953), working at $\lambda = 1.20$ m with the helical antenna of The Ohio State University, detected radio waves from a region of the sky extending approximately along $\alpha = 12^h$ between $15° > \delta > -15°$. The axis of this region lies very close to the supergalactic equator, and its width is about $15°$. In the same year Hanbury Brown and Hazard (1953*d*), using the 218-ft reflector at $\lambda = 1.89$ m, found the prolongation of this region in the Northern Hemisphere, between $+40° < \delta < +70°$. In Fig. 180 the broken line delimits the emission region detected by Kraus and Ko, and the full line the region detected by Hanbury Brown and Hazard. The intensity observed in the northern part of the Supergalaxy at $\nu = 158$ Mc/s is 3×10^{-26} w m^{-2} (c/s)$^{-1}$ deg^{-2}, and the intensity in the southern part at $\nu =$

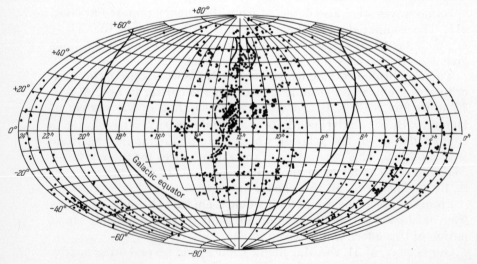

Fig. 180. Distribution of bright galaxies over the sky (Hanbury Brown and Hazard 1953*d*).

250 Mc/s (at $\delta = -5°$) is 10^{-25} w m^{-2} (c/s)$^{-1}$ deg^{-2}. Preliminary observations carried out at Jodrell Bank at $\nu = 72$ Mc/s seem to indicate that the radio spectrum of the Supergalaxy obeys the law $I \propto \nu^{-0.5}$. Extrapolating by means of this relation to $\nu = 250$ Mc/s, we may assume that the intensity at $\delta = -5°$ exceeds by a factor of five the intensity measured in the northern regions. This effect can easily be understood by taking into account that the center of the Supergalaxy, where the largest number of galaxies concentrates, is in the constellation of Virgo.

Bolton, Westfold, Stanley, and Slee (1954) detected an extended radio source in Virgo (source H in Table 11, Sec. 7) with approximate coordinates $\alpha = 12^{\rm h}40^{\rm m}$, $\delta = +7°$ (the probable error was several degrees). The position of this source agrees to within observational error with the direction of the supergalactic center. The angular extent of the source is about $4°$, and its flux at $\lambda = 3$ m is 2.5×10^{-23} w m^{-2} (c/s)$^{-1}$. The intensity of the radiation is 1.5×10^{-24} w m^{-2} (c/s)$^{-1}$ deg^{-2}, and at $\lambda = 1.20$ m it is 10^{-24} w m^{-2} (c/s)$^{-1}$ deg^{-2}, which is one order of magnitude greater than the value given by Kraus. As reasons for the discrepancy we should keep in mind that, apparently, the intensity at $\delta = -5°$ is less than at $\delta = +7°$ to $+10°$, where the concentration of galaxies is considerably greater. Furthermore, the spectrum of the Supergalaxy may fall off on the high-frequency side more steeply than was assumed by Hanbury Brown and Hazard. Finally, Kraus and Ko used an antenna whose beamwidth in declination was $8°$.

The brightness temperature of the Supergalaxy at $\lambda = 1.20$ m in the region observed by Kraus and Ko is $17°$K, which represents approximately 25 percent of the background in this region of the sky. According to Bolton *et al.*, $T_{\rm b} \approx 1500°$K at the supergalactic center ($\lambda = 3$ m); that is, it exceeds the brightness of the local background.

A crude estimate, based on the assumption that the optical radiation of the central part of the Supergalaxy is equivalent to 100 twelfth-magnitude galaxies (apparently the actual radiation is less), gives a radio index $m_{\rm r} - m_{\rm pg} \approx -3^m$, the same value as for other clusters of galaxies.

Further investigations of the variations in the cosmic radio-wave background at high galactic latitudes will undoubtedly reveal new peculiarities of the "radio universe."

25. THE NATURE OF THE RADIO GALAXIES

If a large fraction of the radio emission from normal galaxies such as our own has a nonequilibrium character, the same should be true to an even greater extent for the radio galaxies. In these the sources are undoubtedly relativistic electrons moving in magnetic fields. If the radio-star hypothesis has exceptional difficulties in explaining the nonequilibrium radiation even in our own stellar system, in the case of radio galaxies it becomes simply absurd. The relative radio emitting power of radio galaxies is 10^3 and even 10^5 times as great as for normal ones. Therefore the number of radio stars in these galaxies should be correspondingly larger than in normal systems, where the number required is already excessively large (Sec. 11).

In the radio galaxies the sources of radio waves must be continuously distributed through the interstellar medium. Under such conditions the emission of relativistic electrons in interstellar magnetic fields is the only mechanism that can account for the radio flux emitted by peculiar galaxies. Further evidence favoring this mechanism is the great similarity between the spectra of the radio galaxies and the nonequilibrium radiation in our own stellar system. Figures 68 and 69 (Sec. 6), for example, show how the radio galaxy Cygnus A resembles the nebula Cassiopeia A over an enormous frequency interval.

The differences in intensity and, probably, the small differences in spectra among various galaxies may be explained by different cosmic conditions present in them. The great radio emitting power of some galaxies (within the frame of the assumed emission mechanism) may be explained either by a large number of relativistic electrons, or by a high intensity of the random magnetic fields in which the electrons move, or by both.

If we assume that a radio galaxy emits at the present high level during a period of time long compared to the time of deceleration of the relativistic electrons in the magnetic fields, then we require the existence of strong sources of relativistic electrons that would continuously compensate for those that lose their energy by radiation. Accordingly, the source of electrons in Centaurus A should be several tens of times as strong as in our Galaxy, and that in Cygnus A about 10^5 times as strong. However, it does not necessarily follow that radio galaxies have been emitting at the present rate during their entire evolution, that is, during several billion years. In principle it is quite possible that for some reason, either internal or external, a normal galaxy undergoes a sudden change either in the number of relativistic electrons or in the intensity of the magnetic fields present in the galaxy. As a consequence, the radio emission from this galaxy will be increased sharply to a very high level; the electrons will continually be slowed down, and the decrease in their energy will not be compensated by the sources of relativistic electrons, which are not very strong under normal conditions. As a result, the flux from the radio galaxy will gradually diminish until it becomes once again normal.

Radio-astronomical and astrophysical data now show that for

several radio galaxies the second possibility obtains. In these galaxies there is no dynamic equilibrium between the energy-losing and energy-gaining relativistic electrons, as there is in our Galaxy; rather, the emission process has a sharply nonsteady-state character. However, the question still remains open whether there are radio galaxies in a steady state.

Cygnus A seems to be a typical example of a radio galaxy in an extreme nonsteady state. In Sec. 8 we described in detail the optical observations of this strong discrete source, which is identified with two late spirals colliding face-to-face. There we also showed that the radio flux integrated over frequency is several times as great as the optical flux. The relative velocity of the colliding galaxies is 1000 to 3000 km/s, while the internal velocities of the interstellar gas, computed from the width of the spectral lines, is 1000 km/s. The intensity distribution of the radio waves over the source is very peculiar: according to the most plausible interpretation of the data (see Sec. 7), the source appears to be double, in agreement with the colliding-galaxies hypothesis.

However, the interpretation of the radio emission from Cygnus A presents two difficulties. First of all, the radio regions clearly do not agree with the optical regions. Figure 181 depicts the relative positions of the colliding galaxies, observed in visible light, together with the two centers of radio emission deduced from interferometric observations. Obviously the optical and radio emission regions do not coincide; however, the complete symmetry of the picture leaves no doubt that we are observing two almost identical interacting objects.

Notice the enormous linear dimensions of the emitting region in Cygnus A. The extent of the region in the east–west direction is at least 2′; since the colliding galaxies are about 100 Mpc away, the linear dimension becomes 60 kpc, and the extent of each of the two radio regions is about 20 kpc.

In order to interpret the observed picture (Shklovsky 1954), we must assume that at the time of the collision each galaxy was surrounded by a spherical corona of highly rarefied gas. A similar corona encircles our Galaxy and seems to give rise to the spherical distribu-

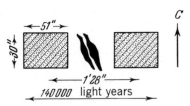

Fig. 181. Distribution of radio intensity (shaded) and optical intensity (filled) in Cygnus A (Shklovsky 1954).

tion of sources of nonequilibrium radiation (relativistic electrons). Such coronas have also been observed for other galaxies and we may expect that they are a feature common to all giant spirals of late type, to which the colliding galaxies in Cygnus also belong. Since in spite of the extremely low densities the mean free path in the corona is 10^{-5} times the extent of the corona itself, the latter can be considered a continuous medium. Thus, the face-on collision of the two galaxies in Cygnus is fundamentally a collision of two continuous gas masses, which we can assume to be to some extent elastic. The enormous relative velocity (≈ 3000 km/s) and the magnetization of the interstellar medium speak in favor of this assumption. Each corona is an extremely complicated sphere of magnetic lines of force that may retain their identity during the collision, while being more or less deformed. Meanwhile, at such high collisional velocity the stars of one galaxy traverse the other galaxy without suffering appreciable perturbations. We must keep in mind that the interstellar gas is not, of course, a resisting medium for stellar bodies.

Figure 182 shows schematically the positions of the two colliding galaxies at different instants of time. The drawings will help us to understand the observed picture. Because of the collision a shock wave passes through the two gaseous coronas, and afterward the coronas interpenetrate because of the partial elasticity of the gas. Meanwhile the stars and the relatively dense clouds of interstellar gas belonging to the flat systems of the two galaxies continue to approach; at the present time they have already passed through each other. Since the separation of the stars and interstellar gas of the two galaxies, the coronas have traveled about 25 kpc. With a relative velocity

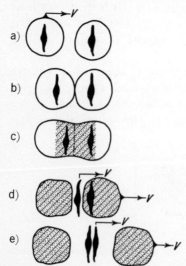

Fig. 182. Successive stages in the collision of galaxies (Shklovsky 1954).

v of 3000 km/s, the time taken is about 2.5×10^{14} sec or 8×10^6 yr. Until the gaseous coronas had separated, a shock wave traveled through the gas with a velocity $\frac{4}{3}v = 4000$ km/s, so that the time required for the wave to traverse each galaxy is 4.4×10^6 yr. This means that we can assume that about 12 million years have elapsed since the beginning of the collision.

The source of energy in Cygnus A can only be the kinetic energy of the gaseous masses of the colliding galaxies (see Sec. 22). Assuming that the gas density in the corona is approximately the same as in our Galaxy (10^{-25} gm/cm^3), at a velocity $v = 3 \times 10^8$ cm/s the kinetic energy of the colliding gas masses is 5×10^{59} erg. The radio luminosity of Cygnus A is given by

$$L = 4\pi R^2 \int F_\nu d\nu.$$

Since the integrated radio flux is 1.3×10^{-10} erg cm^{-2} sec^{-1} and the distance R to the source is 3×10^{26} cm, it follows that $L = 1.6 \times 10^{44}$ erg/s, or about ten times the optical luminosity. Over the time $t = 10^7$ yr $= 3 \times 10^{14}$ sec the energy emitted has been $Lt = 5 \times 10^{58}$ erg, or approximately 10 percent of the available kinetic energy.

What is the immediate reason for the strong radio emission from Cygnus A: is it the amplification of the magnetic field or is it an increase in the number of relativistic particles after the collision? It is easy to show that only the second alternative can be of real importance. Indeed, the growth of the field increases the energy lost by the electrons in the form of radiation, since $dE/dt \propto H^2E^2$. Furthermore, the energy of the electrons satisfies $E \propto H^{-1/2}$, so that $dE/dt \propto H$. If the reason for the strong radio emission from Cygnus A were the increase in the strength of the interstellar magnetic field, and the number of relativistic electrons remained approximately the same as in our Galaxy, then the intensity of the field should become 10^3 times its normal value, that is, of the order of 10^{-2} gauss. However, if that were the case the relativistic electrons would lose their energy in the form of radiation in 10^4 to 10^5 yr, while the collision in Cygnus took place about 10^7 yr ago and the source still emits strongly. In other words, the relativistic electrons are losing their energy quite slowly and this can occur only when the magnetic field is not very large. Notice that after the collision we cannot expect any mechanism that will replenish the energy lost by the relativistic electrons.

We can cite one further argument against the assumption of an increase in the magnetic field. Since the magnetic field is built up at the expense of the kinetic energy of the colliding galaxies, we can write the relation

$$VH^2/8\pi \lesssim \tfrac{1}{2}\mathfrak{M}v^2, \tag{25-1}$$

where $V \approx 10^{68}$ cm^3 is the volume of the corona, $\mathfrak{M} \approx 10^{44}$ gm is its mass, and $v \approx 3 \times 10^8$ cm/s is the relative velocity of collision. Therefore $H \lesssim 8 \times 10^{-4}$ gauss. From Eq. (25-1), if $\rho \approx 10^{-25}$ gm/cm^3 and $v \approx 10^8$ cm/s (which corresponds to the width of the emission lines in the spectrum of Cygnus A), then $H \approx 10^{-4}$ gauss; this is the value that we shall adopt.

The intensity of the source is given by $I_\nu = F_\nu/\Omega$, where $\Omega = 0.72$ min^2 or 6.1×10^{-8} sterad. For $\lambda = 3$ m, $F_\nu = 1.3 \times 10^{-22}$ w m^{-2} (c/s)$^{-1}$ so that

$$I_\nu = 2.1 \times 10^{-12} \text{ erg sec}^{-1} \text{cm}^{-2} \text{ (c/s)}^{-1} \text{ sterad}^{-1},$$

approximately 10^3 times as bright as the quiet sun at the same frequency, or 400 times as bright as the Crab Nebula. Such large surface brightness is explained by the enormous extent of this source along the line of sight.

According to Eq. (13-7), and taking into account the spectrum of Cygnus A ($F \propto \nu^n$, where $n = -1$),

$$I_\nu = 3.1 \times 10^{-15} K H^2 \langle R \rangle \lambda,$$

provided the differential energy spectrum of the relativistic electrons in this source follows the law

$$N(E) = K E^{-\gamma};$$

according to the radio spectrum, $\gamma = 3$. With $H = 10^{-4}$ gauss, $\lambda = 3 \times 10^2$ cm, $\langle R \rangle = 6 \times 10^{22}$ cm, we have $K = 3 \times 10^{-15}$. The concentration of relativistic electrons with energies greater than $E_0 = 3 \times 10^7$ ev is given by

$$\langle n \rangle = \int_0^E K E^{-3} \approx 3 \times 10^{-7} \text{ cm}^{-3}. \tag{25-2}$$

The total energy stored in the relativistic electrons is $E_1 \approx 10^{58}$ erg, which corresponds to approximately 2 percent of the kinetic energy of the gas in the corona. In 10^7 yr the energy emitted is approximately five times as great as the energy now stored in the relativistic electrons; in other words, the electrons have had time to radiate a large share of the energy which they received at the time of collision. Undoubtedly our discussion is very rough, but apparently it reveals a basic energetic relation.

The fact that under certain conditions the collision of two galaxies gives rise to strong radio waves whose spectral distribution is approximately the same as for the nebular ejecta of novae and supernovae clearly shows that the reasons for the formation of relativistic particles in these objects are macroscopic gas-dynamical processes which occur in the presence of the magnetic field associated with the material. Although in the case of supernova remnants there is a connection between the relativistic particles and the cause of their formation, apparently the outburst, such a tie does not exist for colliding galaxies.

Notice, however, that Ambartsumian assumes that in the case of Cygnus A (and a number of other radio galaxies) what we observe is not the collision of two galaxies but, on the contrary, the breakup of one (this view was presented at the 1955 Dublin congress of the I. A. U.). In this case the formation of relativistic electrons would be connected with several processes whose nature is not well understood at the present time. It is doubtful, however, whether such a viewpoint is tenable.

We can assume that in the case of colliding galaxies the electrons are accelerated by a statistical mechanism. The characteristic parameter α of this mechanism can be obtained from the relation $\alpha T \approx 10$, where T is the acceleration time and is approximately equal to the time of travel of the shock wave through the galaxy, that is, $T \approx 10^{14}$ sec. Therefore $\alpha \approx 10^{-13}$ sec^{-1}. We also have (Sec. 22)

$$\alpha = u^2/cl.$$

Taking $u = 10^8$ cm/s and $l \approx 3 \times 10^{18}$ cm (or about 20 times the mean free path) we find that $\alpha \approx 10^{-13}$ sec^{-1}, in agreement with the previous estimate.

If the accelerating mechanism is indeed statistical, the energy of the relativistic electrons will be limited by the condition

$$\alpha E_1 = \left(\frac{dE}{dt}\right)_{\text{mag}}^{E=E_1}, \tag{25-3}$$

where $(dE/dt)_{\text{mag}}$ is the rate of dissipation of energy by the relativistic electrons in the form of radiation in the magnetic field. If $H \approx 10^{-4}$ and $\alpha \approx 10^{-13}$, then $E_1 \approx 2 \times 10^9$ ev. Therefore we can assume that the energy spectrum of the relativistic electrons in the range $5 \times 10^8 < E < 2 \times 10^9$ ev slopes very gently, and hence the radio spectrum at relatively high frequencies will be equally gentle. If the energy spectrum of the electrons has a sharp cutoff at 2×10^9 ev, we would expect a similar cutoff on the high-frequency side of the radio spectrum. With $H = 10^{-4}$ gauss this would occur at $\nu \approx 2 \times 10^{10}$ c/s ($\lambda = 1.5$ cm). It would be very desirable to make an attempt to observe this effect.

Thus, the theory of the emission of relativistic electrons in a magnetic field provides a qualitative and even quantitative explanation of the unusual phenomena observed in Cygnus A. From the point of view of this theory, emission sources of the type of the colliding galaxies in Cygnus are a completely new kind of cosmic body: they are enormous rarefied clouds of gas of galactic dimensions, very hot, and devoid of stars. Their surface brightness in optical light should be very small since their emission measure is about 100. We can assume that in the course of time these gaseous coronas will gradually disperse into the intergalactic medium.

Besides Cygnus A several other colliding galaxies have been detected which are also sources of radio waves. In Sec. 8 we pointed out the peculiar galaxy NGC 1275 which, according to Baade and Minkowski, is also a double galaxy in collision. However, in all other well-known cases the relative radio emitting power is far smaller than for Cygnus A. Apparently this fact can be explained by the exceptional circumstance that in Cygnus A we have a face-on collision, plus the fact that both objects are giant galaxies with well-developed coronas; also the relative collision velocity in Cygnus A is anomalously large. Statistically, the simultaneous occurrence of all these favorable factors should be very rare and we would not expect to observe many such instances (Sec. 26). Apparently in all other cases we observe the much more probable sideways collisions in which the conditions for the formation of a large number of relativistic electrons are considerably less favorable.

Baade and Minkowski were inclined to assume that all radio galaxies are colliding pairs. For example, they interpreted NGC 5128 as being a late-type spiral in collision with an elliptical galaxy (Baade and Minkowski 1954b), the spiral being oriented edgewise with respect to the observer. At about the same time, however, Shklovsky and Kholopov identified Fornax A with NGC 1316, which has a very similar appearance to NGC 5128. De Vaucouleurs showed that several other galaxies of the same type seem to be radio sources. It is not possible to assume that in every one of these cases we observe the collision of a spiral with an elliptical galaxy; it is even less likely that they are always identically oriented with respect to the observer.

Thus we are led to the conclusion that there are anomalously strong radio galaxies which do not appear to be in collision. In such cases the high radio emitting power is due only to internal sources. A typical example is NGC 4486 (M 87), identified with Virgo A, one of the strongest sources (Shklovsky 1955a).

The spheroidal galaxy NGC 4486 is one of the brightest members of the well-known cluster in Virgo. According to the Shapley–Ames catalogue, its visual magnitude is 10^m, its angular extent is about $5'$, and, as with other spheroidal galaxies, its brightness is strongly concentrated toward the center. Its radio flux at $\lambda = 3$ m is 1.2×10^{-23} w m^{-2} (c/s)$^{-1}$ and the radio spectrum is characterized by a decrease in intensity with increasing frequency; in this respect it resembles Cygnus A, Cassiopeia A, and several other nonequilibrium sources.

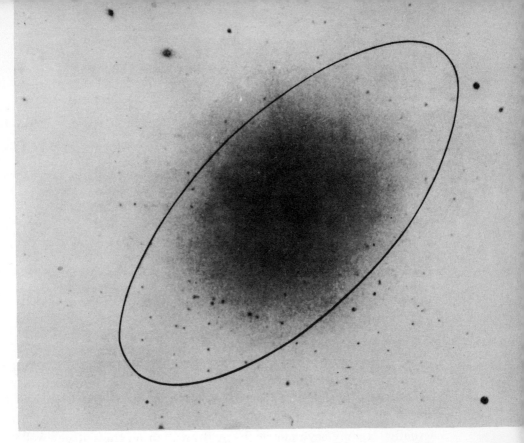

Fig. 183. The radio galaxy NGC 4486, whose "radio shape" is represented by the ellipse.

The brightness distribution over the source is very interesting: the "radio shape" of NGC 4486 is an ellipse whose axes are 6′ and 2′.5 (see Fig. 183). It is very striking that in optical light this galaxy is spherical while in radio light it is strongly elliptical; as a rule the radio emitting region encircles the region of optical emission (for example M 31, Cygnus A), but in NGC 4486 we observe the opposite situation.

Baldwin and Smith (1956), using one of the elements of the Cambridge interferometer at $\lambda = 3.7$ m, detected an extended (50′) weak corona surrounding NGC 4486. Its flux was about 20 percent of the total flux from Virgo A, while its surface brightness was about 1/500 that of the central regions of the source. Apparently the presence of an extended corona is a characteristic feature of radio galaxies; the reason for its formation is, evidently, the pressure exerted by relativistic particles on the interstellar medium (Sec. 23). Figure 184 gives an idea of the extent of the halo around NGC 4486.

Recently Mills made an important contribution to the question of the halo in NGC 4486. From observations at $\nu = 85.5$ Mc/s he deduced that the halo does not surround the nebula uniformly but is

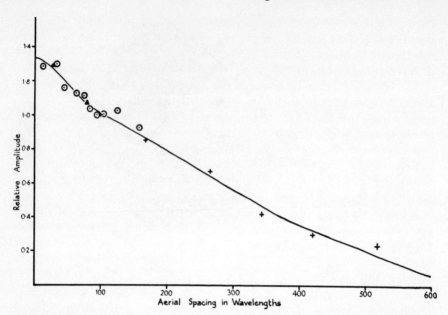

Fig. 184. Amplitude as a function of antenna spacing for NGC 4486 (Baldwin and Smith 1956).

actually strongly asymmetric, being drawn out on the side of a well-known peculiarity first detected by Curtis (1918), the so-called "jet."

At first sight NGC 4486 neither differs from other spheroidal galaxies nor appears to be a radio source capable of strong emission. However, in the central and brightest region of the galaxy there is a striking jet about 20″ long and of 2″ average width (Fig. 185), containing several condensations. Superposed on the usual type G spectrum is an emission line at λ 3727 displaced with respect to the Fraunhofer lines. The displacement indicates a velocity of the emitting gas of -295 ± 100 km/s, which is quite significant. Baade and Minkowski assume that the jet is formed by matter moving away from the central regions and that it is responsible for the emission at λ 3727.

The spectrum of the jet was photographed by Humason (see Baade and Minkowski 1954b); it is bluer than the integrated spectrum of the galaxy and there are no lines visible.

Baade and Minkowski pointed out the very significant fact that the major axis of the "radio ellipse" in NGC 4486 is perpendicular to the axis of the jet. This phenomenon indicates a basic connection between the jet and the radio waves. It is difficult to explain, however, the difference in size between the radio emitting region and the jet, the volume of the former exceeding that of the latter by a factor of 10^5.

It seems to us that the key to an understanding of the nature of the radio waves from NGC 4486 lies in the strictly continuous spectrum of the jet. In 1954 we put forward the suggestion that the optical emission from the jet is due, as in the case of the Crab Nebula, to relativistic electrons moving in a weak magnetic field. One important difference between the two objects is the fact that in NGC 4486 the extent of the region of continuous optical emission is significantly smaller than the radio region, while in the Crab Nebula the two regions are practically identical. In order to understand this peculiarity of NGC 4486 we must look into the quantitative aspects of the problem in more detail.

First of all we need to estimate the surface brightness of the jet in visible light. Unfortunately no special photometric investigation of NGC 4486 has yet been made, and hence our estimate must necessarily be very rough. As we can see from the photographs of the jet published by Baade and Minkowski (1954b), the brightness of its knots exceeds by at least several times the brightness of the central region of the galaxy. Further, this region is about 10 times as bright as points 1' from the center. We can roughly estimate that the optical flux from the jet is about ten times as small as from the whole galaxy.

The total area of the four brightest knots is about 10 sec^2. If the visual magnitude of NGC 4486 is about 10^m then, according to our estimate, the visual magnitude of the jet is between 12^m.5 and 13^m. On the other hand, for the Crab Nebula $m_{vis} \approx 9^m$ (computed mainly from its emission in the continuum) and its area is about 5 $\times$ 10^4 sec^2. Therefore, the surface brightness of the knots is approximately 100 times that of the Crab Nebula. Since the diameters of the knots are

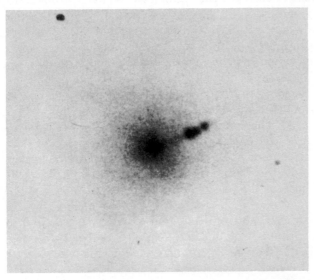

Fig. 185. Central part of the radio galaxy NGC 4486.

about 2″ and since NGC 4486, as a member of the Virgo cluster, is at a distance of about 10 Mpc, the linear dimensions of the knots are of the order of 100 pc (the diameter of the Crab Nebula is about 1 pc). Thus we have shown that the emission in the continuum per unit volume for both objects is of the same order of magnitude. We can assume that the strength of the magnetic field in the jet is $H \approx 10^{-4}$ gauss, which is approximately the same field as in the amorphous outer parts of the Crab Nebula.

With a field $H = 10^{-4}$ gauss we require $E \approx 5 \times 10^{11}$ ev if the frequency of the radiation is to fall in the optical region. Let us now compute the concentration of relativistic electrons, which seem to produce the optical emission in NGC 4486.

Let us take the visual magnitude of one of the knots to be 14^m. Then it can easily be shown that its total flux is $F \approx 10^{-10}$ erg cm^{-2} sec^{-1}, and the flux per unit frequency $F_\nu \approx 3 \times 10^{-25}$ erg cm^{-2} sec^{-1} (c/s)$^{-1}$. The angular diameter of the jet is 2″, and therefore it subtends an angle $\Omega = 7 \times 10^{-11}$; since $F_\nu = \langle I_\nu \rangle \Omega$, the mean intensity of the optical emission from a knot is $\langle I_\nu \rangle = 5 \times 10^{-15}$ erg cm^{-2} sec^{-1} (c/s)$^{-1}$ sterad^{-1}.

Assuming that the intensity computed above does not depend on the wavelength (which follows from the fact that the spectrum of the jet is bluer than that of a G-type star), we find that $\gamma = 1, u(\gamma) = 0.37$. Taking $H = 10^{-4}$ gauss and $R = 1.8 \times 10^{20}$ cm, we obtain $K = 5 \times 10^{-9}$, from which it follows that the concentration of relativistic electrons in the energy interval $10^{11} \leq E \leq 10^{12}$ is about 10^{-8} cm^{-3}.

The energy density of the "luminous" relativistic electrons in a knot is approximately 3×10^{-9} erg/cm^3, which is of the same order of magnitude as the kinetic and magnetic energies. This circumstance seems to indicate that the energy of the relativistic electrons may ultimately be due to the kinetic energy of the random macroscopic motions in the jet.

Let us assume that the acceleration of the electrons in the jet is due to some statistical mechanism. How long will the acceleration process take? The maximum time obviously would be the time of formation of the entire jet. With a linear extent of 600 pc and a velocity of 300 km/s (the velocity must actually be larger, since 300 km/s is only the radial velocity of the gas emitting λ 3727 in the neighborhood of the jet), we find that the time is $t \approx 6 \times 10^{13}$ sec or 2×10^6 yr. Then, if $E = 5 \times 10^{11}$ ev and $\alpha t = 15, \alpha \geq 2.5 \times 10^{-13}$ sec^{-1}. With this value of α the gain of energy per second by the electrons is about 0.1 ev/s; but in a field $H = 10^{-4}$ gauss the loss by radiation is about 10 ev/s. To compensate for this loss we need $\alpha \geq 2 \times 10^{-11}$ sec^{-1}. From the relation

$$\alpha = u^2 c^{-1} \langle l \rangle^{-1}$$

we find that, if $u = 3 \times 10^7$ cm/s and $\langle l \rangle = 10^{15}$ cm, $\alpha = 3 \times 10^{-11}$ sec^{-1}. Notice that the mean free path determined by the Coulomb interaction of electrons and ions is of the order of 10^{12} to 10^{13} cm. Also, at different points in the jet, and in particular in the knots, the density and the turbulent velocities may be higher than those adopted by us.

Thus a statistical accelerating mechanism may, in principle, account for the very energetic relativistic electrons observed in the knots of the jet in NGC 4486; in this case the characteristics of the turbulent gas in the vicinity of the jet are not in contradiction with the observations. The jet, then, seems to be a powerful generator of relativistic particles. The relativistic electrons formed in the jet will diffuse into the surrounding space and in one or two million years (the time of formation of the jet) they will fill a considerable fraction of the galaxy. While wandering in the interstellar magnetic field these electrons will radiate at radio frequencies and cause the anomalously high radio emitting power of NGC 4486.

The argument presented above gives a qualitative description of the nature of the radio waves from NGC 4486. It also explains the extent of the radio region over the whole galaxy, while the original cause of the emission—the jet—takes up a very small fraction of the total volume of NGC 4486.

Let us now estimate the mean concentration of relativistic electrons responsible for the radio emission. According to observations in the meter and decimeter ranges, the flux is inversely proportional to the first power of the frequency, and hence $\gamma = 3$. The flux at 3 m is $F_\nu = 1.2 \times 10^{-20}$ erg cm^{-2} sec^{-1} (c/s)$^{-1}$ and the solid angle subtended by the source is $\Omega \approx 10^{-6}$ sterad. The average extent of NGC 4486 along the line of sight is $R \approx 10\ 000$ pc or 3×10^{22} cm. Adopting an interstellar magnetic field of $H = 10^{-5}$ gauss we find, using Eq. (13-7), $K = 7 \times 10^{-15}$. The concentration of relativistic electrons with energies $E > E_0 = 5 \times 10^7$ ev is

$$N = \tfrac{1}{2} K / E_0^{\,2} = 3 \times 10^{-8} \text{ cm}^{-3},$$

whence their energy density is

$$\epsilon = K/E_0 \approx 10^{-10} \text{ erg/cm}^3.$$

On the basis of Eq. (13-7) we can show that the concentration of relativistic electrons is almost independent of the magnitude of H, while their energy density is given by $\epsilon \propto H^{-1}$ (when changing H it is necessary simultaneously to change E_0).

Since the volume of the radio region of NGC 4486 is $V \approx 10^{67}$ cm^3, the total number of relativistic electrons is $NV \approx 3 \times 10^{59}$, and their energy $\epsilon V \approx 6 \times 10^{56}$ erg. On the other hand, the volume of the jet is of the order of 10^{61} cm³ and the number of *Leuchtelektronen* (electrons emitting in the visible range) about 10^{53}, that is, several million times fewer than in the whole galaxy. If $E = 5 \times 10^{11}$ ev and $H = 10^{-4}$ gauss, a *Leuchtelektron* loses a considerable fraction of its energy by radiation in about 1000 yr, unless it is continually accelerated. This, in particular, shows that we cannot assume that the relativistic electrons responsible for the radio emission are former *Leuchtelektronen* which abandoned the jet and were decelerated. We need to postulate that in the jet, along with the *Leuchtelektronen* with $E \approx 5 \times 10^{11}$ ev, there is a considerably larger number of electrons with $E \approx 10^9$ to 10^{10} ev. This is a very natural result since the formation of the less energetic electrons does not require such special conditions. Apparently the region of formation of relativistic electrons is not restricted to the jet but comprises a more or less significant volume in its vicinity. It would be very desirable to photograph the central region of NGC 4486 in the light of λ 3727. Such a photograph would, in particular, allow us to estimate more accurately the extent of the region where the relativistic electrons are generated.

Now we shall look into the conditions for the diffusion of the relativistic electrons. According to diffusion theory the mean square distance traversed by a relativistic electron in time t is

$$\langle L \rangle^2 = 2\langle l \rangle ct.$$

Taking $t = 2 \times 10^6$ yr, $\langle L \rangle = 5 \times 10^3$ pc, we find that $\langle l \rangle \approx 20$ pc, which is approximately the length of the clouds of interstellar gas in our Galaxy.

If we assume that the optical emission continues at the observed level for 10^6 yr, we find that in this length of time the *Leuchtelektronen* have lost about 10^{55} erg of their energy. This is about one and a half orders of magnitude less than the estimated total energy of the relativistic electrons in NGC 4486 and should be regarded as the lower limit of the energy of the particles in the nebula.

We shall compute now the integrated radio emission from NGC 4486 after 10^6 yr. The radio luminosity over the entire radio range is given by

$$L = 4\pi R^2 F_\nu \Delta\nu,$$

where $R \approx 10^7$ pc — the distance to NGC 4486. At $\nu = 100$ Mc/s, $F_\nu = 1.2 \times 10^{-20}$ erg cm⁻² sec⁻¹ (c/s)⁻¹, and $\Delta\nu = 10^9$ c/s, whence

$$L \approx 10^{41} \text{ erg/s},$$

so that after 10^6 yr the total emission is 5×10^{54} erg. It follows that the relativistic electrons in the interstellar medium lose their energy very slowly and will be emitting radio waves for at least 5×10^8 yr.

During this time the jet will disappear, leaving no visual traces. This raises an interesting point: it would be possible to observe radio galaxies with the same absolute and relative radio luminosity as NGC 4486 but without any optically noticeable peculiarities!

What is the nature of the phenomenon causing the formation of a jet which in turn generates an enormous number of relativistic particles and other remarkable peculiarities of the radio galaxy NGC 4486? It would be natural to think of some explosion of grandiose proportions, exceeding by far even such exceptional phenomena as supernova outbursts (the energy of this explosion would exceed that of a supernova by a factor of about 10^7; the mass equivalent of the energy released would be about $100\mathfrak{M}_\odot$).

As we have pointed out before, the morphological peculiarities of NGC 4486 (for example, the striking proximity of the very small knots in the jet to the nucleus of the galaxy, the distribution of these knots along a straight line) indicate that the observed jet is genetically tied to the galaxy and is not the result of a collision with another galaxy. If we adopt this position, we are compelled to account for the nature of the jet or jets. In principle, of course, we can assume that the observed phenomenon is due to a process very different from those known at the present time and that unknown laws of nature are in operation.

Such an approach is naturally fruitless. Only when we are convinced of the impossibility of explaining some new phenomenon by means of the well-known laws of nature can we appeal to the unknown and derive new laws from the analysis of the new facts. Thus, for example, G. R. Burbidge (1956), who also constructed a model for a radio galaxy of the type of NGC 4486, assumes that the formation of the large number of relativistic particles in such galaxies can be explained on the basis of interactions between matter and antimatter. On the strength of the considerations presented above, we believe such a viewpoint to be unsatisfactory. Notice, also, that it is difficult to imagine that two aggregates of matter and antimatter may fully annihilate each other, for, as soon as they come into contact, the energy released in their vicinity by the annihilation will separate them again and thus not allow the process to continue.

The aggregate of all facts concerning the nature of the most different discrete sources of nonequilibrium cosmic radio waves indicates that favorable conditions for the emission seem to be connected with large-scale motions of ionized gas inside the cosmic objects. Sources of completely different scale—such as Cygnus A and Taurus A—apparently owe their large radio emitting powers to the same cause: in both cases the energy of the radio waves is eventually obtained at the expense of the kinetic energy of random motions of gaseous masses.

It is only natural to assume that this is also the case in NGC 4486.

At first sight we see no reason for the presence of such a large amount of kinetic energy, since collisions with other galaxies (which may produce energies up to 10^{59} erg) in this case are absent. We can invoke, however, an internal source of energy: collisions among its own globular clusters or similar aggregates, which are so numerous in NGC 4486.

The mass of a large globular cluster in our Galaxy is about $10^6 \mathfrak{M}_\odot$. Notice, however, that in elliptical galaxies such as NGC 4486 the masses of some globular clusters can be considerably larger. If the mass of gas in such aggregates is of the order of $5 \times 10^6 \mathfrak{M}_\odot$, then in principle their mutual collisions may explain the formation of the jet from the energetic point of view.

There exists still another mechanism which in principle can explain the radio properties of NGC 4486. The absolute magnitude of this galaxy is exceptionally large: -20^M; at the same time it is an elliptical galaxy of fairly late spectral class. It is well known that elliptical systems are characterized by a very large ratio of mass to luminosity (Sec. 24); if we adopt $\mathfrak{M}/L = 300$, the mass of NGC 4486 has the enormous value $3 \times 10^{12} \mathfrak{M}_\odot$.

Recently some more evidence has been obtained which points to a very large mass. Morgan and Mayall (1957) discovered that the line widths in the integrated spectrum of the galaxy are anomalously large; this may indicate a larger than normal dispersion in stellar velocities due to the huge mass of the galaxy. Furthermore, the mass concentration toward the center should be very pronounced. If, for example, we assume that 10 percent of the total mass of the galaxy is found within a region 300 pc in radius at its center, then the stellar concentration (mainly subdwarfs) will be about 5×10^3 pc^{-3}, or 5×10^4 times as great as in the neighborhood of the sun.

We can now estimate the number of head-on collisions in the nucleus of NGC 4486: assuming that the collisional cross section is equal to the geometric cross section $\pi(r_1 + r_2)^2 \approx 10^{-15}$ pc^2 (actually, of course, it is larger), we find that per cubic parsec in 10^6 yr there are $(5 \times 10^3)^2 \times 10^3 \times 10^{-15} = 2.5 \times 10^{-5}$ collisions. The relative collisional velocity will be of the order of the velocity dispersion, which we can take as 10^3 km/s or 10^3 pc/10^6 yr. In the entire central region there will be 2.5×10^3 collisions in 10^6 yr, or one every 400 yr.

What takes place during such a collision? A strong shock wave will heat the two stars to a temperature of about 10^8 °K, and hence produce nuclear reactions which may release about 10^{50} to 10^{51} erg, mostly in the form of relativistic particles. If such collisions are a permanent feature of NGC 4486, then on the average relativistic particles will be formed at a rate of 10^{40} to 10^{41} erg/s, which is sufficient to support the radio emission of NGC 4486 at its observed level.

As regards the jet itself, it may be formed by a break-through of relativistic particles from the center, carrying with them the magnetic field and gas.

According to the concepts expressed above, the elliptical form of the radio region is explained by the fact that the interstellar gas (and associated magnetic field) into which the electrons diffuse does not form a spherical but rather an ellipsoidal system. The spherical form of the galaxy as a whole is due to the globular clusters and other type II objects.

It is hardly necessary to stress that all the preceding considerations on the formation of the jet are of a very preliminary and hypothetical character. This important problem is very difficult, and unexpected developments will probably occur in the near future.

We can expect that, as is the case in the Crab Nebula, the emission from the jet will be polarized in a direction parallel to its axis (Shklovsky 1955a). As far as we know, the first attempt to observe polarization was made at the Byurakan Observatory. According to a private communication of Ambartsumian, no polarization was detected. The next attempt was made by Oort and Walraven (1956), who photographed the central region of NGC 4486 through a doubly refracting prism. In this way they obtained two pictures polarized in mutually perpendicular directions; in one they found evidence of polarization, with the electric vector parallel to the axis of the jet. We must point out, however, that the small scale of the Byurakan and Leiden photographs renders these observations very unreliable, and only by means of the largest instruments will it be possible to investigate this important effect predicted by theory.

In March 1956, Baade (1956b) detected with the 200-in. reflector strong polarization in each of the three knots in the jet. The large scale of the photographs allowed the knots to be studied individually. Baade calls the three knots A, B, and C, going outward from the center (Fig. 186), and he finds that the direction of vibration of the electric vector in B lies at a considerable angle to that in A and C. The degree of polarization reaches 30 percent. It is interesting to notice that a fourth, very weak knot close to the nucleus of the galaxy is also polarized approximately in the same direction as B. This alternation in the polarization angle is extremely interesting and must be due to some cause connected with the formation of the jet.

With the discovery of polarization in the jet of NGC 4486 we can consider as an established fact that the mechanism of synchrotron radiation is responsible for the radio waves in both galactic and ex-

Fig. 186. A 200-in. photograph of NGC 4486 in polarized light, with the direction of the electric vector normal to the jet (Baade 1956*b*). Knots *A* and *B* form a close bright pair. Compare with the unpolarized photograph of the knots in Fig. 185.

tragalactic objects. Now we need very precise radio-interferometric observations of the brightness distribution over NGC 4486; we cannot exclude the possibility that the intensity may have a sharp maximum at the very center of the galaxy. Finally, we must attempt to detect interstellar hydrogen in globular clusters (see Chapter IV).

To conclude this section we shall pause on the question of the nature of the emission from clusters of galaxies. In Sec. 24 we remarked that the radio indices of clusters are negative, indicating that their radio emission is abnormally high. The same is also true of the Supergalaxy; in this case it is particularly clear that the phenomenon cannot be explained by the presence of radio galaxies in the cluster, since, with the exception of NGC 4486, none are found within the volume of the Supergalaxy among objects down to the twelfth magnitude.

We may put forward the hypothesis that the cause of the anomalous radio waves is the retarded emission from relativistic electrons which move in intergalactic magnetic fields within the cluster.

The presence of relativistic particles in intergalactic space would not be surprising. First, these particles may reach intergalactic space from the galaxies: very energetic relativistic protons escaping from galaxies may produce relativistic electrons by colliding with the nuclei of intergalactic atoms. Second, galactic collisions may give rise to a large number of relativistic particles which subsequently diffuse gradually into intergalactic space. If the relativistic particles formed during the collision in Cygnus diffuse into a volume of radius 3 Mpc, their mean concentration will be of the order of 10^{-13} to 10^{-14} cm^{-3} (remember that in our Galaxy the concentration of relativistic particles is about 10^{-10} cm^{-3}).

We can also expect the presence of very weak extragalactic magnetic fields. This necessarily follows from the existence of intergalactic gas. In a cluster of galaxies almost every member has undergone a collision at some time during its evolution; these collisions lead to a loss of galactic interstellar gas which, in the end, diffuses into the intergalactic medium. The expected density of intergalactic gas can be estimated by distributing the interstellar gas contained in the galaxies uniformly over the entire space occupied by the cluster: the density thus obtained is of the order of 10^{-29} gm/cm^3. The intergalactic medium will be in a state of turbulent motion and the velocities of the largest eddies will be of the order of the peculiar velocities of the galaxies, that is, 3×10^7 cm/s. From the familiar relation

$$\rho v^2/2 = H^2/8\pi,$$

we find the intensity of the intergalactic magnetic field to be $H \approx$ 10^{-7} gauss, higher than expected.

From the equations of synchrotron radiation we find

$$n = \int_0^\infty k E^{-\gamma}\, dE \approx 10^{-13} \text{ cm}^{-3},$$

in agreement with the previous estimate.

Thus the emission by relativistic electrons in a magnetic field may, in principle, explain the basic observational data concerning radio waves from extragalactic objects.

26. RADIO ASTRONOMY AND COSMOLOGY

Discrete sources of radio waves are abundantly represented in extragalactic space; hence radio investigations must play a great and ever-increasing role in cosmology. We are here confronted above all with the problem of the ability of large modern radio telescopes to penetrate to great distances, for successful research in cosmology depends strongly upon this property of the observational resources of radio astronomy.

We might expect the penetration of a telescope to depend primarily on its radiation-collecting power. In optical astronomy the latter is of course determined by the diameter of the objective of a refractor or the mirror of a reflector. And this is also true in radio astronomy: the greater the effective area of the antenna, the weaker the flux that can be measured and, as a result, the greater the distance to which sources can be studied. But there is a further circumstance which influences the penetration of a telescope to no small degree—the power of the sources of radiation. Obviously, the more powerful sources can be observed to a greater distance, for apparatus of given sensitivity.

The development of radio astronomy over the past few years has shown that certain extragalactic radio sources possess a power so enormous that the penetration of large radio telescopes already *exceeds* that of the greatest optical telescopes.

We pointed out in Sec. 3 that cosmic objects differ sharply in their relative flux of radiation at optical and at radio wavelengths. Let us consider this matter more carefully. The nearest extragalactic objects, the Magellanic Clouds, are 46 kpc away; we receive an optical flux from them which is 10^{-11} times that from the sun. And these are the nearest objects! In optical astronomy, the solar radiation absolutely dominates the flux from all other cosmic objects, for it is 10^8 times the combined radiation from all other bodies in the universe except the moon. Naturally this results from the close proximity of the sun; and yet we find an entirely different situation in radio astronomy. For example, the flux of cosmic radio waves at a wavelength ≈ 10 m exceeds the corresponding solar radiation by a hundred thousand times!

As indicated in Sec. 8, the radio flux received from the colliding galaxies in Cygnus, at a distance of 300 million light years from us, is approximately the same as that from the quiet sun, which is 8 light minutes distant. Thus the radio luminosity of Cygnus A exceeds that of the sun by 26 orders of magnitude, while the optical luminosity is only 10 orders greater than that of the sun.

The source Cygnus A can hardly be unique in the universe. Although extremely rare, such objects must occur in considerable number in the aggregate. Many of the weakest discrete sources observed at the present time may well be objects of the type of Cygnus A. At their limit, the most powerful modern radio telescopes can detect, at 3 m, a flux $F_{min} \approx 2 \times 10^{-26}$ w m^{-2} (c/s)$^{-1}$. This minimal flux is 1/6500 of that from Cygnus A. If a source identical with Cygnus A were to be responsible for so weak a flux, it would be located at a distance 80 times that of Cygnus A—a distance of 8×10^9 parsecs!

At these extreme distances, the effects predicted by relativity theory, such as the curvature of space and the red shift, will have become prominent; they will lead to a considerable diminution of flux (see below), but we shall not take them into account for the present. What is the penetrating power of the greatest optical telescopes in existence today (aside from the effects of relativistic cosmology)? The detection of a 23^m object represents the frontier of photoelectric techniques. If such an object corresponds to a giant galaxy of absolute magnitude -20^M, it will lie at a distance of 4×10^9 pc. This means that the penetration of existing radio telescopes is already twice that of the unique 200-in. Palomar reflector. And not only are ever more powerful radio telescopes coming into operation, but the sensitivity of receiving apparatus is rapidly being improved as well. The vast penetration of the radio telescope will in the near future ensure a substantial broadening of the observational basis of cosmology.

Another advantage of radio astronomy over optical astronomy is the opportunity of measuring the integrated radio brightness of the entire universe. Olbers' photometric paradox is an effect of optical astronomy. This states that, if stars were distributed with constant density over the universe, the surface brightness of the sky would be the same as that of the sun, a typical star; and this is not observed. The removal of the photometric paradox was one of the basic problems of cosmology. It was resolved in the framework of classical cosmology through the introduction of a "hierarchic" structure for the universe (the Charlier scheme). Relativistic cosmology removed the paradox by proceeding from the existence of the red shift. However, in order to understand the nature of the universe it is not enough to explain why the celestial sphere is not dazzlingly bright. We must find the true intensity as determined by the radiation from all the objects in the universe. In optical astronomy it is virtually impossible to derive this quantity: the extragalactic component of the night-sky brightness is completely overwhelmed by the galactic, zodiacal, and atmospheric components. Optical methods can therefore provide only an upper limit on the intensity of extragalactic radiation.

The situation is just the reverse in radio astronomy. At meter wavelengths the zodiacal and atmospheric components of the observed radiation are negligible. Cosmic radio emission consists of galactic and extragalactic components only. Until 1955 it was widely believed that at high galactic latitudes the intensity of cosmic radio emission is almost entirely governed by extragalactic sources (see Sec. 4). It is now clear, however, that this is not so; thus a supplementary investigation is required in order to determine the share of the extragalactic component. As early as 1953 we had found, by an analysis of the distribution of the intensity of cosmic radio waves over the sky, that at $\lambda = 3$ m the brightness temperature T_b of extragalactic space is between 100 and 200 °K, corresponding to an intensity between one-third and one-fourth of that observed at the galactic poles (Shklovsky 1953e). English investigators subsequently obtained the same result by a similar method, but with more qualitative observational material (Baldwin 1955c). For the brightness temperature of extragalactic space at $\lambda = 3$ m we may therefore take $T_b = 150\,°K$, with a probable error of several tenths of this value.

Let us now assume that the observed intensity of extragalactic radio emission is given by the radiation from radio galaxies distributed more or less uniformly through space. We may then formulate the following problem. Let a sphere of radius R be filled with radio sources at a constant spatial density equal to the observed density of radio galaxies. What should the radius R be in order that the intensity of the radiation from all the sources may equal the observed extragalactic intensity? To answer this question we have employed (Shklovsky 1953e) the Oort–Westerhout formulas introduced in Sec. 11. The desired radius is given by a formula readily deduced from Eqs. (11-2) and (11-3):

$$R = \frac{r_N}{3N} \frac{8\pi\nu^2 kT_b}{P_N c^2}, \tag{26-1}$$

where P_N denotes the radiation flux from the Nth extragalactic source in order of decreasing intensity, r_N the most probable distance to the Nth source, and T_b the extragalactic brightness temperature. On the basis of the scanty data then available with regard to the sources, we obtained for R a value close to 250 Mpc on the old distance scale, or ≈ 750 Mpc on the new scale.

The quantity R may be called the "photometric" or "equivalent" radius of the Metagalaxy. It must be emphasized that the "equivalent" radius turns out to be considerably less than the distance to the most remote galaxies observable with present optical and radio sensitivities. We must at once conclude that the classical hierarchic scheme, the Charlier universe, cannot account for the photometric paradox: the observed dimension of the second tier of the hierarchic structure, the Metagalaxy (the first tier was our own Galaxy), is found to exceed the equivalent radius of the Metagalaxy by a significant margin. There remains but one possible explanation for this "shortened" equivalent radius—the *red shift*.

We shall regard the red shift as a Doppler effect. The red-shift law for the domain of the universe observable optically may be written in the form

$$\frac{\nu - \nu_1}{\nu} = \frac{\Delta\nu}{\nu} = -kr, \tag{26-2}$$

where ν_1 is the observed frequency of the radiation, ν is the frequency at the source, which is in motion with a velocity $v(r) = ckr$, and k is the red-shift constant. The best available data give a value for k of ≈ 150 km sec^{-1} Mpc^{-1}.

The relativistic formula for the Doppler effect is

$$\nu_1 = \frac{\nu(1 - v^2/c^2)^{1/2}}{1 + v/c}. \tag{26-3}$$

Let the spectrum of a radio source assume the form

$$\epsilon(\nu) = (B/\nu)\,d\nu,$$

where $\epsilon(\nu)$ represents the radiation per unit frequency interval from a single source, over all directions. The total intensity of extragalactic radio emission will then be given by

$$I_\nu = \frac{Bn}{4\pi\nu_1}\int_0^R \left[\frac{\nu}{\nu(r)}\right]^2 dr = \frac{Bn}{4\pi\nu_1}R = \frac{Bn}{4\pi\nu_1}\int \frac{dr}{(1+kr)^2} = \frac{Bn}{4\pi\nu_1 k}, \quad (26\text{-}4)$$

whence

$$R = 1/k = 2000 \text{ Mpc}, \quad (26\text{-}5)$$

a value 2.5 to 3 times that found empirically. The primary reason for this discrepancy between the theoretical and empirical values for the photometric radius is simply the unreliability of our present data on the distances to radio galaxies. For example, we had made an estimate of 2.3 Mpc for the distance to NGC 4486 (Shklovsky 1953e), while more recent determinations give a distance of 10 Mpc for this giant galaxy. The empirical equivalent radius of the Metagalaxy forthwith increases to 1200 Mpc. Furthermore, we have taken a linear law, Eq. (26-2), for the red shift. In general, this is not correct; a second-order term in Eq. (26-2) may decrease the value of R given by Eq. (26-5). Finally, we have adopted a Euclidean metric for space, which may lead to error at the great distances involved. Thus on the whole we may regard the red shift, if interpreted as an effect of the recession of the galaxies, as a full explanation of the observed intensity of extragalactic radio emission.

Shakeshaft (1954) has also called attention to the value of radioastronomical methods for the study of cosmology. He attempted to estimate the intensity of radio emission per unit volume of extragalactic space through an analysis of the observational material. Assuming that the sources of extragalactic radio emission are as a rule colliding galaxies, Shakeshaft, proceeding from elementary laws in kinetic theory, derived a formula for the number of collisions of galaxies per cubic parsec:

$$Z = 2^{1/2}\pi a^2 \rho^2 v,$$

where a is the radius of a galaxy, v the velocity, and ρ the mean concentration of galaxies. By taking $\rho = 4 \times 10^{-18}$ pc^{-3}, he found that $Z = 7 \times 10^{-23}$. If a source of radio waves is created in such a collision, and if its power is the same as that of Cygnus A, then according to Shakeshaft about 10^5 sources of flux $> 10^{-24}$ w m^{-2} (c/s)$^{-1}$ could be observed over the sky. Actually, however, they are incomparably rarer. Shakeshaft therefore assumed that collisions ordinarily lead to the formation of radio galaxies such as NGC 1275, whose radio power is several orders of magnitude less than that of Cygnus A. Arbitrarily increasing the frequency of galaxy collisions per unit volume by an order of magnitude, and allowing for radio emission from "normal" galaxies as well (Shakeshaft regards the latter as significant, although this is debatable, to say the least), he obtains a mean value for the emission per unit volume of extragalactic space of from 400 to 1200 w pc^{-3} (c/s)$^{-1}$ sterad^{-1}.

If we take a uniform distribution of sources in extragalactic space, and neglect both the curvature of space and any possible variations in the red-shift law, we may compute the extragalactic radio flux observed at the Earth by means of the formula

$$B(\nu)\Delta\nu = \int_0^\infty \epsilon(\nu')\Delta\nu'(1 + Z)^{-2-\alpha}dr. \tag{26-6}$$

Here ν is the frequency of the radiation in the "natural" coordinate system, and $Z = r/T$, where $T = 1/k$ is Hubble's constant on the new distance scale. Now the parameter α vanishes in relativistic cosmology; whereas in the "steady-state" theory of the universe, as developed by certain theoreticians but by no means well established at the present time, $\alpha = 3$. We also have

$$\nu'/\nu = \Delta\nu'/\Delta\nu = 1 + Z = 1 + r/T, \quad \epsilon(\nu) = \nu^{-0.5},$$

so that

$$B(\nu) = \int_0^\infty \frac{\epsilon(\nu)dr}{(1 + r/T)^{1.5+\alpha}} = \frac{\epsilon(\nu)T}{0.5 + \alpha}. \tag{26-7}$$

With $\alpha = 0$, this formula is equivalent to our Eq. (26-4), but the emission $\epsilon(\nu)$ is here proportional to $\nu^{-0.5}$ rather than to ν^{-1}. It appears that in fact $\epsilon(\nu) \propto \nu^{-0.8}$.

We shall now derive the brightness temperature T_b:

$$T_b = \frac{B(\nu)c^2}{2k\nu^2} = \frac{\epsilon(\nu)T\lambda^2}{2k(0.5 + \alpha)}.$$

For $\epsilon(\nu) = 400$ to 1200 w pc^{-3} (c/s)$^{-1}$ sterad^{-1}, $T_b = 315$ to $940°$K if $\alpha = 0$, and $T_b = 45$ to $135°$K if $\alpha = 3$. If we take the more accurate spectrum $\epsilon(\nu) \propto \nu^{-0.8}$, then if $\alpha = 3$ we have the very small value $T_b = 28$ to $85°$K. Nevertheless, the indeterminacy in the spectrum $\epsilon(\nu)$ has not yet permitted us to make an unequivocal choice between the two cosmological theories. Yet the possibility still exists of analyzing the radio observations in the hope of reaching a conclusion as to the validity of one cosmological hypothesis or another. If a choice between different cosmological concepts cannot be made at the present level of radio techniques, it will soon become possible to decide the question.

Ryle and Scheuer (1955) have conducted a statistical analysis of the distribution of discrete radio sources, based on the material in the Cambridge catalogue. The sky was divided into seven large areas, in each of which was determined the number N of sources per unit solid angle whose radiation was greater than F_ν. We shall suppose that the sources have the same radio luminosity L and are distributed through space with a constant concentration ρ. Then the flux from a source at distance r will be equal to

$$F = L/r^2,$$

and the number of sources per unit solid angle of flux $> F_\nu$ will be

$$N = \tfrac{1}{3}\rho r^3 = \tfrac{1}{3}\rho L^{2/3} F^{-3/2}. \tag{26-8}$$

If $\log N$ is now plotted against $\log F$, we obtain a straight line of slope 1.5. It can be shown that this result holds even when there is a dispersion in the radio luminosity of the sources (see Sec. 8).

Figure 187 gives the results of the reduction of the seven areas in the Cambridge catalogue. In every case a striking deviation is observed from the law

$$2 \log N \sim -3 \log F.$$

This law is satisfied only for relatively powerful sources; for weak sources the curve is considerably steeper than the theoretical straight line (broken in Fig. 187) corresponding to a uniform source distribution. The latter part of the curve can be approximated by a straight line of slope ≈ -3. At very small values of F_ν the curve slopes gently, approaching the horizontal asymptotically. This is purely an instrumental effect: sources whose intensity is below the threshold of sensitivity are not detected, so that for some F_1, $N(F < F_1) = \text{const}$ (see Sec. 8).

From the curve in Fig. 187 we may conclude that the sun is at the center of a region in which the concentration of sources increases with distance. If the radio luminosity L is independent of distance, then the curve indicates that $\rho \propto r^3$. If on the other hand ρ were constant, we would obtain the peculiar result $L \propto r^3$, which can hardly be accepted. If the increase of ρ with distance were to extend to arbitrarily great r, then the brightness temperature of the sky could evidently be arbitrarily high. From an analysis of the probability distribution of the amplitudes of interference patterns, Ryle and Scheuer found that the radius of the region within which the law $\rho \propto r^3$ holds must be $1.2r_0$ or $1.3r_0$, where r_0 is the distance to the weakest detectable sources.

Were we to suppose that the sources are localized in the Galaxy, we would encounter serious difficulties. By applying a formula analo-

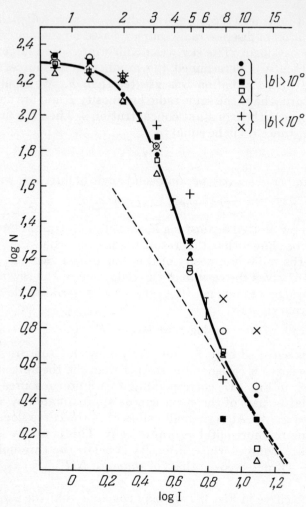

Fig. 187. Results of the reduction of selected areas in the Cambridge catalogue (Ryle and Scheuer 1955).

gous to the Oort–Westerhout formula (Sec. 11), we may obtain a *lower* bound on the dimensions of the region containing the sources. This is found to be 2 kpc. It is highly unlikely that the sun is just in the center of so large a system of sources, a system in which the spatial density increases monotonically in all directions.

These isotropically distributed sources must therefore be regarded as extragalactic objects. To be sure, this conclusion has already been indicated in Sec. 8, through analysis of the distribution of Class II sources, although that treatment employed less extensive material. Inasmuch as the sources are localized beyond the Galaxy, the radius of the spherical region within which they are distributed becomes at least 100 Mpc.

If all the above considerations are correct, then any cosmological theory must account for two facts: (a) the increase of concentration (or luminosity) of the sources with distance; (b) the breaking off of this increase at distances only slightly in excess of r_0. In Ryle and Scheuer's opinion the observational data sharply contradict the steady-state theory of the universe, according to which the concentration and luminosity of the galaxies remains constant, in the mean, throughout the universe. On the other hand, the data are found to be in good agreement with fundamental consequences of relativistic cosmology, which implies that both ρ and L vary with time. If, for example, we assume that most of the weak sources observed are colliding galaxies of the type of Cygnus A, then the number of such objects must clearly be proportional to the square of the concentration of the galaxies. At the earliest epoch of the expansion of the portion of the universe now accessible to observation, this concentration must have been considerably greater than today, and collisions between galaxies must have occurred much more frequently. Furthermore, Ryle and Scheuer also admitted the possibility that the galaxies may have been powerful radio sources at the epoch of their condensation from a diffuse medium. This would explain the cutoff in ρ at $r \approx 1.3r_0$. At that point we observe galaxies several billion light years away, so that we observe them as they were several billion years ago. At an epoch even earlier than this, galaxies would in general not yet have been formed.

Taking 4×10^9 pc as the distance to the weakest sources observed, and adopting a formula similar to Eq. (26-4) for the attenuation of the flux by virtue of the red shift, Ryle and Scheuer found that the radio luminosity of these sources is approximately four times that of Cygnus A. They concluded from this that the majority of the weak discrete sources observed are in fact objects of the Cygnus A type. To interpret the curve of Fig. 187 we must adopt for the concentration of these sources in extragalactic space the very small value $\rho \approx 10^{-26}$ pc^{-3}. Ryle and Scheuer believe that this explains why so few of the sources in the Cambridge catalogue can be identified with optical objects; because of their vast distances only a few such galaxies can be detected by modern optical telescopes.

If the results of Ryle and Scheuer prove correct, they will be of extreme significance for cosmology. Unfortunately, however, more recent investigations do not support the results of the Cambridge radio astronomers. In Sec. 7 we discussed the outstanding investigation of discrete sources conducted at Sydney with the Mills cross at 3.5 m (Mills and Slee 1957). As we pointed out, the Sydney observers brought to light serious inadequacies in the Cambridge catalogue of discrete sources, arising from various causes of an instrumental nature. Statistical analysis of the Sydney results, carried out by Mills, showed no appreciable increase in the concentration of extragalactic sources with respect to distance from the sun.

Figure 188 provides a diagram analogous to that of Fig. 187, prepared for the Sydney and Cambridge observations. The slope of the line representing the Sydney observations is -1.7. Mills ascribes the deviation from the value -1.5, which would have been expected for a homogeneous distribution of sources in space, to the influence of Class I sources. By employing sources for which $|b| > 12°.5$, and which therefore belong mainly to Class II, Mills secured the diagram given as Fig. 189. The broken line has slope -1.5. It is evident from Fig. 189 that the spatial distribution of Class II sources departs very little from a homogeneous one. The remaining discrepancy between the observed and theoretical lines can be interpreted as an instru-

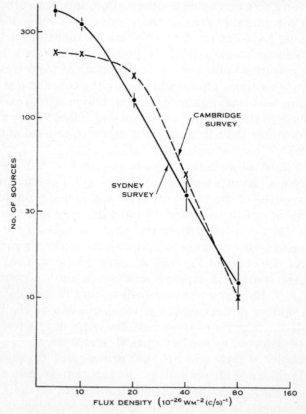

Fig. 188. A comparison of the Sydney and Cambridge source counts (Mills and Slee 1957).

mental effect. Thus the Cambridge results, which had created a sensation, now appear to be in error.

We now turn to the following important question: with what type of radio galaxy are the majority of the Class II sources to be identified? Ryle and Scheuer treated these sources as colliding galaxies of the Cygnus A type, in general. At first glance such an assumption seems quite plausible. Indeed, one can hardly believe that the Cygnus A source is unique in the universe; there must be other sources of the same character. These must be distinguished by their immense surface brightness. Morris, Palmer, and Thompson (1957) have recently estimated the angular dimensions of three rather intense sources $[F_\nu \approx 5 \times 10^{-25} \text{ w m}^{-2} \text{ (c/s)}^{-1}]$, finding their extent to be $< 12''$. It follows that their brightness temperature $T_b > 2.4 \times 10^7 \text{ °K}$, while for Cygnus A, $T_b = 10^8 \text{ °K}$. These sources are very likely analogous in nature to Cygnus A.

However, without a special discussion we cannot exclude another

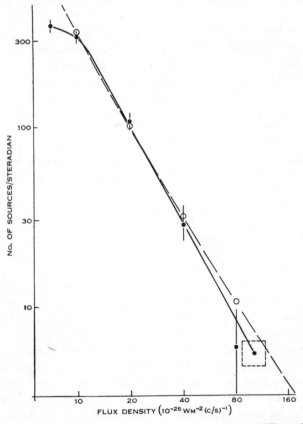

Fig. 189. A count of Class II sources observed at Sydney (Mills and Slee 1957).

possibility, namely, that most Class II sources are to be identified
with considerably weaker, but more abundant, radio galaxies, such
as NGC 4486 and 5128. Let us investigate this matter in some detail.

In stellar astronomy the following equation is proved:

$$\frac{N_m(M_1)}{N_m(M_2)} = \frac{d_1}{d_2} \times 10^{0.6(M_2 - M_1)}, \tag{26-9}$$

where $N_m(M)$ is the number of radiating objects, uniformly distributed in space, whose apparent stellar magnitude lies within the
interval $(m - \frac{1}{2}, m + \frac{1}{2})$, and which have the absolute magnitude M
and the concentration d in space. Assume that the majority of the
Class II sources are objects of the Cygnus A type. Over the entire
sky there are about 10 sources identified with "normal" galaxies and
having a radio magnitude between 9 and 10^m, whereas the number of
Class II sources in the same magnitude interval is ≈ 1000. In the following discussion we shall assign the subscript *1* to symbols referring
to objects of the Cygnus A type, and the subscript *2* to "normal"
galaxies.

Since the radio luminosity of Cygnus A is some 10^6 times that of
"normal" galaxies, $M_2 - M_1 \approx 15^m$. We have seen that $N(M_1)/N(M_2)$
≈ 100; hence $d_2/d_1 \approx 10^7$. This means that for every 10 million "normal" galaxies there will occur only one colliding pair such as Cygnus
A. The mean distance between such rare objects would have to be
100 to 200 Mpc, so that the more distant objects would already lie
billions of parsecs away.

If we assume that the majority of the weak sources of radio waves
are objects similar to NGC 4486, we will be dealing with galaxies
whose radio magnitude is 7 or 8^m greater than that of Cygnus A.
Thus $M_2 - M_1 = 7^m.5$ and $d_2/d_1 \approx 300$. Finally, if the majority of
the weak discrete sources correspond to objects of still lower radio
luminosity, such as NGC 5128, then $M_2 - M_1 \approx 5^m$ and $d_2/d_1 \approx 10$.

The following argument supports the assertion that most of the
weak discrete sources are not the exceptionally rare objects of the
Cygnus A type, but the far more widely distributed radio galaxies of
the NGC 4486 and 5128 types. About 20 galaxies may be found in the
sky with an apparent photographic magnitude less than 10^m. Of these,
three galaxies are radio galaxies: NGC 5128, 4486, and 1316. The first
to point this out was V. A. Ambartsumian, who noted that in optical
light the radio galaxies are supergiants; their absolute photographic
magnitude is about -20^M. We must take into consideration the fact
that one-seventh of the brightest galaxies observed are radio galaxies.
This cannot be fortuitous, and we may suppose that as we pass to
fainter galaxies this proportion of radio galaxies must be preserved to
some extent. Since the radiation flux from the weakest objects in

Mills's catalogue falls below that from NGC 4486 by a factor of ≈ 200, if we identify these objects with radio galaxies of the NGC 4486 type we find that their apparent photographic magnitude must be 15 or 16^m. In the entire sky the total number of galaxies of this photographic magnitude is of the order of 40 000. Even if the fraction of radio galaxies among these is $\approx 1/20$, a value one-third of that for the bright galaxies, the number of radio galaxies among the objects of 15 to 16^m will be ≈ 2000; this is of the same order as the total count of sources in the Cambridge and Sydney catalogues.

Galaxies of 15^m have still received very little study. Attempts to identify such galaxies with faint sources demand a rather elaborate investigation.

There is one further argument against an identification of the majority of the Class II sources with objects of the Cygnus A type. If this were the case, then sources with $F_\nu = 10^{-25}$ w m^{-2} (c/s)$^{-1}$ would be 35 times as far from us as Cygnus A, that is, at a distance of 3.5×10^9 pc. At such extreme distances cosmological effects such as the red shift, will have come strongly into play, and this will undoubtedly be reflected in the apparent distribution of Class II sources. But it is clear from Fig. 189 that the Sydney observations reveal practically no departure from homogeneity in the distribution of sources having a flux up to $F_\nu = 10^{-25}$ w m^{-2} (c/s)$^{-1}$. This is perfectly understandable if the weak Class II sources represent radio galaxies of the NGC 4486 type. Their distance would then be ≈ 100 Mpc, and no cosmological effects would distort the distribution of Class II sources.

To close this section, we arrive at the conclusion that radio astronomy has encroached upon the domain of cosmological studies, and will shortly occupy a leading position in this field.

27. MONOCHROMATIC EMISSION

As we have demonstrated in Chapter IV, the analysis of observations of the hydrogen radio line has become a powerful tool for studying the physical properties of the interstellar medium and the dynamics of the Galaxy. At the present writing it has become abundantly clear that this method offers rich perspectives for extragalactic study as well.

The Australian radio astronomers Kerr, Hindman, and Robinson (1954) have carried out a particularly valuable investigation of the monochromatic hydrogen radio emission from the Magellanic Clouds. Their pioneering study was the first attempt to apply the new technique to extragalactic sources. Larger reflectors are now being exploited, and with these many other investigations will certainly be devoted to the analysis of the hydrogen radio emission from various extragalactic objects.

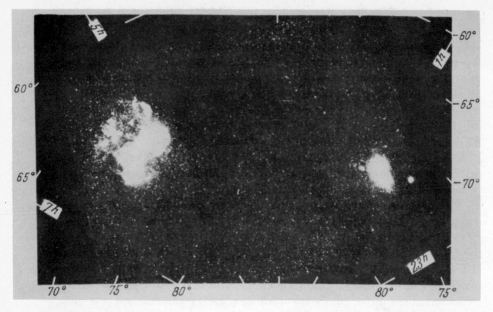

Fig. 190. The Magellanic Clouds.

The Magellanic Clouds (Fig. 190) are the nearest external galaxies. Their distance is 46 kpc; the two galaxies are almost equally far from the sun (Shapley 1953). Since the dimension of our own Galaxy is about 25 kpc, the Clouds are actually very close to us. Our Galaxy and the two Clouds may, in fact, be regarded as comprising a multiple system whose components are united dynamically. De Vaucouleurs (1954) has demonstrated that the Large Magellanic Cloud is joined to the Galaxy by a "bridge" resembling that observed by Zwicky (1953) in other galaxies.

When observed with the unaided eye, the Large and Small Clouds exhibit diameters of 7° and 3°.5 respectively. However, on photographic plates traces of the Large Cloud can be detected over a region 20° in extent; the Small Cloud is also found to be much larger than it appears visually.

On the accepted system for the classification of galaxies, the Magellanic Clouds are usually grouped with the irregular galaxies. But de Vaucouleurs has convincingly shown that the outer portion of the Large Cloud exhibits spiral structure. According to Shapley (1950), the Small Cloud is a typical irregular dwarf galaxy. The Large Cloud contains stars predominantly of population I, while the Small Cloud presents a far more complex picture. The content of cosmic dust in the two galaxies is entirely different; dust is abundant in the Large Cloud, while the Small Cloud is practically free of it.

In any study of the dynamics of these galaxies, it is essential to know the radial velocities of the various objects within them. Optical astronomy is circumscribed here by the very scanty data. Radial velocities are known for only 17 bright nebulosities in the Large Cloud and for one in the Small Cloud. The values range from $+251$ to $+309$ km/s for the Large Cloud, with $+168$ km/s for the Small Cloud. The Magellanic Clouds are exceptionally useful objects for the investigation of the hydrogen radio line. The great radial velocities in these galaxies, as well as their high galactic latitude, enable us to separate their monochromatic radio emission from that of the Galaxy. Furthermore, the proximity of the Magellanic Clouds to our own stellar system permits the study of details which cannot be resolved in other galaxies.

The Australian observers employed a 12-m paraboloid with $1°.5$ beamwidth, mounted as a transit instrument with axis east and west. An interesting feature was that these investigators did not use the frequency-modulation technique which is customary in such cases. Observations were made on a fixed frequency with a 40-kc/s bandwidth. The reflector was stationary, the axis being directed at a certain altitude. Because of diurnal rotation, the directional diagram of the telescope swept out a zone in the sky; the inclination of the reflector was then changed, and another zone was swept out. By reducing the tracings obtained, it was possible to derive isophotes of the region of the sky under investigation, at a fixed frequency. The frequency was then varied and the procedure repeated. Profiles of the radio line were thereby obtained for every point of the region.

Figure 191 presents isophotes, at several definite frequencies, for that part of the southern sky where the Small Magellanic Cloud is located. These frequencies are given in terms of Doppler velocities with respect to the unperturbed position of the radio line ($v = 0$ at $\nu = 1420.4$ Mc/s). Evidently the monochromatic radio emission of this galaxy corresponds to a basic radial velocity of $\approx +150$ km/s, in agreement with the optical observations (see above).

From isophotes such as these we may construct profiles of the radio line at various points. A number of these are given in Fig. 192 for the 21-cm line in the Small Cloud. Generally speaking, the profiles are rather intricate. The brightness temperature attains a maximum of $30°$K, which is one-fourth of the maximum temperature of the monochromatic galactic radio emission in the directions of the center or the anticenter.

The mean width of the hydrogen radio line in the Magellanic Clouds is about 250 kc/s, corresponding to a velocity dispersion in the radiating gas of 50 km/s. The velocity dispersion is somewhat smaller for the Large than for the Small Cloud.

The total intensity of the radio line (integrated over the spectrum) in the Magellanic Clouds may be determined from the expression

$$I = 2k\lambda^{-2} \int T_b(\nu) \, d\nu. \tag{27-1}$$

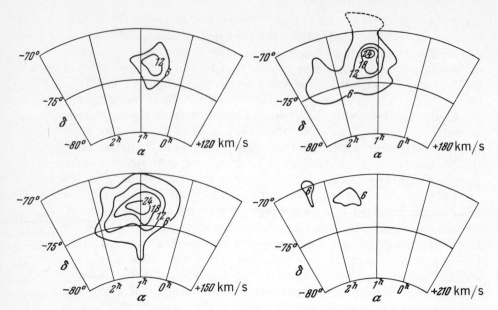

Fig. 191. Isophotes of radio emission from the Magellanic Clouds at four frequencies (Kerr, Hindman, and Robinson 1954).

Isophotes of total intensity, derived in this way, appear in Fig. 193. Here the dotted curve delimits the area within which measurable radio emission can still be detected at the sensitivity of the apparatus employed. Figure 194 shows the mean radial velocities derived from an analysis of the 21-cm profiles throughout the Magellanic-Cloud region.

It is noteworthy that the extent of the Clouds, as observed in the radio line, is so much greater than that observed optically (compare Figs. 190 and 193). Another interesting fact is that the monochromatic radio fluxes from the Large and Small Clouds are almost the same, whereas in optical radiation the Small Cloud is much the fainter of the two. It is important to note one further circumstance. If the apparatus had been two or three times as sensitive, isophotes embracing *both* Clouds would evidently have been obtained. This simply means that these two nearby galaxies are enveloped in a single overlying corona of rarefied hydrogen. We shall return to this important finding presently.

The monochromatic radio brightness of the Magellanic Clouds falls off with distance from their centers considerably more slowly than in visible light. Interstellar gas is therefore concentrated to the centers of these galaxies less strongly than the stars of high luminosity, which are primarily responsible for the optical radiation of the Clouds. Figure 195 illustrates the distribution of brightness along sections in right ascension and declination for optical radiation (broken lines) and radio

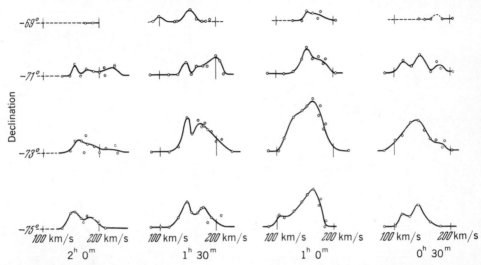

Fig. 192. Profiles of the hydrogen radio line at a number of points in the Small Magellanic Cloud (Kerr, Hindman, and Robinson 1954).

waves, thus demonstrating the difference in the spatial distribution of interstellar gas and bright stars in the Clouds. It is quite evident that the interstellar gas may be regarded as an extended envelope surrounding the system of bright stars in these galaxies.

In this connection we recall that in S. B. Pikelner's opinion the interstellar gas in our Galaxy forms a spherical system of vast extent,

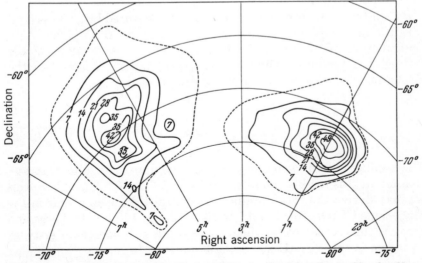

Fig. 193. Isophotes of total radio intensity in the Magellanic Clouds (Kerr, Hindman, and Robinson 1954).

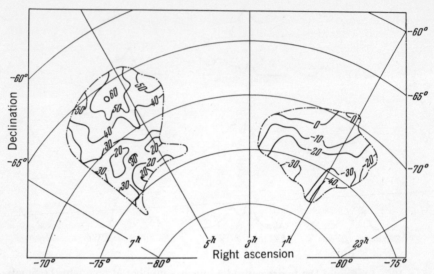

Fig. 194. Mean radial velocities in the Magellanic Clouds (Kerr, Hindman, and Robinson 1954).

while at the same time a relatively dense gas cloud is concentrated to the galactic plane (Sec. 12). Thus the observations of monochromatic radio emission from the Magellanic Clouds indirectly support the new ideas, as developed in the U.S.S.R., concerning the distribution of interstellar matter. The radio observations indicate that the interstellar gas in these "coronas" is basically neutral, as Pikelner (1953) had also considered it to be.

By applying the formulae of Sec. 14 to a measured brightness temperature and radio-line profile, we may compute the number N of neutral hydrogen atoms in a column of unit cross section passing through the portions of the Clouds investigated. If we introduce the total intensity, we obtain the following expression:

$$N = \frac{\int_0^\infty T_b(\nu)\, d\nu}{2.6 \times 10^{-15}} = 6.2 \times 10^{35}\, I. \tag{27-2}$$

The number of hydrogen atoms within a column of angular cross section δA deg^2 is readily found to be $\delta N = 3.8 \times 10^{78}\, I\, \delta A$. From this we may derive the total number of neutral hydrogen atoms in the Magellanic Clouds. We obtain 7×10^{65} and 5×10^{65} for the Large and Small Clouds respectively, which corresponds to a mass of interstellar hydrogen in these galaxies of 6×10^8 and $4 \times 10^8 \mathfrak{M}_\odot$.

We see, then, that the new method for investigating the interstellar medium has yielded a brilliant demonstration of its power; we can literally count the number of hydrogen atoms in each of these galaxies. It is particularly interesting that, while the quantity of interstellar hydrogen in the Large and in the Small Cloud appears to be almost the same, the quantity of interstellar dust in the two objects is entirely different; the Large Cloud is extremely dusty, the Small Cloud almost transparent. This proves that the ratio of the masses of interstellar gas and dust can differ widely for different stellar systems. For comparison we note that within our own stellar system a general correlation exists between interstellar gas and dust (Sec. 16).

In the external portions of the Magellanic Clouds, where the total intensity $I < 10 \times 10^{-16}$ w m^{-2} sterad^{-1}, we are evidently dealing with a highly rarefied spherical corona. The number of hydrogen atoms in a column of unit cross section, for $I = 7 \times 10^{-16}$ w m^{-2}

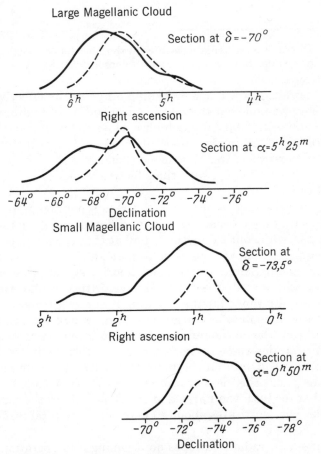

Fig. 195. Distribution of radio and optical brightness along two sections in each of the Magellanic Clouds (Kerr, Hindman, and Robinson 1954).

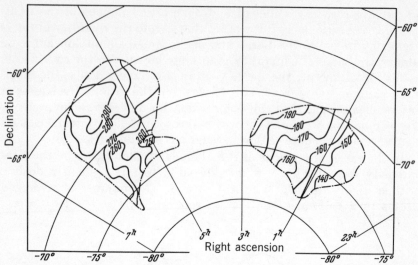

Fig. 196. Distribution of radial velocities in the Magellanic Clouds, corrected for galactic rotation (Kerr, Hindman, and Robinson 1954).

sterad^{-1}, is $N \approx 4 \times 10^{20}$ cm^{-2}. Taking 5 kpc as the effective length of a column, we find that the mean concentration of hydrogen atoms in the corona is ≈ 0.03 cm^{-3}, which is close to the estimates of the concentration of gas in the spherical corona of our own Galaxy. For the central portions of the Magellanic Clouds the mean concentration is found to be ≈ 0.7 cm^{-3}, with the effective thickness of the galaxies assumed to be 1 kpc.

It is extremely important that with the new method more than 200 radial velocities have been obtained for different points in the Magellanic Clouds, while optical methods had at that time yielded only 18. This circumstance now enables us to make a thorough study of the kinematics and dynamics of these neighboring stellar systems. Figure 196 charts the distribution of radial velocities in the Magellanic Clouds after allowance for the rotation of our own Galaxy. Analysis of this material enables us to conclude that each of the galaxies is in rotation. From the characteristics of the rotation, we may derive in elementary fashion the *total* mass for each of these galaxies. We have just seen that the data on the monochromatic radio emission from the Clouds yields the mass of interstellar gas. This leads to the conclusion that for both the Large and the Small Cloud the mass of the interstellar gas constitutes about 20 percent of the total mass of each galaxy.

The large new radio telescopes now coming into operation make it possible for us to investigate the monochromatic radio emission from galaxies more distant than the Magellanic Clouds.

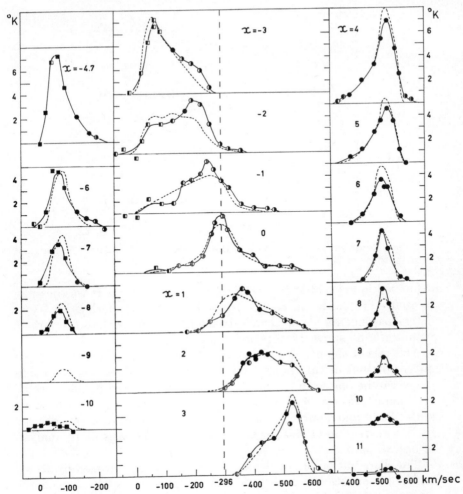

Fig. 197. Profiles of the hydrogen radio line at a number of points along the major axis of M 31 (van de Hulst, Raimond, and van Woerden 1957). Broken curves give profiles computed for a model galaxy.

Van de Hulst, Raimond, and van Woerden (1957) have published the results of a study of the monochromatic emission from the Andromeda Nebula (M 31), the largest of the galaxies in the vicinity of our own stellar system. The observations were carried out with the new 25-m reflector of the radio-astronomy observatory at Dwingeloo in The Netherlands. The main lobe was quite narrow, about 0°.6. Both the observations and the reduction were handled with the exemplary care and accuracy for which the Dutch investigators are noted.

In Fig. 197 profiles of the hydrogen radio line are given for a number of regions in M 31 along its major axis. The coordinate x is

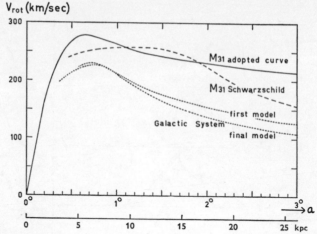

Fig. 198. Rotation laws for M 31 and the Galaxy (van de Hulst, Raimond, and van Woerden 1957).

expressed in units of $\frac{1}{4}°$ from the center of the galaxy. It was possible to secure profiles as far as 2°.5 from the center, on both sides. Typical measurements of the central frequency of the profiles for various values of x all show the rotation of M 31.

Analysis of the profiles yields a fairly reliable rotation law for the outer portions of this galaxy, but for the interior regions less confidence can be placed in the rotation law obtained. Figure 198 presents rotation curves for M 31 and for our own Galaxy, drawn to the same scale. The broken curve for M 31 represents the rotation law derived by Schwarzschild (1954) for that system on the basis of an analysis of optical data.

The distribution of the density of interstellar hydrogen projected on the major axis can be derived from a curve giving the dependence upon x of the profiles integrated over frequency (Fig. 199). This dis-

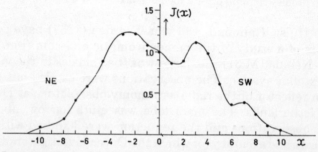

Fig. 199. Integrals of the observed profiles of the hydrogen radio line (van de Hulst, Raimond, and van Woerden 1957).

tribution is given as Fig. 200, where on the same scale the corresponding distribution for our own Galaxy has also been included. Circles and triangles represent the results of the computations for the southwest and northeast sectors of M 31. The shaded graph gives the mean value of the density distribution for these two parts of the Andromeda Nebula.

Notice the sharp maximum in the projected density distribution of the hydrogen at $x = 4$, corresponding to a distance of 8.7 kpc from the center of M 31 if we take the galaxy to be $\approx$500 kpc away. There is a similar, though less sharply defined maximum in our own Galaxy at a distance of about 6.5 kpc from the center.

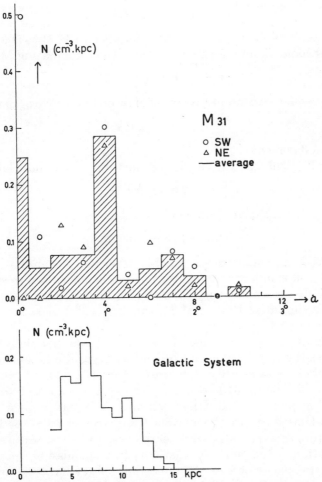

Fig. 200. Distribution of interstellar hydrogen in M 31 and the Galaxy (van de Hulst, Raimond, and van Woerden 1957).

From an analysis of the profiles of the hydrogen radio line in M 31, such as those given in Fig. 197, we may determine the component of the peculiar velocity of clouds of interstellar hydrogen along the line of sight. The mean value of this component is ≈ 8 km/s, approximately the same as in our own Galaxy.

On the basis of the radial-velocity observations just described, Schmidt (1957b) has constructed a model for the distribution of the total mass of M 31. The mass is found to be $3.4 \times 10^{11} \mathfrak{M}_\odot$. Schmidt has here assumed a distance of 630 kpc for the Andromeda Nebula.

It is of great interest to compare the two giant stellar systems, M 31 and the Galaxy. The mass of interstellar hydrogen in M 31, according to calculations by van de Hulst and his colleagues, amounts to

$$\mathfrak{M}_H = 0.25 \times 10^{10}\, c^2\, \mathfrak{M}_\odot,$$

where c denotes the ratio of the true distance to M 31 to 500 kpc. On the other hand, Schmidt's model gives as the total mass of M 31,

$$\mathfrak{M} = 26.8 \times 10^{10}\, c\, \mathfrak{M}_\odot.$$

Hence the relative content (by mass) of interstellar hydrogen is

$$\mathfrak{M}_H / \mathfrak{M} = 0.01\, c,$$

or 1.0 to 1.3 percent.

Schmidt (1956) has also calculated the total mass of our Galaxy:

$$\mathfrak{M} = 7.0 \times 10^{10}\, \mathfrak{M}_\odot;$$

the mass of interstellar hydrogen is

$$\mathfrak{M}_H = 0.15 \times 10^{10}\, \mathfrak{M}_\odot.$$

The mass of our Galaxy is less reliably determined than that of M 31, since the rotation law in the outer portion is not known to sufficient accuracy. If we assume that the rotation law in that portion of the Galaxy is similar to that in M 31, then the total mass increases to $10^{11}\, \mathfrak{M}_\odot$.

At any rate, the total mass of the Andromeda Nebula is three to four times the mass of the Galaxy, and its mass of interstellar hydrogen is about twice as great. The result that the mass of interstellar hydrogen in both M 31 and in the Galaxy is only 1 to 2 percent of the total masses of these stellar systems is of great interest.

In the United States, the new 60-ft radio telescope of the Harvard Observatory has recently been employed to initiate a systematic investigation of the monochromatic hydrogen emission from a number of extragalactic objects. An important survey has been completed by Heeschen (1957a), who has investigated several nearby galaxies. These include the elliptical system M 32, one of the companions of

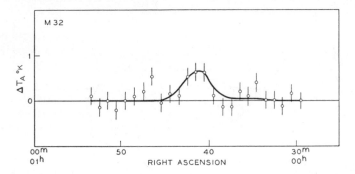

Fig. 201. The mean of 20 drift curves across M 32 (Heeschen 1957a).

the Andromeda Nebula; the Sc spiral M 51; and the Sb spiral M 81. The beamwidth of the antenna was 49′. Following the customary frequency-modulation methods, two frequency bands were used, separated by 3 Mc/s. Normally the antenna was held stationary. As the galaxies under investigation passed through the main lobe, the difference ΔT_a in antenna temperature was measured as a function of right ascension for each frequency. Analogous observations were then made at another declination. The bandwidth of the receiver corresponded to $\Delta v = 42$ km/s; the accuracy of a single tracing averaged about ± 1 K deg.

Figure 201 gives the mean of 20 of these drift curves across M 32 at a frequency corresponding to a radial velocity of -214 km/s. Since this velocity is equal to the radial velocity of M 32 with respect to the sun, ΔT_a must reach a maximum at this frequency. However, the radial velocity of that part of the great Andromeda Nebula which is adjacent to M 32 is approximately -300 km/s, so that it is possible to separate the line emission of M 32 from that of M 31.

In Fig. 202 are reproduced the averaged results of observations of the well-known spiral galaxy M 51 in Canes Venatici. The upper curve gives the mean profile of the radio line, while the lower gives the dependence of ΔT_a on right ascension at a frequency corresponding to a radial velocity of $+421$ km/s.

Finally, the results of the M 81 observations are given in Fig. 203. Isophotes of the monochromatic radio emission from this galaxy are shown in Fig. 204, where the small shaded area represents the portion of the galaxy observable optically. The region in which hydrogen is localized is of remarkably great extent: it far exceeds the visual portion of M 81. We have already met a similar situation in the case of the Magellanic Clouds. Evidently this is a widespread phenomenon in the realm of the galaxies.

On the basis of his observations Heeschen also derived, by the

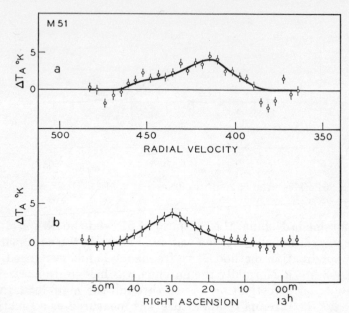

Fig. 202. The mean profile of the hydrogen radio line in M 51 (top), and the mean of six drift curves across M 51 (Heeschen 1957a).

usual procedure, the mass of interstellar hydrogen in the galaxies investigated by him.

The plane of symmetry of M 51 is almost perpendicular to the line of sight. If we take 1200 kpc as the distance to this galaxy, and 2° as the angular size of the region within which the monochromatic radio emission is found (the optical dimensions of this galaxy are 14′ × 10′), we find that the diameter of the system of interstellar gas in M 51 is about 40 kpc. The extent of this system along the line of sight is not known. Both M 81 and M 51 are thus surrounded by a gigantic gaseous corona, analogous to that in the Magellanic Clouds.

The Dutch observations have not yet revealed the corresponding phenomena in M 31 and in the Galaxy; the research program at Dwingeloo has been devoted to other problems. It is possible that, with special preparations, gigantic gaseous coronas could in fact be observed surrounding these systems. We should keep in mind that the gas density in the corona of our Galaxy and, we may suppose, in the corona of M 31 as well, is considerably lower than in M 51. For example, according to Sec. 23, $n_H \approx 0.01$ in the halo. If we also take into account the greater velocity dispersion of the gas clouds in the halo, we may expect that the brightness temperature resulting from the interstellar hydrogen in the halo of our Galaxy (and also in M 31) can hardly exceed 1 or 2°K. Only specially arranged observations with

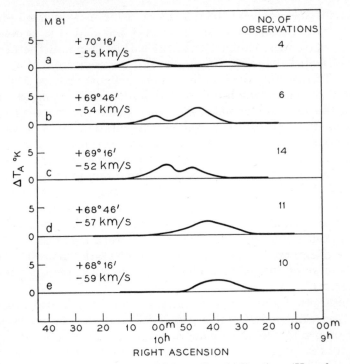

Fig. 203. Mean drift curves across M 81 at five declinations (Heeschen 1957a).

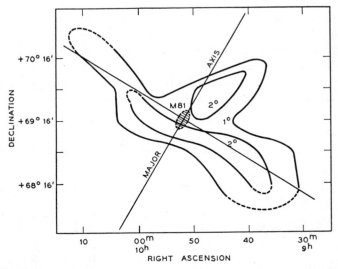

Fig. 204. Contours of equal antenna temperature in the 21-cm line for the region of M 81 (Heeschen 1957a).

high-sensitivity receivers (such as those using molecular amplifiers) could enable a hydrogen halo in the Galaxy and in M 31 to be detected.

At Harvard, Mrs. Dieter (1957, 1958; Heeschen and Dieter 1958) has investigated the monochromatic radio emission from M 33, an Sc spiral in Triangulum. Just as for the Magellanic Clouds, M 51, and M 81, the region in which interstellar hydrogen is localized greatly exceeds the optical dimensions of this galaxy. In particular, this affords us an opportunity to follow the rotation of M 33 to a far greater distance from the center than formerly. From an analysis of the rotation of M 33, Mrs. Dieter has derived the value $1.6 \times 10^{10}\mathfrak{M}_\odot$ for the mass of this system; this is considerably in excess of earlier estimates. The total mass of interstellar hydrogen in the galaxy was found to be $2.8 \times 10^9\mathfrak{M}_\odot$, or 15 percent of the entire mass.

A very important result of this investigation is that, in addition to the interstellar hydrogen partaking in the rotation of M 33, a large quantity of interstellar gas was found to be stationary with respect to the center of the galaxy. This gas forms an extended source of diameter $\approx 1°$.

The line profiles of this stationary gas have a half-width of ≈ 50 km/s. The total mass of this gas is $5.2 \times 10^8\mathfrak{M}_\odot$. If we assume a spherical form for this stationary gas, then its mean concentration is no greater than 9×10^{-2} cm^{-3}. In all its characteristics the stationary gas appears identical with a gaseous corona or "hydrogen halo" in M 33.

The demonstration by Mrs. Dieter of a very small velocity of galactic rotation in the halo is an especially important finding. This result must have great cosmogonic significance.

At the present time, then, data on the monochromatic radio emission of several galaxies have already been secured. The crucial point is that the observations yield very different results for these galaxies.

Table 26 summarizes the new information on masses. For M 31, a distance of 500 kpc has been adopted. M 81 has been excluded from this list because its radial velocity is so small (≈ 45 km/s) that the

TABLE 26. Masses of galaxies.

Galaxy	Type	$\mathfrak{M}_H$ $(10^9\mathfrak{M}_\odot)$	$\mathfrak{M}$ $(10^9\mathfrak{M}_\odot)$	$\mathfrak{M}_H/\mathfrak{M}$
M 51	Sc	4.6	—	—
M 31	Sb	2.5	270	0.01
Galaxy	Sb	1.5	100	0.015
M 32	E2	0.008–0.04	25	0.0003–0.0015
M 33	Sc	2.8	16	0.15
Large Mag. Cl.	Ir	0.6	3	0.2
Small Mag. Cl.	Ir	0.4	2	0.2

measurements may be heavily influenced by hydrogen emission from our own Galaxy.

It is clear from Table 26 that the relative abundance of interstellar gas can vary within very wide limits. Elliptical galaxies are found to be severely wanting in interstellar gas. There is a dependence of relative abundance of interstellar gas on the class of galaxy. The interstellar gas content of Sb systems is considerably more meager than for Sc systems, as optical observations also indicate. Sc galaxies are prolific in population I stars, genetically related to the interstellar gas. There are many associations of stars and H II emission regions. This holds even more strongly for galaxies of the Magellanic-Cloud type. The systematic variation of $\mathfrak{M}_H/\mathfrak{M}$ with the type of galaxy is undoubtedly related closely to the evolution of the galaxies. However, in order to analyze this very difficult and critical problem it is necessary to bring to bear far more extensive observational material than is now at hand.

The radio-line emission from a number of galaxies can be observed with the 250-ft reflector. At $\lambda = 21$ cm the width of the directional diagram of this reflector is about $10'$; the maximum sensitivity of the receiving apparatus is of the order of $\Delta T = 1$ K deg, which is readily attained with a bandwidth of 100 to 200 kc/s. If we assume that the brightness temperature of the radiation from a galaxy is $\approx 10°$K, we find that the minimum angular dimensions of galaxies (more precisely, of their gaseous coronas) which can be detected are $\approx 3'$. If the effective linear size of the corona for a galaxy such as M 33 is taken to be ≈ 10 kpc, then the most remote of these galaxies observable will lie at a distance of ≈ 10 Mpc. There are not a few such objects. In fact, it is entirely possible that galaxies will be detected which are anomalously abundant in interstellar gas, or even extragalactic objects of a new character, devoid or almost devoid of stars and composed only of gas. But this is still unexplored territory.

We have emphasized that there are weighty reasons for believing that the Large and Small Magellanic Clouds are surrounded by a general gaseous corona. At $\lambda = 21$ cm, the Clouds occupy a region in the sky extending for about $35°$, which corresponds to 30 kpc at the distance of these galaxies. It is natural to suppose that this rarefied corona also stretches in the direction of our Galaxy, a distance of 46 kpc. We might regard all three galaxies—the Magellanic Clouds and our own system—as enveloped by one common gaseous corona.

This introduces the possibility of observing intergalactic gaseous matter, a problem of fundamental significance. Although a common isophote surrounding both Clouds has not yet been well observed, its existence is highly probable. Let us take the brightness temperature along such an isophote of monochromatic emission to be $1°$K; then by formulae given earlier the number of hydrogen atoms in a column of unit cross section will be $N = 6 \times 10^{19}$ cm^{-3}. For a column of length 50 kpc $= 1.5 \times 10^{23}$ cm, we find that the mean concentration of inter-galactic hydrogen surrounding the Galaxy and the Magellanic Clouds will be $\approx 4 \times 10^{-4}$ cm^{-3}.

There are, then, strong grounds for believing that the space between the galaxies is filled with neutral hydrogen. On the average its density must, of course, be less than in the neighborhood of our Galaxy and the Magellanic Clouds. The greatest concentrations of the intergalactic gas will be in the region of clusters of galaxies. In this connection, we mention that according to Baade and Spitzer (1951) collisions between the galaxies in clusters must, to a considerable extent, have swept out interstellar gas into intergalactic space. It was assumed that in an average cluster of galaxies each galaxy will suffer at least one collision over a period of $\approx 10^{9}$ years which tends to deprive it of gas. As a result, we would expect gas to be present in the space between the galaxies, within the cluster. On the other hand, clusters of galaxies may contain "primitive" intergalactic gas, present since the "birth" of the galaxies.

Consequently we may well expect clusters of galaxies to be sources of monochromatic hydrogen radio emission. This proposal was put forward in 1954 by the author (see, for example, Shklovsky 1955b). Other authors had independently arrived at the same conclusion (Stone 1955). As a concrete example, we shall consider the great cluster of galaxies in Coma Berenices, which occupies a $4°$ region in the sky. It is 25 Mpc away from us, and is composed of about 1000 galaxies of 16 to 17^{m}. The radial velocity of the cluster, obtained from its red shift, is well known and is equal to 6680 km/s, with a dispersion in the velocities of individual galaxies in the cluster amounting to 800 km/s. The total mass of the cluster, $3 \times 10^{14} \mathfrak{M}_{\odot}$, is estimated from the velocity dispersion by means of the virial theorem. A notable feature is the anomalously high mass/luminosity ratio for this cluster: $\mathfrak{M}/L \approx 1000$, a value much greater than for other galaxies of very different types. May so anomalous a ratio be explained by a concentration of the basic mass of this cluster in intergalactic neutral hydrogen, which is not luminous because of the absence of sources of excitation, such as hot stars?

A natural assumption is that the mass of intergalactic hydrogen in the Coma cluster is of the same order as the mass of the entire cluster, namely $\mathfrak{M}_{\text{gas}} \approx 10^{14} \mathfrak{M}_{\odot} = 2 \times 10^{47}$ gm, corresponding to a population of hydrogen atoms of $N_{\text{H}} \approx 10^{71}$. This cluster is well concentrated toward its center; we shall take $\approx 1°$ as its effective angular size. Then the radius R of the cluster is $\approx 7 \times 10^{23}$ cm, and its volume $V = 4\pi R^{3}/3 \approx 1.4 \times 10^{72}$ cm^{3}, whence the mean concentration of inter-galactic hydrogen atoms is 7×10^{-2} cm^{-3}. There are $\approx 10^{23}$ atoms in a column of unit cross section passing through the cluster. It should be kept in mind that the velocity dispersion of intergalactic atoms will be ≈ 800 km/s. Entering this value in the formulae of Sec. 14, we find that the brightness temperature of the cluster at $\lambda = 21$ cm will be $\approx 50°$K, assuming that the kinetic temperature of intergalactic hydrogen is $\approx 100°$K.

This estimate provides an upper limit on the brightness temperature of the monochromatic radio emission from the Coma cluster of galaxies. The brightness temperature may actually be a dozen times as small. For example, if the volume occupied by the intergalactic hydrogen is 4 or 5 times as great as that assumed, the brightness temperature will be 16 to 25 times as small as we have computed. It is similarly possible that the mass of intergalactic hydrogen in this cluster of galaxies is less than we have adopted. But even in this event the sensitivity of modern apparatus would allow us to set the problem of detecting radio-line emission from the Coma cluster. Note that extremely large antennas are not required for such observations, since the angular dimensions of the cluster are quite large.

In 1956 the detection of hydrogen radio emission from this cluster of galaxies in Coma Berenices was reported by Heeschen (1956), at the Agassiz Station of the Harvard Observatory. The peak emission was located at the frequency 1386.8 Mc/s. The receiver bandwidth was 0.2 Mc/s, the maximum antenna temperature $2°$K, the angular dimension of the radio source about $5°$, the beamwidth of the antenna $1°.7$. A 500-km/s velocity dispersion was observed in the radiating gas. Assuming no self-absorption, the mass of hydrogen in the cluster was estimated in the usual way, and was found to be $\approx 10^{14} \mathfrak{M}_{\odot}$.

The same author has subsequently employed the new 60-ft Harvard reflector to observe monochromatic radio emission from two other clusters of galaxies, in Hercules and Corona Borealis (Heeschen 1957b). The angular diameter measured for the Hercules cluster of galaxies was $4°.5$. From the radio observations the radial velocity was 10 040 km/s; optical observations yielded 10 277 km/s. The Corona cluster of galaxies has an angular size less than $1°$, and a radial velocity of 21 024 km/s (optically, 21 530 km/s).

The 21-cm radio line from clusters of galaxies must be strongly shifted toward low frequencies because of the great velocities of recession which prevail in the Metagalaxy. For example, in the Coma Berenices cluster of galaxies, this shift will be equal to 31.6 Mc/s for the hydrogen line. Thus the center of gravity of the radio line will not be at 1420.4 Mc/s, but at 1388.8 Mc/s. The shift will assume greater values for more distant clusters of galaxies. For example, if a cluster of galaxies identical in character to the Coma cluster were at a distance so great that its angular dimension was $10'$, the beamwidth of the

250-ft reflector at $\lambda = 21$ cm, then the shift would amount to 200 Mc/s, or even more.

Since the expected velocity dispersion for the intergalactic gas must be of the same order of magnitude as the velocity dispersion of the galaxies in a cluster, the width of the radio line from clusters of galaxies must be very great. If, for example, the dispersion in the velocities of the Coma galaxies and gas is 800 km/s, the Doppler half-width of the hydrogen radio line from the cluster will be about 4 Mc/s.

For observations of this type, the receivers used for the study of monochromatic galactic radio emission are not suitable. In the first place, such receivers are not capable of a large shift in their transmission band; they are designed only to handle shifts (1 or 2 Mc/s) arising from the motions of hydrogen clouds within our Galaxy. Secondly, their bandwidth is too narrow, since this is again dictated by the specific requirements of observation of galactic clouds of interstellar hydrogen. In order to observe monochromatic radio emission from clusters of galaxies, it is necessary to exploit specialized receiving apparatus which can accomplish a shift of the passband by tens of megacycles per second toward lower frequencies from the basic frequency of 1420.4 Mc/s. A wide-band receiver is required. This latter circumstance materially raises the sensitivity of the apparatus, an especially valuable result. For example, if the most sensitive of modern receivers has a sensitivity $\Delta T = 3$ K deg for a 5-kc/s bandwidth at the hydrogen radio line, then a 2-Mc/s bandwidth will imply a theoretical sensitivity as great as 0.15 K deg. As molecular amplifiers come into application, the sensitivity of the receiving apparatus will increase by at least an order of magnitude beyond this.

At the present time a receiver is being developed at the Sternberg Astronomical Institute in the U.S.S.R., designed for the detection of extragalactic monochromatic radio emission. The receiver will enable a 40-Mc/s shift in the passband to be realized.

Let us turn now to the question whether hydrogen radio emission may be detected from the Metagalaxy as a whole. Although the mean density of the gas over all intergalactic space is extremely small, the tremendous extent of the Metagalaxy leads us to expect, in principle, some observable effect. If the red shift were absent, then the 21-cm radio line would necessarily be observable, since the number of hydrogen atoms could be rendered sufficient by taking a unit column extending as far as desired. However, the red shift will tend to smear out the hydrogen radio line into a very broad band of exceedingly low intensity. This band must extend many hundreds of megacycles per second from 1420.4 Mc/s toward lower frequencies.

In a fixed frequency interval from ν to $\nu + d\nu$ radiation will be received from intergalactic hydrogen atoms at a distance from us between R and $R + \Delta R$, where, according to the red shift law, $dv \propto \Delta R$. With the current value of the red-shift constant, $k = 150$ km sec^{-1} Mpc^{-1}, we find that the number of intergalactic hydrogen atoms in a 1-Mpc column of unit cross section will be

$$N_H = 3 \times 10^{24} \, n_H \, \text{cm}^{-2},$$

where n_H is the concentration of atoms. The radiation from this column will be spread over a bandwidth corresponding to $\Delta v = 150$ km/s, or $\Delta\nu \approx 0.75$ Mc/s. If we suppose that $n_H \approx 3 \times 10^{-5}$ and the kinetic temperature is $\approx 100°$K, the formulas of Sec. 14 imply a brightness temperature of $\approx 0°.25$. Radiation from the more distant parts of intergalactic space will be found at lower frequencies. To observe the integrated effect, it is necessary to employ a receiver of maximal bandwidth, scores of megacycles per second; this may, for example, be attained with a traveling-wave tube.

In connection with this general problem we should also note the following circumstance. We have assumed that the intensity of the intergalactic radio line (both from clusters of galaxies and from the entire Metagalaxy) is proportional to the number of hydrogen atoms in a column of unit cross section, namely $\int n_H ds$. This is valid only when the number of hydrogen atoms in the initial excited level of the hyperfine structure of the ground state is proportional to the total number of atoms (Boltzmann distribution, Sec. 14). The condition will be fulfilled provided that the frequency of the collisions populating the initial excited hyperfine level is greater than the probability of spontaneous downward cascades:

$$\nu_{coll} \approx n_H v\sigma > A_{21} = 2.85 \times 10^{-15} \, \text{sec}^{-1}.$$

In the Galaxy, this condition is always satisfied. The effective cross section for collisions accompanied by exchange of electrons of the colliding atoms must be quite large (Sec. 14). We shall adopt $\sigma = 5 \times 10^{-15}$ cm^2. Let us further suppose that $v = 10^5$ cm/s. Then the condition formulated above will be fulfilled if $n_H > 0.6 \times 10^{-5}$, which holds for all cases of monochromatic extragalactic radio emission that we have considered. If n_H is less than some limiting value, the intensity of the radio emission will not be proportional to $\int n_H ds$, but to $\int n_H{}^2 ds$, since each unit of radiation will arise in the collision of two hydrogen atoms. In this event the intensity of the monochromatic radio emission will *fall off* with decreasing n_H. It is not precluded that this "unfavorable" situation is the very one which is realized in the case of monochromatic intergalactic emission. The answer will be given by observations which will probably be undertaken before long.

We shall consider one further circumstance, of a rather curious character. Along with radiative transitions between the components of the hyperfine structure leading to the emission of a photon with frequency $\nu = 1420.4$ Mc/s, we may also expect those radiative transitions in which two photons are emitted, the sum of whose frequencies is equal to 1420.4 Mc/s. As Kipper has shown, two-photon transitions of this type between the 2^2S and 1^2S levels are responsible for the continuous spectrum of ionized gaseous nebulae. The probability of this forbidden two-photon process must in any case be comparable with the probability of the transition (also forbidden) which gives rise to the hydrogen radio line. The radiation excited will form a continuous spectrum extending toward the low-frequency side of 1420.4 Mc/s, and attaining a maximum several hundred megacycles per second away from the radio line at 1420.4 Mc/s. For galactic radio emission, this effect is quite negligible. It can readily be shown that the brightness temperature in the direction of the galactic center will be increased by barely 0.1 K deg at $\nu = 1420$ Mc/s. However, under intergalactic conditions this radiation may in principle yield an intensity comparable with the intensity of the 21-cm radio line. In fact, the red shift also tends to spread the radio line out into a wide frequency band, as we have seen. In a given frequency interval, radio-line emission will therefore appear which has arisen in the limited layer of extragalactic space between R and $R + \Delta R$, together with two-photon radio emission from the vast extragalactic domain $R_2 - R_1$, where $R_2 - R_1 \gg \Delta R$. This circumstance must be kept in mind in all succeeding investigations. It is therefore of interest to carry out a computation for such a two-photon process. In the U.S.S.R., this radiation has recently been calculated; it now appears that the probability of the two-photon process is in fact very low.

In general, one factor interfering with the study of monochromatic extragalactic radio emission is the background of continuous radio emission from our Galaxy. Galactic emission is mainly thermal in this frequency range. On the basis of the data available, we may assume a brightness temperature for the Galaxy of roughly $1°K$ at $\nu = 1420$ Mc/s in the continuum, and at high galactic latitudes. However, in observations with frequency-modulation techniques the relevant quantity is not the absolute value of the brightness temperature of the background, but its difference ΔT for sufficiently close frequencies. The latter quantity, as may easily be seen, will be proportional to $\Delta \nu / \nu$, where $\Delta \nu$ is the difference in frequencies. It follows that ΔT will be a small quantity if $\Delta \nu$ is not too great. Even then, however, the galactic background should be taken into account.

Extragalactic hydrogen can be observed not only in emission, but also in absorption. The first work in this direction was carried out by Lilley and McClain (1956) with the 50-ft reflector of the Naval Research Laboratory in Washington. The idea behind these observations is the following. If hydrogen exists in the neighborhood of the collid-ing galaxies in Cygnus, then a band of absorption would be anticipated in the radio spectrum of the Cygnus A source. In Sec. 17 we have described the results of observations of an absorption band in this source caused by *galactic* hydrogen. If the red shift is a Doppler effect, the absorption band due to *extragalactic* hydrogen must be strongly shifted toward the low-frequency end of the spectrum. Optical obser-vations yield a velocity of recession for the colliding galaxies of $\approx 17\,000$ km/s, so that we would expect a shift in the absorption band of ≈ 80 Mc/s from the normal position at 1420.4 Mc/s.

Lilley and McClain's observations were conducted as follows. The antenna temperatures of Cygnus A and Cassiopeia A were alternately observed at several frequencies between 1334 and 1348 Mc/s. Observa-tions at wavelengths of 9.4 and 21 cm had indicated that the energy distribution in the spectra of these sources was almost the same over the range from 9.4 to 21 cm. Therefore a measurement of the ratio of the antenna temperatures of the two sources for various frequencies would, in the *absence* of the absorption band in the Cygnus A spec-trum, yield an almost constant value; at most, this quantity would vary slowly and monotonically. But if an absorption band were pres-ent in the spectrum of the Cygnus A source, it would at once be reflected in the ratio of the two antenna temperatures.

Figure 205 gives the results of a set of very careful measurements

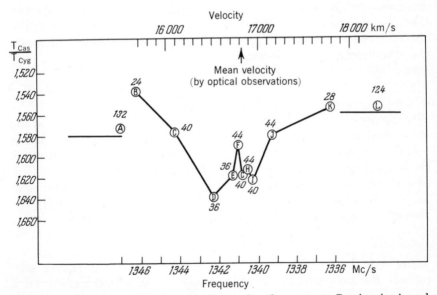

Fig. 205. Ratio of antenna temperatures for the sources Cassiopeia A and Cygnus A (Lilley and McClain 1956).

of the ratio of the antenna temperatures for the sources Cassiopeia A and Cygnus A, at a number of different frequencies. The numerals above the circles refer to the number of observations which were combined to yield the indicated ratios of antenna temperature. Circles *A* and *L,* and the corresponding horizontal lines, give the mean value for the ratio of antenna temperatures in the spectral intervals 1348 to 1358 Mc/s and 1324 to 1334 Mc/s.

It is evident from Fig. 205 that in the frequency interval 1336 to 1346 Mc/s the ratio of the antenna temperature of Cassiopeia A to that of Cygnus A is *greater* than on either side of this interval. It follows that the source Cygnus A has an absorption line in this part of the spectrum. The depth of the absorption is 6.2 percent. Since for so small an absorption the percentage by which the radiation flux is diminished is practically equal to the optical depth, we conclude that $\tau = 0.06$ or 0.07. The weighted mean of the absorption reaches a maximum at $\nu = 1341$ Mc/s, corresponding to a velocity of recession of 16 700 km/s. This value practically coincides with the mean velocity of recession derived from optical observations. The half-width of the absorption band corresponds to a dispersion in radial velocity of 850 km/s.

The above result demonstrates that the red-shift ratio $\Delta\nu/\nu$ remains constant over the immense wavelength range from 4000 A to 21 cm. This is equivalent to a change in wavelength of 500 000 times, or a spectral interval of 19 octaves. When we discussed the problem of the integrated brightness from extragalactic space in Sec. 26, we showed that the red shift must also take place in the radio range. But the observations just described provide a more convincing demonstration that the fundamental relation

$$\Delta\nu/\nu = \Delta\lambda/\lambda = \text{const}$$

is valid over a vast gamut of electromagnetic waves.

If the optical depth of intergalactic hydrogen reaches 0.06, and the velocity dispersion for the internal motions of the absorbing gases is ≈ 850 km/s, then by assigning a kinetic temperature we can determine the mass of the absorbing hydrogen. Let us take the kinetic temperature of intergalactic hydrogen to be $\approx 100°$K, about the same as in galactic H I clouds. Then we obtain the following expression for the absorbing mass of intergalactic gas, in terms of the solar mass:

$$\mathfrak{M} = 7 \times 10^7 A \, \mathfrak{M}_\odot,$$

where A is the surface area of the Cygnus A source in square kiloparsecs. For a distance of 100 Mpc to this source, with dimensions and form taken from the interferometric observations by Jennison and Das Gupta (see Sec. 7), we find $A = 625$ kpc², whence $\mathfrak{M} = 4 \times 10^{10} \mathfrak{M}_\odot$.

If we consider that roughly the same quantity of intergalactic hydrogen also lies beyond the source, then the mass of hydrogen will be $\approx 10^{11} \mathfrak{M}_\odot \approx 10^{44}$ gm.

Two hypotheses may be made as to the nature of the absorbing intergalactic hydrogen: (a) it is found in the neighborhood of colliding galaxies; (b) it is localized in intergalactic space in the neighborhood of clusters of galaxies, such as the cluster of which the Cygnus A source is a member.

The former possibility does not seem to us to be a probable one. To begin with, it is difficult to imagine that interstellar hydrogen could remain neutral during so highly effective a collision of galaxies. Optical observations show that in these colliding galaxies the interstellar medium is strongly ionized (Sec. 8). Furthermore, the mass of interstellar hydrogen in these galaxies would in any event be too great. Finally —and this is perhaps the most important argument—there is no clear indication in Fig. 205 that the absorption is produced by two masses of gas moving at a high velocity with respect to each other. In fact, according to the optical observations the relative radial velocity of the colliding galaxies is ≈ 3000 km/s. But if the two minima in the curve drawn in Fig. 205 referred to the two galaxies, then their relative radial velocity would be only ≈ 400 km/s, a value far less than that derived optically.

If we regard the absorption as produced by intergalactic hydrogen associated with the cluster, and if we adopt 1 Mpc as a diameter of the cluster, $T \approx 100°$ K, and the velocity dispersion ≈ 850 km/s, then we find that the mean concentration of intergalactic hydrogen in the cluster is $\approx 5 \times 10^{-3}$ cm^{-3}. And if the kinetic temperature of the intergalactic gas is significantly less than $100°$K, as is entirely possible, the mean concentration of hydrogen atoms will be correspondingly lower.

Further observations will undoubtedly assist us in clarifying the nature of the intergalactic hydrogen responsible for the absorption in the Cygnus A spectrum.

Radio methods have opened such exceptional perspectives for us in the investigation of extragalactic space that we eagerly look forward to continued efforts to survey the radio-line emission from additional extragalactic objects. Along with the study of extragalactic radio emission at meter wavelengths from clusters of galaxies and from radio galaxies, the analysis of monochromatic emission enriches cosmology with much of the basic new observational data which it so acutely needs.

Appendix

This volume endeavors to reflect the progress in the field up to the 1958 Paris symposium on radio astronomy. Dr. Shklovsky has, however, expressed the desire that attention be called to selected papers published since that time. We are summarizing these briefly here, keying each note to the corresponding page of the text.

R. B. R.

April 1960

Pages 105 and **127.**   Hindman and Wade (1959) have mapped the brightness distribution of Centaurus A at $\nu = 1400$ Mc/s. They confirm the position of the bright central nucleus (identified with the peculiar E0 galaxy NGC 5128), but merely report it as a point source, contributing 23 percent of the total flux of 13×10^{-24} w m^{-2} (c/s)$^{-1}$. The elongated component extends for 7° in declination, and exhibits a maximum 2° south of the nucleus. In agreement with Sheridan (1958), Shain (1958b) models the object as an open spiral nebula, with the internal magnetic-field distribution accounting for the spectral dependence of the contours of radio luminosity. The role of the dark equatorial band is still open to speculation.

Page 166.   Hindman and Wade (1959) have also observed the η Carinae nebula at $\lambda \approx 21$ cm, finding a brightness distribution even more symmetrical than that observed by Mills, Little, and Sheridan (1956c) at 3.5 m (Fig. 114); there is a strong central concentration.

Page 167.   The extraordinary emission nebulosity in the Vela–Puppis sector of the Milky Way has been surveyed by Rishbeth (1958). Both Sydney Mills crosses were used—one designed for 3.5 m (Mills *et al.* 1958), the other for 15.2 m (Shain 1958c); at the latter wavelength, terrestrial and ionospheric effects complicate the technique. Rishbeth presents contour maps of T_a for both wavelengths, and tabulates T_a and the flux density for some 90 discrete sources distributed over the Vela–Puppis region. Of those sources observed at the longer wavelength, some represent discrete *minima* in the galactic background.

Major optical H II regions, especially that excited by γ Vel, appear in emission and absorption at 3.5 and 15 m respectively; however, the correspondence between the principal optical and radio features is not complete, presumably because of interstellar absorption. Emission measures for the H II zones are consistent with thermal processes. But three intense nonthermal sources, several degrees in diameter, have been found at $l \approx 232°$, $b \approx -1°$; these are apparently unrelated to the H II regions. At 15 m, $T_a > 200\,000°$K; since dark nebulosity is present, these extremely high temperatures suggest emission by high-velocity particles in magnetic fields.

Nonthermal emission is also intense along the galactic equator, with T_b following the law $T_b \propto \nu^{-2.5}$, and attaining $\approx 100\,000°$K at 15 m. The gradients of the background T_b seem to correlate with the known spiral structure, in view of the path lengths through the spiral arms in the appropriate directions.

Page 232. At Sydney, the 21-cm survey of the southern galactic equator has now been completed by Kerr, Hindman, and Gum (1959). Preliminary reduction of this material had yielded the data on hydrogen distribution incorporated into the left-hand sides of Figs. 139 and 140. These diagrams combine the Dutch survey for the northern Milky Way with the Australian results, which cover the range $175° < l < 365°$ in scans crossing the equator at 41 longitudes. A 36-ft paraboloid was employed (beamwidth, $1°.4$; bandwidth, 40 kc/s). For each crossing, a line profile is given, with T_b as a function of radial velocity. Contour diagrams were then prepared for the 41 scans to illustrate the additional dependence of T_b on the latitude b near the equator. These diagrams will form the basis of a study of large-scale galactic structure.

Page 253. At the National Academy of Sciences' 1959 Bloomington symposium on radio astronomy, Rougoor and Oort (1960) discussed further the observations of the inner Galaxy made at $\lambda = 21$ cm with the 25-m Dwingeloo paraboloid. They stress that the expanding arm between the sun and the galactic center, at $R \approx 3$ kpc, is remarkably homogeneous both in gas density and in its motion. This is demonstrated not only by the sharp emission and absorption profiles, but by the smooth variation of the approach velocity over the $\approx 25°$ longitude interval within which the arm is observable. The observations also suggest another structure, on the opposite side of the galactic center, whose mass and orientation are rather similar to the expanding 3-kpc arm. The form of this feature is less regular, but it too is probably expanding outward from the center (*away* from the sun). Both structures seem to lie entirely within the galactic plane (thickness between half-density layers, 120 pc). Contour diagrams of T_b in an $(l, \Delta v)$ system were prepared for several b.

Page 254. At 21 cm, Rougoor and Oort (1960) find the following detail at the nucleus itself. A dense core ($R < 100$ pc) is surrounded by a disk (hydrogen concentration 1 cm^{-3}) extending to $R \approx 300$ pc. The disk spins very rapidly, at ≈ 200 km/s. Beyond the disk, a well-defined ring (again with $n_H = 1$ cm^{-3}) appears from $R = 500$ to 590 pc, and rotates at 265 km/s. The absorption and emission line profiles reveal no expansions within $R = 600$ pc.

A tentative curve of rotational velocity as a function of R is derived by combining the foregoing results with normalized data for the Andromeda Nebula, and with other studies of the Galaxy. Rougoor and Oort then construct Table 27 on the basis of a purely gravitational explanation for the rotation; here T is the period of revolution. Notice especially the strong central condensation in the density ρ and the great preponderance of stars over gas.

High-frequency observations ($\lambda = 3.75$ cm) with the 85-ft radio telescope of the National Radio Astronomy Observatory have revealed a pair of intense thermal emission regions (presumably H II) between $R = 10$ and 20 pc, each with mass $\approx 50\,000$ $\mathfrak{M}_\odot$ (Drake 1959). A ring of nonthermal (synchrotron) emission, with spectral index 0.7, appears to envelop the thermal regions at $R = 90$ pc. In the opinion of the author of this volume, the absence of synchrotron radiation in the inner nucleus, $R < 80$ pc, is due to the high gas density there. As a result, collisional ionization leads to a quite rapid destruction of

TABLE 27. Mass distribution and rotation in the galactic nucleus.

R (pc)	$\log \mathfrak{M}$ ($\mathfrak{M}_\odot$)	$\log \rho$ (gm/cm^3)		$\log T$ (yr)
		stars	gas	
10	7.5	−18.6	−21	5.7
20	7.8	−19.1	−22	6.0
100	9.0	−20.1	−23.3	6.5
500	9.9	−21.4	−23.7	7.1
3000	10.2	−22.2	−24.0	8.0
8200	10.8	−23.0	−23.7	8.36

relativistic electrons. This accounts for the surrounding ring of nonthermal radio emission discovered by Drake.

It follows from Rougoor and Oort's discussion that the source Sagittarius A arises from an intensification of the interstellar magnetic field by a factor of ≈ 10 at the galactic nucleus. This is indicated by the rapid motions of the interstellar gas clouds found there, well in excess of 100 km/s.

Page 290. The author of this volume has investigated further the dependence of the flux and brightness of an expanding radio nebula on its radius r. It is now assumed that the magnetic field falls off with r according to the natural law $H \propto r^{-2}$; that is, the magnetic flux is conserved. In calculating the continuous decrease in the energy of relativistic particles in an expanding nebula under the assumption that the synchrotron mechanism is responsible for the radio emission (compare Sec. 22), we then find that

$$F_\nu \propto r^{-\beta}, \qquad I_\nu \propto r^{-(\beta+2)}, \tag{20-7}$$

where $\beta = 2\gamma$. These considerations supersede the proposals on page 327 for the behavior of F_ν and I_ν.

In particular, Eq. (20-7) will imply that when the filamentary nebulosity in Cygnus had the same radius as the Cassiopeia A source has now, its radio luminosity was ≈ 2000 times as great as now, close to the present radio luminosity of Cassiopeia A. This computation suggests at once that the Cygnus Loop and Cassiopeia A are objects of essentially the same character, but at different stages of evolution. They are the remnants of type II supernova explosions.

Another interesting consequence of Eq. (20-7) concerns the possibility of observing a secular decrease in the flux of radio waves from Cassiopeia A. If the age of this radio nebula is ≈ 260 yr, as follows from Minkowski's observations, then the relative annual attenuation in the radio flux is evidently equal to

$$-\Delta F_\nu / F_\nu = \beta/260 = 2 \text{ percent.}$$

It would be of great value to have an observational demonstration of this prediction of theory.

Page 352. The Andromeda Nebula has recently been surveyed with the 250-ft Jodrell Bank paraboloid at 158 and 237 Mc/s (Hanbury Brown and Hazard 1959) and at 408 Mc/s (Large *et al.* 1959*a*). The emission is resolved at the lower frequencies into two components: a disk, strongly concentrated to the galactic plane and corresponding roughly to the optical emission, and a large spheroidal corona. Along its major axis, the corona extends for fully $10°.5$, or over 110 kpc; its axial ratio, corrected for projection, is 0.57, about the same as that believed to obtain in the corona of our own Galaxy. The corona contributes 90 percent to the integrated radio magnitude $m_r = +5.8$ at 158 Mc/s; the radio index $m_r - m_{pg} = +1.5$. An analogous corona was observed in M 33.

Distortions of the corona's elliptical isophotes have been noted at 408 Mc/s. Extensive wings protrude at an inclination of $30°$ to the minor axis, perhaps reflecting the magnetic-field structure. At this frequency, the coronal contribution is 96 percent.

Page 355. In an attempt to verify Kraus and Ko's report (1953) of supergalactic radio emission, Hill (1958) has inspected tracings obtained at $\lambda = 3.5$ m with one of the large Sydney crosses. This antenna array, described recently by Mills *et al.* (1958), provided higher resolution than the earlier survey (beamwidth $\approx 1°$). After correcting for galactic and discrete-source radiation, Hill indeed found a band of emission in the region investigated by Kraus and Ko at 1.2 m. However, the band does not correspond to the area within the broken curve in Fig. 180. The feature extends from $\alpha = 11^h 30^m$, $\delta = -15°$ to $\alpha = 14^h$, $\delta = +22°$, but Fig. 180 shows no concentration of galaxies along this line. The new feature would presumably mask any radio emission from intergalactic space that may be present.

At Cambridge, Shakeshaft and Baldwin (1959) also find emission bands in this region, at $\lambda = 1.9$ m; but again the correlation with galaxy counts is poor.

Page 404. Radio emission from the Coma cluster has been observed at $\lambda = 75$ cm with the 25-m Dwingeloo paraboloid (Seeger, Westerhout, and Conway 1957); the total flux density $F_\nu = 11 \times 10^{-26}$ w m^{-2} (c/s)$^{-1}$. More recently, the 250-ft Jodrell Bank instrument has yielded $F_\nu = (26 \pm 6) \times 10^{-26}$ w m^{-2} (c/s)$^{-1}$ at $\lambda = 73.5$ cm (Large *et al.* 1959b). Three-fourths of this radiation arises in several peculiar discrete sources, two of which are $< 30'$ in diameter; thus the contribution of normal galaxies or intergalactic matter is relatively small. Isophotes of T_b are given.

On the other hand, Muller (1959) has failed to detect the 21-cm line in the Coma cluster. Observing with the Dwingeloo telescope at various frequencies near the red-shifted $\nu = 1389$ Mc/s, he finds no significant signal ΔT_a due to the cluster, and concludes that the maximum possible value for T_a at 21 cm is $0.05°$K. (It is assumed that the dispersions in the velocities of the galaxies and of any H I in the field are both 1000 km/s.) In a column of unit cross-sectional area, there are $< 6 \times 10^{20}$ H atoms. Furthermore, Oort has estimated the total mass of the cluster under the assumption that $\langle \mathfrak{M}/L \rangle = 50$. Using this value, Muller finds $\mathfrak{M}_H/\mathfrak{M} < 0.03$ for the cluster, an entirely acceptable ratio for *individual* galaxies (see Table 26), especially since most Coma galaxies are of types E and S0. If we retain the distance 25 Mpc for the cluster, the mean intergalactic hydrogen concentration seems to be $< 10^{-4}$ cm^{-3}.

References

1846   Biot, Eduard. "Catalogue des Comètes observées en Chine depuis l'an 1230 jusqu'à l'an 1640 de notre ère. Catalogue des Etoiles extraordinaires observées en Chine, depuis les temps anciens jusqu'à l'an 1203 de notre ère," *Connaissance des Temps* **1846**, 44–68.

1850   Humboldt, Alexander von. "Neu erschienene und verschwundene Sterne ...," *Kosmos* **3** (Cotta, Stuttgart), 150–186.

1871   Williams, John. *Observations of comets, from* B.C. *611 to* A.D. *1640. Extracted from the Chinese annals. Translated, with introductory remarks, and an appendix, comprising the tables necessary for reducing Chinese time to European reckoning; and a Chinese celestial atlas* (Strangeways and Walden, London), 34 + 124 pp. + atlas.

1891   Schönfeld, E. "Über den neuen Stern von 1006," *A.N.* **127**, 153–154.

1912   Schott, G. A. *Electromagnetic radiation and the mechanical reactions arising from it* ... (Cambridge University Press, Cambridge).

1918   Curtis, H. D. "Descriptions of 762 nebulae and clusters photographed with the Crossley reflector," *Pub. Lick Obs.* **13**, 9–42.

1919   Zinner, E. "Die neuen Sterne," *Sirius* **52**, 25–35, 127–134, 152–165.

1921   Lampland, C. O. "Observed changes in the structure of the 'Crab' Nebula (NGC 1952)," *Pub. A. S. P.* **33**, 79–84.

         Lundmark, K. "Suspected new stars recorded in old chronicles and among recent meridian observations," *Pub. A. S. P.* **33**, 225–238.

1929   Van Vleck, J. H. "On σ-type doubling and electron spin in the spectra of diatomic molecules," *Phys. Rev.* **33**, 467–506.

1931   Mulliken, R. S., and A. Christy. "Λ-type doubling and electronic configurations in diatomic molecules," *Phys. Rev.* **38**, 87–119.

1932   Jansky, K. G. "Directional studies of atmospherics at high frequencies," *Proc. I. R. E.* **20**, 1920–1932.

         Shapley, H., and A. Ames. "A survey of the external galaxies brighter than the thirteenth magnitude," *Harv. Ann.* **88**, 41–75.

1933   Bottlinger, K. F. "Beiträge zur Theorie der Rotation des Sternsystems," *Veröff. Berlin-Babelsberg* **10**, Nr. 2.

         Jansky, K. G. "Electrical disturbances apparently of extraterrestrial origin," *Proc. I. R. E.* **21**, 1387–1398.

         Swann, W. F. G. "A mechanism of acquirement of cosmic-ray energies by electrons," *Phys. Rev.* **43**, 217–220.

1934   Iba, Y. "Fragmentary notes on astronomy in Japan," *Pop. Astr.* **42**, 243–252.

1935   Condon, E. U., and G. H. Shortley. *The theory of atomic spectra* (Cambridge University Press, Cambridge).

         Jansky, K. G. "A note on the source of interstellar interference," *Proc. I. R. E.* **23**, 1158–1163.

1937   Menzel, D. H. "Physical processes in gaseous nebulae I. Absorption and emission of radiation," *Ap. J.* **85**, 330–339.

1938   Baker, J. G., and D. H. Menzel. "The Balmer decrement," *Ap. J.* **88**, 52–64.

1939   Alfvén, H. "Atomic nuclei in primary cosmic radiation," *Nature* **143**, 435.
        Herzberg, G. *Molecular spectra and molecular structure* 1, *Diatomic molecules,* trans. J. W. T. Spinks (Prentice-Hall, New York).

1940   Dunham, T. "The concentration of interstellar molecules," *Pub. A. A. S.* **10**, 123–124.
        Pasternack, S. "Transition probabilities of forbidden lines," *Ap. J.* **92**, 129–155.
        Reber, G. "Cosmic static," *Ap. J.* **91**, 621–624.
        Shapley, H. "An extension of the Small Magellanic Cloud," *Harv. Bull.* **914**, 8–9.

1942   Baade, W. "The Crab Nebula," *Ap. J.* **96**, 188–198.
        Mayall, N. U., and J. H. Oort. "Further data bearing on the identification of the Crab Nebula with the supernova of A.D. 1054," *Pub. A. S. P.* **54**, 95–104.
        Minkowski, R. "The Crab Nebula," *Ap. J.* **96**, 199–213.
        Zwicky, F. "On the frequency of supernovae II, *Ap. J.* **96**, 28–36.

1943   Baade, W. "Nova Ophiuchi of 1604 as a supernova," *Ap. J.* **97**, 119–127.
        Minkowski, R. "The spectrum of the nebulosity near Kepler's nova of 1604," *Ap. J.* **97**, 128–129.
        Seyfert, C. K. "Nuclear emission in spiral nebulae," *Ap. J.* **97**, 28–40.

1944   Reber, G. "Cosmic static," *Ap. J.* **100**, 279–287.

1945   Baade, W. "B Cassiopeiae as a supernova of type I," *Ap. J.* **102**, 309–317.
        Bakker, C. J., and H. C. van de Hulst. "Radiogolven uit het wereldruim," *Nederl. Tijds. v. Natuurkunde* **11**, 201–221. (Bakker, "Ontvangst [reception] der radiogolven"; van de Hulst, "Herkomst [origin] der radiogolven.")
        Barbier, D. "Le spectre continu de la Crab Nebula dans la région des courtes longueurs d'onde," *Ann. d'Ap.* **8**, 35–42.

1946   Artsimovich, L. A., and I. Ya. Pomeranchuk. "The radiation of fast electrons in a magnetic field" [in Russian], *Zhur. Eksp. Teor. Fiz.* **16**, 379–389.
        Dicke, R. H. "The measurement of thermal radiation at microwave frequencies," *Rev. Sci. Instruments* **17**, 268–275.
        Hey, J. S., S. J. Parsons, and J. W. Phillips. "Fluctuations in cosmic radiation at radio frequencies," *Nature* **158**, 234.
        Hey, J. S., J. W. Phillips, and S. J. Parsons. "Cosmic radiations at 5 metres wavelength," *Nature* **157**, 296–297.
        Moxon, L. A. "Variation of cosmic radiation with frequency," *Nature* **158**, 758–759.
        Oort, J. H. "Some phenomena connected with interstellar matter," *M. N.* **106**, 159–179.
        Sander, K. F. "Measurements of galactic noise at 60 Mc/s," *J. Inst. Elec. Eng.* **93**, 1487–1496.
        Shklovsky, I. S. "On the radiation of radio waves by the Galaxy and by the upper layers of the solar atmosphere" [in Russian], *Astr. Zhur. SSSR* **23**, 333–347.
        Terletsky, Ya. P. "On the induction of a flow of fast charged particles by rotating magnetized cosmic bodies" [in Russian], *Zhur. Eksp. Teor. Fiz.* **16**, 403–409.

1947   Nafe, J. E., E. B. Nelson, and I. I. Rabi. "The hyperfine structure of atomic hydrogen and deuterium," *Phys. Rev.* **71**, 914–916.

1947 Nagle, D. E., R. S. Julian, and J. R. Zacharias. "The hyperfine structure of atomic hydrogen and deuterium," *Phys. Rev.* **72,** 971.

Reber, G., and J. L. Greenstein. "Radio-frequency investigations of astronomical interest," *Observatory* **67,** 15–26.

1948 Bolton, J. G. "Discrete sources of galactic radio-frequency noise," *Nature* **162,** 141–142.

Bolton, J. G., and G. J. Stanley. "Variable source of radio-frequency radiation in the constellation of Cygnus," *Nature* **161,** 312–313.

Herbstreit, J. W., and J. R. Johler. "Frequency variation of the intensity of cosmic radio noise," *Nature* **161,** 515–516.

Ivanenko, D. I., and A. A. Sokolov. "On the theory of the 'luminous' electron" [in Russian], *Dokl. Akad. Nauk SSSR* **59,** 1551–1554.

Nafe, J. E., and E. B. Nelson. "The hyperfine structure of hydrogen and deuterium," *Phys. Rev.* **73,** 718–728.

Reber, G. "Cosmic static," *Proc. I. R. E.* **36,** 1215–1218.

Ryle, M., and F. G. Smith. "A new intense source of radio-frequency radiation in the constellation of Cassiopeia," *Nature* **162,** 462–463.

Spitzer, L., Jr. "The distribution of interstellar sodium," *Ap. J.* **108,** 276–309.

Terletsky, Ya. P. "Stars with magnetic moments as possible sources of cosmic rays" [in Russian], *Vestnik Mosk. Gos. Univ.* **3,** No. 1, 75–82.

Vladimirsky, V. V. "Influence of the terrestrial magnetic field on large Auger showers" [in Russian], *Zhur. Eksp. Teor. Fiz.* **18,** 392–401.

1949 Bolton, J. G., and G. J. Stanley. "The position and probable identification of the source of galactic radio-frequency radiation Taurus A," *Austral. J. Sci. Res.* **2A,** 139–148.

Bolton, J. G., G. J. Stanley, and O. B. Slee. "Positions of three discrete sources of galactic radio-frequency radiation," *Nature* **164,** 101–102.

Fermi, E. "On the origin of the cosmic radiation," *Phys. Rev.* **75,** 1169–1174.

Ginzburg, V. L. *Teoriya rasprostraneniya radiovoln v ionosfere* [Theory of the distribution of radio waves in the ionosphere] (Gostekhizdat, Moscow).

Hall, J. S. "Observations of the polarized light from stars," *Science* **109,** 166.

Hall, J. S., and A. H. Mikesell. "Observations of polarized light from stars," *A. J.* **54,** 187–188.

Hiltner, W. A. "On the presence of polarization in the continuous radiation of stars II," *Ap. J.* **109,** 471–478.

Ryle, M. "Evidence for the stellar origin of cosmic rays," *Proc. Phys. Soc.* **A62,** 491–499.

Schwinger, J. "On the classical radiation of accelerated electrons," *Phys. Rev.* **75,** 1912–1925.

Shklovsky, I. S. "Monochromatic radio emission from the Galaxy and the possibility of its observation" [in Russian], *Astr. Zhur. SSSR* **26,** 10–14.

Terletsky, Ya. P. "The origin of cosmic rays" [in Russian], *Zhur. Eksp. Teor. Fiz.* **19,** 1059–1075.

Unsöld, A. "Über den Ursprung der Radiofrequenzstrahlung und der Ultrastrahlung in der Milchstrasse," *Zs. f. Ap.* **26,** 176–199.

1950 Alfvén, H., and N. Herlofson. "Cosmic radiation and radio stars," *Phys. Rev.* **78,** 616.

1950   Allen, C. W., and C. S. Gum. "Survey of galactic radio noise at 200 Mc/s," *Austral. J. Sci. Res.* **3A,** 224–233.

Bolton, J. G., and K. C. Westfold. "Galactic radiation at radio frequencies I. 100-Mc/s survey," *Austral. J. Sci. Res.* **3A,** 19–33.

Hall, J. S., and A. H. Mikesell. "Polarization of light in the Galaxy . . . ," *Pub. U. S. Naval Obs.* **17,** 1–62.

Kiepenheuer, K. O. "Cosmic rays as the source of general galactic radio emission," *Phys. Rev.* **79,** 738–739.

Little, C. G., and A. C. B. Lovell. "Origin of the fluctuations in the intensity of radio waves from galactic sources—Jodrell Bank observations," *Nature* **165,** 423–424.

Pikelner, S. B. "The dissipation of the corona and its significance" [in Russian], *Dokl. Akad. Nauk SSSR* **72,** 255–258.

Prodell, A. G., and P. Kusch. "On the hyperfine structure of hydrogen and deuterium," *Phys. Rev.* **79,** 1009–1010.

Ryle, M., F. G. Smith, and B. Elsmore. "A preliminary survey of the radio stars in the northern hemisphere," *M. N.* **110,** 508–523.

Shapley, H. "Comparison of the Magellanic Clouds with the galactic system," *Pub. Obs. Univ. Mich.* **10,** 79–84.

Smith, F. G. "Origin of the fluctuations in the intensity of radio waves from galactic sources," *Nature* **165,** 422–423.

Spitzer, L., Jr., and M. P. Savedoff. "The temperature of interstellar matter III," *Ap. J.* **111,** 593–608.

Stanley, G. J., and O. B. Slee. "Galactic radiation at radio frequencies II. The discrete sources," *Austral. J. Sci. Res.* **3A,** 234–250.

1951   Baade, W., and L. Spitzer, Jr. "Stellar populations and collisions of galaxies," *Ap. J.* **113,** 413–418.

Davis, L., Jr., and J. L. Greenstein. "The polarization of starlight by aligned dust grains," *Ap. J.* **114,** 206–240.

Ewen, H. I., and E. M. Purcell. "Radiation from galactic hydrogen at 1420 Mc/s," *Nature* **168,** 356.

Ginzburg, V. L. "Cosmic rays as a source of galactic radio radiation" [in Russian], *Dokl. Akad. Nauk SSSR* **76,** 377–380.

Hanbury Brown, R., and C. Hazard. "Radio emission from the Andromeda Nebula," *M. N.* **111,** 357–367.

Hiltner, W. A. "The polarization of 841 stars," *Ap. J.* **114,** 241–271.

Hulst, H. C. van de. *A course in radio astronomy,* Harvard lectures (Leiden Obs., Leiden), mimeographed.

Kalinyak, A. A., V. I. Krasovsky, and V. B. Nikonov. "Observation of the region of the galactic center in radiation of about 10 000 A" [in Russian], *Izv. Krym. Astrofiz. Obs.* **6,** 119–129.

Muller, C. A., and J. H. Oort. "The interstellar hydrogen line at 1420 Mc/s, and an estimate of galactic rotation," *Nature* **168,** 357–358.

Oort, J. H. "Interaction of nova and supernova shells with the interstellar medium," *Problems of cosmical aerodynamics* (Central Air Documents Office, Dayton), 118–124.

Piddington, J. H. "The origin of galactic radio-frequency radiation," *M. N.* **111,** 45–63.

Piddington, J. H., and H. C. Minnett. "Observations of galactic radiation at frequencies of 1210 and 3000 Mc/s," *Austral. J. Sci. Res.* **4A,** 459–475.

1951   Ryle, M., and B. Elsmore. "A search for long-period variations in the intensity of radio stars," *Nature* **168,** 555–556.

Shain, C. A. "Galactic radiation at 18.3 Mc/s," *Austral. J. Sci. Res.* **4A,** 258–267.

Smith, F. G. (*a*) "An accurate determination of the positions of four radio stars," *Nature* **168,** 555.

————. (*b*) "An attempt to measure the annual parallax or proper motion of four radio stars," *Nature* **168,** 962–963.

Spitzer, L., Jr., and D. R. Bates. "The density of molecules in interstellar space," *Ap. J.* **113,** 441–463.

Vitkevich, V. V. "A new method for investigating the solar corona" [in Russian], *Dokl. Akad. Nauk SSSR* **77,** 585–588.

Westerhout, G., and J. H. Oort. "A comparison of the intensity distribution of radio-frequency radiation with a model of the galactic system," *B. A. N.* **11,** 323–333 (No. 426).

1952   Blaauw, A. "The velocity distribution of the interstellar calcium clouds," *B. A. N.* **11,** 459–473 (No. 436).

Christiansen, W. N., and J. V. Hindman. "A preliminary survey of 1420 Mc/s line emission from galactic hydrogen," *Austral. J. Sci. Res.* **5A,** 437–455.

Critchfield, C. L., E. P. Ney, and S. Oleksa. "Soft radiation at balloon altitudes," *Phys. Rev.* **85,** 461–467.

Getmantsev, G. G. "Cosmic electrons as a source of the radio emission from the Galaxy" [in Russian], *Dokl. Akad. Nauk SSSR* **83,** 557–560.

Gum, C. S. "A large H II region at galactic longitude 226°," *Observatory* **72,** 151–154.

Hanbury Brown, R., and C. Hazard. (*a*) "Radio-frequency radiation from Tycho Brahé's supernova (A.D. 1572)," *Nature* **170,** 364–365.

————. (*b*) "Extragalactic radio-frequency radiation," *Phil. Mag.* (7), **43,** 137–152.

Hanbury Brown, R., R. C. Jennison, and M. K. Das Gupta. "Apparent angular sizes of discrete radio sources," *Nature* **170,** 1061–1063.

Link, F. "Eclipses coronales des radioétoiles," *Bull. Astr. Inst. Czech.* **3,** 6–12 (No. 1).

Machin, K. E., and F. G. Smith. "Occultation of a radio star by the solar corona," *Nature* **170,** 319–320.

Mills, B. Y. (*a*) "The distribution of the discrete sources of cosmic radio radiation," *Austral. J. Sci. Res.* **5A,** 266–287.

————. (*b*) "Apparent angular sizes of discrete radio sources—observations at Sydney," *Nature* **170,** 1063–1064.

————. (*c*) "The positions of 6 discrete sources of cosmic radio radiation," *Austral. J. Sci. Res.* **5A,** 456–463.

Morgan, W. W., S. Sharpless, and D. Osterbrock. "Some features of galactic structure in the neighborhood of the sun," *A. J.* **57,** 3.

Piddington, J. H., and H. C. Minnett. "Radio-frequency radiation from the constellation of Cygnus," *Austral. J. Sci. Res.* **5A,** 17–31.

Ryle, M. "A new radio interferometer and its application to the observation of weak radio stars," *Proc. Roy. Soc.* **A211,** 351–375.

Shajn, G. A., and V. F. Gaze. "The structure and masses of the diffuse gaseous nebulae NGC 6523, 6618, and 2237" [in Russian], *Izv. Krym. Astrofiz. Obs.* **8,** 80–92.

1952   Shklovsky, I. S. (*a*) "On the nature of the radio emission from the Galaxy" [in Russian], *Astr. Zhur. SSSR* **29,** 418–449.
——————. (*b*) 8th general assembly of the I. A. U., Roma.
——————. (*c*) "Radio spectroscopy of the Galaxy" [in Russian], *Astr. Zhur. SSSR* **29,** 144–153.
Shklovsky, I. S., and P. N. Kholopov. "Identification of the nebula NGC 1316 with the radio star in the constellation of Fornax" [in Russian], *Astr. Tsirk.* **131,** 2–3.
Shklovsky, I. S., and P. P. Parenago. "Identification of the supernova of the year 369 with an intense radio star in Cassiopeia" [in Russian], *Astr. Tsirk.* **131,** 1–2.
Smith, F. G. "Apparent angular sizes of discrete radio sources—observations at Cambridge," *Nature* **170,** 1065.
Vitkevich, V. V. "An experimental possibility of observing monochromatic galactic radio emission" [in Russian], *Astr. Zhur. SSSR* **29,** 532–537.
Wild, J. P. "The radio-frequency line spectrum of atomic hydrogen and its applications in astronomy," *Ap. J.* **115,** 206–221.

1953   Babcock, H. W. "The solar magnetograph," *Ap. J.* **118,** 387–396.
Dewhirst, D. W. "The identification of radio sources," *J. B. A. A.* **63,** 263–266.
Ginzburg, V. L. (*a*) "The origin of cosmic rays and radio astronomy" [in Russian], *Uspekhi Fiz. Nauk* **51,** 343–392.
——————. (*b*) "The statistical mechanism of the acceleration of particles on the sun and in stellar shells" [in Russian], *Dokl. Akad. Nauk SSSR* **92,** 727–730.
——————. (*c*) "Supernovae and novae as sources of cosmic and radio radiation" [in Russian], *Dokl. Akad. Nauk SSSR* **92,** 1133–1136.
Ginzburg, V. L., and M. I. Fradkin. "The electron component and the origin of cosmic rays" [in Russian], *Dokl. Akad. Nauk SSSR* **92,** 531–534.
Greenstein, J. L., and R. Minkowski. "The Crab Nebula as a radio source," *Ap. J.* **118,** 1–15.
Hanbury Brown, R., and C. Hazard. (*a*) "A radio survey of the Milky Way in Cygnus, Cassiopeia, and Perseus," *M. N.* **113,** 109–122.
——————. (*b*) "A survey of 23 localized radio sources in the Northern Hemisphere," *M. N.* **113,** 123–133.
——————. (*c*) "Radio-frequency radiation from the spiral nebula Messier 81," *Nature* **172,** 853.
——————. (*d*) "An extended radio-frequency source of extragalactic origin," *Nature* **172,** 997–998.
Jennison, R. C., and M. K. Das Gupta. "Fine structure of the extraterrestrial radio source Cygnus 1," *Nature* **172,** 996–997.
Kraus, J. D., and H. C. Ko. "Radio radiation from the supergalaxy," *Nature* **172,** 538–539.
Lew Hin. "The hyperfine structure and nuclear moments of Pr[141]," *Phys. Rev.* **91,** 619–630.
Mills, B. Y. "The radio brightness distributions over four discrete sources of cosmic noise," *Austral. J. Phys.* **6,** 452–470.
Morgan, W. W., A. E. Whitford, and A. D. Code. "A preliminary determination of the space distribution of the blue giants," *Ap. J.* **118,** 318–322.
Pikelner, S. B. "The kinematic properties of the interstellar gas in connec-

tion with the isotropy of cosmic rays" [in Russian], *Dokl. Akad. Nauk SSSR* **88**, 229–232.

1953 Sanders, T. M., Jr., A. L. Schawlow, G. C. Dousmanis, and C. H. Townes. "A microwave spectrum of the free OH radical," *Phys. Rev.* **89**, 1158–1159.

Scheuer, P. A. G., and M. Ryle. "An investigation of the H II regions by a radio method," *M. N.* **113**, 3–17.

Shajn, G. A. "On the system of the emission nebulae" [in Russian], *Dokl. Akad. Nauk SSSR* **93**, 993–996.

Shapley, H. "Magellanic Clouds VI. Revised distances and luminosities," *Proc. Nat. Acad. Sci.* **39**, 349–357.

Shklovsky, I. S. (*a*) "The problem of cosmic radio waves" [in Russian], *Astr. Zhur. SSSR* **30**, 15–36.

———. (*b*) "An attempt to compare the brightness of the Milky Way in infrared and photographic light" [in Russian], *Izv. Krym. Astrofiz. Obs.* **10**, 169–182.

———. (*c*) "The possibility of observing the monochromatic radio radiation of interstellar molecules" [in Russian], *Dokl. Akad. Nauk SSSR* **92**, 25–28.

———. (*d*) "On the origin of cosmic radiation" [in Russian], *Dokl. Akad. Nauk SSSR* **91**, 475–478.

———. (*e*) "The problem of the origin of cosmic rays and radio astronomy" [in Russian], *Astr. Zhur. SSSR* **30**, 577–592.

———. (*f*) "On the nature of the radiation from the Crab Nebula" [in Russian], *Dokl. Akad. Nauk SSSR* **90**, 983–986.

Vaucouleurs, G. de. (*a*) "Normal and abnormal galaxies as radio sources," *Observatory* **73**, 252–254.

———. (*b*) "Evidence for a local supergalaxy," *A. J.* **58**, 30–32.

Wild, J. P., J. D. Murray, and W. C. Rowe. "Evidence of harmonics in the spectrum of a solar radio outburst," *Nature* **172**, 533–534.

Zwicky, F. "Luminous and dark formations of intergalactic matter," *Phys. Today* **6**, 7–11.

1954 Baade, W., and R. Minkowski. (*a*) "Identification of the radio sources in Cassiopeia, Cygnus A, and Puppis A," *Ap. J.* **119**, 206–214.

———. (*b*) "On the identification of radio sources," *Ap. J.* **119**, 215–231.

Baldwin, J. E. (*a*) "The distribution of radio brightness across the Crab Nebula," *Observatory* **74**, 120–123.

———. (*b*) "Radio emission from the Andromeda Nebula," *Nature* **174**, 320–321.

Baldwin, J. E., and D. W. Dewhirst. "Position and identification of a bright extended radio source in Gemini," *Nature* **173**, 164–165.

Baldwin, J. E., and B. Elsmore. "Radio emission from the Perseus cluster," *Nature* **173**, 818.

Bolton, J. G., G. J. Stanley, and O. B. Slee. "Discrete sources at 100 Mc/s between declinations $+50°$ and $-50°$," *Austral. J. Phys.* **7**, 110–129.

Bolton, J. G., K. C. Westfold, G. J. Stanley, and O. B. Slee. "Discrete sources with large angular widths," *Austral. J. Phys.* **7**, 96–109.

Cowling, T. G. "Nuclear processes and stellar structure: general survey," *Mém. Soc. Roy. Sci. Liège* **14**, 75–87.

Dombrovsky, V. A. "On the nature of the radiation from the Crab Nebula" [in Russian], *Dokl. Akad. Nauk SSSR* **94**, 1021–1024.

1954   Gaze, V. F., and G. A. Shajn. "The nebula IC 443 and a source of radio emission in Gemini" [in Russian], *Astr. Zhur. SSSR* **31,** 409–412.

Ginzburg, V. L. "On the generation of cosmic rays by nova and supernova explosions" [in Russian], *Trudy Soveshch. Vopr. Kosmog.* **3,** 258–264.

Gordon, I. M. "The outbursts of supernovae and the problem of the origin of the electron component of cosmic rays" [in Russian], *Dokl. Akad. Nauk SSSR* **94,** 413–416.

Haddock, F. T., C. H. Mayer, and R. M. Sloanaker. (*a*) "Radio observations of ionized hydrogen nebulae and other discrete sources at a wavelength of 9.4 cm," *Nature* **174,** 176–177.

————. (*b*) "Radio emission from the Orion Nebula and other sources at a wavelength of 9.4 cm," *Ap. J.* **119,** 456–459.

Hagen, J. P., E. F. McClain, and N. Hepburn. "Radio discrete sources and galactic absorption at 21-cm wavelength," *A. J.* **59,** 323.

Hanbury Brown, R., H. P. Palmer, and A. R. Thompson. "Galactic radio sources of large angular diameter," *Nature* **173,** 945–946.

Heald, M. A., and R. Beringer. "Hyperfine structure of nitrogen," *Phys. Rev.* **96,** 645–648.

Hey, J. S., and V. A. Hughes. "Intensities of discrete radio sources in Cygnus and Cassiopeia at 22.6 Mc/s," *Nature* **173,** 819–820.

Hulst, H. C. van de, C. A. Muller, and J. H. Oort. "The spiral structure of the outer part of the galactic system derived from the hydrogen emission at 21-cm wavelength," *B. A. N.* **12,** 117–149 (No. 452).

Kerr, F. J., J. V. Hindman, and B. J. Robinson. "Observations of the 21-cm line from the Magellanic Clouds," *Austral. J. Phys.* **7,** 297–314.

Kraus, J. D., H. C. Ko, and S. Matt. "Galactic and localized source observations at 250 Mc/s," *A. J.* **59,** 439–443.

Kwee, K. K., C. A. Muller, and G. Westerhout. "The rotation of the inner parts of the galactic system," *B. A. N.* **12,** 211–222 (No. 458).

Schwarzschild, M. "Mass distribution and mass–luminosity ratio in galaxies," *A. J.* **59,** 273–284.

Shain, C. A. "A comparison of the intensities of cosmic noise observed at 18.3 Mc/s and at 100 Mc/s," *Austral. J. Phys.* **7,** 150–164.

Shain, C. A., and C. S. Higgins. "Observations of the general background and discrete sources of 18.3 Mc/s cosmic noise," *Austral. J. Phys.* **7,** 130–149.

Shajn, G. A. "On the groups of emission nebulae" [in Russian], *Astr. Zhur. SSSR* **31,** 217–223.

Shakeshaft, J. R. "The isotropic component of cosmic radio-frequency radiation," *Phil. Mag.* (7), **45,** 1136–1144.

Shklovsky, I. S. "On the nature of the discrete sources of cosmic radio waves" [in Russian], *Dokl. Akad. Nauk SSSR* **98,** 353–356.

Spitzer, L., Jr. "Behavior of matter in space," *Ap. J.* **120,** 1–17.

Townes, C. H. "Microwave spectra of astrophysical interest," *J. Geophys. Res.* **59,** 191.

Vashakidze, M. A. "On the degree of polarization of the light near extragalactic nebulae and the Crab Nebula" [in Russian], *Astr. Tsirk.* **147,** 11–13.

Vaucouleurs, G. de. "The Magellanic Clouds and the Galaxy," *Observatory* **74,** 23–31.

Williams, D. R. W., and R. D. Davies. "A method for the measurement of the distance of radio stars," *Nature* **173,** 1182–1183.

1955   Baldwin, J. E. (a) "A survey of the integrated radio emission at a wave-
       length of 3.7 m," *M. N.* **115,** 684–689.

       ——— . (b) "The distribution of the galactic radio emission," *M. N.* **115,**
       690–700.

       ——— . (c) "A survey of the integrated radio emission at a wavelength
       of 3.7 m," *Observatory* **75,** 229–231.

       Bok, B. J., R. S. Lawrence, and T. K. Menon. "Radio observations
       (21-cm) of dense dark nebulae," *Pub. A. S. P.* **67,** 108–112.

       Carter, A. W. L. "The angular size of the variable radio source Hydra A,"
       Austral. J. Phys. **8,** 564–567.

       Davies, R. D., and D. R. W. Williams. "An alternative identification of the
       radio source in the direction of the galactic centre," *Nature* **175,** 1079–
       1081.

       Denisse, J.-F., E. Leroux, and J.-L. Steinberg. "Observations du rayon-
       nement galactique sur la longueur d'onde de 33 cm," *C. R.* **240,** 278–280.

       Getmantsev, G. G., K. S. Stankevich, and V. S. Troitsky. "The monochro-
       matic radio emission of deuterium at 91.6-cm wavelength from the
       galactic center" [in Russian], *Dokl. Akad. Nauk SSSR* **103,** 783–786.

       Ginzburg, V. L., S. B. Pikelner, and I. S. Shklovsky. "On the mechanism of
       the acceleration of particles in envelopes of novae and supernovae" [in
       Russian], *Astr. Zhur. SSSR* **32,** 503–513.

       Greenstein, J. L. "Theoretical problems of discrete radio sources," in
       Radio astronomy, I. A. U. Symposium **4** (Cambridge University Press,
       Cambridge, 1957), 179–191.

       Gum, C. S. "A survey of southern H II regions," *Mem. R. A. S.* **67,** 155–
       177.

       Hagen, J. P., A. E. Lilley, and E. F. McClain. "Absorption of 21-cm radia-
       tion by interstellar hydrogen," *Ap. J.* **122,** 361–375.

       Heeschen, D. S. "Some features of interstellar hydrogen in the section of
       the galactic center," *Ap. J.* **121,** 569–584.

       Hsi Tsê-tsung. "A new catalogue of ancient novae" [in Chinese], *Acta
       Astr. Sinica* **3,** 196–215.

       Kaidanovsky, N. L., N. S. Kardashev, and I. S. Shklovsky. "Results of
       observations of discrete sources of cosmic radio waves at 3.2 cm" [in
       Russian], *Dokl. Akad. Nauk SSSR* **104,** 517–519.

       Khachikyan, E. E. "On the polarization of the light of the Crab Nebula"
       [in Russian], *Dokl. Akad. Nauk Arm. SSR* **21,** 63–68.

       Kraus, J. D., and H. C. Ko. "A detailed radio map of the sky," *Nature*
       175, 159–161.

       Kraus, J. D., H. C. Ko, and D. V. Stoutenburg. "A fluctuating celestial
       radio source at 242 Mc/s," *Nature* **176,** 304–305.

       Kusch, P. "Redetermination of the hyperfine splittings of hydrogen and
       deuterium in the ground state," *Phys. Rev.* **100,** 1188–1190.

       Lilley, A. E. "A note on the galactic microwave spectrum," *Ap. J.* **122,**
       197–198.

       McClain, E. F. "An approximate distance determination for radio source
       Sagittarius A," *Ap. J.* **122,** 376–384.

       McGee, R. X., O. B. Slee, and G. J. Stanley. "Galactic survey at 400 Mc/s
       between declinations −17° and −49°," *Austral. J. Phys.* **8,** 347–367.

       Menon, T. K. "A 21-cm study of the Orion region," in *Radio astronomy,*
       I. A. U. Symposium **4** (Cambridge University Press, Cambridge, 1957),
       56–65.

1955   Mills, B. Y. "The observation and interpretation of radio emission from some bright galaxies," *Austral. J. Phys.* **8**, 368–389.

Naqvi, A. M., and J. N. Tandon. "The identification of radio stars," *Proc. Nat. Inst. Sci. India* **A21**, 244–251.

National Geographic Society–Palomar Observatory. *Sky Survey Atlas* (Calif. Inst. Technology, Pasadena, 1955–1958).

Pawsey, J. L., and R. N. Bracewell. *Radio astronomy* (Clarendon Press, Oxford).

Ryle, M., and A. Hewish. "The Cambridge radio telescope," *Mem. R. A. S.* **67**, 97–105.

Ryle, M., and P. A. G. Scheuer. "The spatial distribution and the nature of radio stars," *Proc. Roy. Soc.* **A230**, 448–462.

Seaton, M. J. "The kinetic temperature of the interstellar gas in regions of neutral hydrogen," *Ann. d'Ap.* **18**, 188–205.

Shajn, G. A. "On the magnetic fields in interstellar space and in the nebulae" [in Russian], *Astr. Zhur. SSSR* **32**, 381–394.

Shajn, G. A., S. B. Pikelner, and R. N. Ikhsanov. "Measurement of polarization in the Crab Nebula" [in Russian], *Astr. Zhur. SSSR* **32**, 395–400.

Shakeshaft, J. R., M. Ryle, J. E. Baldwin, B. Elsmore, and J. H. Thomson. "A survey of radio sources between declinations − 38° and +83°," *Mem. R. A. S.* **67**, 106–154.

Shklovsky, I. S. (*a*) "On the nature of the radiation from the radio galaxy NGC 4486" [in Russian], *Astr. Zhur. SSSR* **32**, 215–225.

————. (*b*) *Radioastronomiya*, 2nd edition (Gostekhizdat, Moscow).

Slee, O. B. "Apparent intensity variations of the radio source Hydra A," *Austral. J. Phys.* **8**, 498–507.

Stone, S. N. "A possible extragalactic source of 21-cm radio emission," *Pub. A. S. P.* **67**, 183–185.

1956   Adgie, R., and F. G. Smith. "The radio emission spectra of four discrete sources and of the background radiation," *Observatory* **76**, 181–187.

Baade, W. (*a*) "The polarization of the Crab Nebula on plates taken with the 200-inch telescope," *B. A. N.* **12**, 312 (No. 462).

————. (*b*) "Polarization in the jet of Messier 87," *Ap. J.* **123**, 550–551.

Baldwin, J. E., and F. G. Smith. "Radio emission from the extragalactic nebula M 87," *Observatory* **76**, 141–144.

Burbidge, G. R. "On synchrotron radiation from Messier 87," *Ap. J.* **124**, 416–429.

Burbidge, G. R., F. Hoyle, E. M. Burbidge, R. F. Christy, and W. A. Fowler. "Californium-254 and supernovae," *Phys. Rev.* **103**, 1145–1149.

Burke, B. F. "Mills Cross telescopes," *A. J.* **61**, 167–168.

Conway, R. G. "The angular diameter of the radio stars Cygnus A and Cassiopeia A," *Observatory* **76**, 235–237.

Costain, C. H., B. Elsmore, and G. R. Whitfield. "Radio observations of a lunar occultation of the Crab Nebula," *M. N.* **116**, 380–385.

Getmantsev, G. G., and V. A. Razin. "On the problem of the polarization of nonthermal cosmic radio emission" [in Russian], *Trudy Soveshch. Vopr. Kosmog.* **5**, 496–505.

Ginzburg, V. L. "On the origin of cosmic rays" [in Russian], *Izv. Akad. Nauk SSSR, Ser. Fiz.* **20**, 5–16.

Heeschen, D. S. "21-cm line emission from the Coma cluster," *Ap. J.* **124**, 660–662.

1956   Lamden, J. R., and A. C. B. Lovell. "The low-frequency spectrum of the Cygnus (19N4A) and Cassiopeia (23N5A) radio sources," *Phil. Mag.* (8), **1**, 725–737.

Lawrence, R. S. "Radio observations of interstellar neutral hydrogen clouds," *Ap. J.* **123**, 30–33.

Lilley, A. E., and E. F. McClain. "The hydrogen-line red shift of radio source Cygnus A," *Ap. J.* **123**, 172–175.

Mills, B. Y. "The radio source near the galactic centre," *Observatory* **76**, 65–67.

Mills, B. Y., A. G. Little, and K. V. Sheridan. (*a*) "Absorption of 3.5-m radiation in the optical emission nebula, NGC 6357," *Nature* **177**, 178.

————. (*b*) "Radio emission from novae and supernovae," *Austral. J. Phys.* **9**, 84–89.

————. (*c*) "Emission nebulae as radio sources," *Austral. J. Phys.* **9**, 218–227.

Mustel, E. R. "On the magnetic fields of novae" [in Russian], *Astr. Zhur. SSSR* **33**, 182–204.

Oort, J. H., and T. Walraven. "Polarization and composition of the Crab Nebula," *B. A. N.* **12**, 285–308 (No. 462).

Piddington, J. H., and G. H. Trent. "A survey of cosmic radio emission at 600 Mc/s," *Austral. J. Phys.* **9**, 481–493.

Pikelner, S. B. "The magnetic field of the Crab Nebula and of the central star" [in Russian], *Astr. Zhur. SSSR* **33**, 785–799.

Plechkov, V. M., and V. A. Razin. "Results of measurements of the intensity of radio emission from discrete sources at wavelengths of 3.2 and 9.7 cm" [in Russian], *Trudy Soveshch. Vopr. Kosmog.* **5**, 430–435.

Purcell, E. M., and G. B. Field. "Influence of collisions upon population of hyperfine states in hydrogen," *Ap. J.* **124**, 542–549.

Razin, V. A. "Preliminary results of the measurement of the polarization of cosmic radio emission at a wavelength of 1.45 m" [in Russian], *Radiotekhnika i Elektronika* **1**, 846–851.

Reber, G., and G. R. Ellis. "Cosmic radio-frequency radiation near one megacycle," *J. Geophys. Res.* **61**, 1–10.

Rishbeth, H. "An investigation of the radio source 06N2A in Gemini," *Austral. J. Phys.* **9**, 494–504.

Schmidt, M. "A model of the distribution of mass in the galactic system," *B. A. N.* **13**, 15–41 (No. 468).

Severny, A. B. "An investigation of deuterium in the sun" [in Russian], *Izv. Krym. Astrofiz. Obs.* **16**, 12–44.

Spitzer, L., Jr. "On a possible interstellar galactic corona," *Ap. J.* **124**, 20–34.

Stanley, G. J., and R. Price. "An investigation of monochromatic radio emission of deuterium from the Galaxy," *Nature* **177**, 1221–1222.

Tandon, J. N. "A note on the identification of radio stars," *Zs. f. Ap.* **41**, 19–20.

Vaucouleurs, G. de. "Survey of bright galaxies south of −35° declination with the 30-inch Reynolds reflector," *Mem. Commonwealth Obs., Mt. Stromlo* **3**, No. 13.

Wells, H. W. "Flux measurements of discrete sources at frequencies below 30 Mc/s," *J. Geophys. Res.* **61**, 541–545.

Westerhout, G. "Search for polarization of the Crab Nebula and Cassiopeia A at 22-cm wavelength," *B. A. N.* **12**, 309–311 (No. 462).

1957 Baldwin, J. E., and D. O. Edge. "Radio emission from the remnants of the supernovae of 1572 and 1604," *Observatory* **77**, 139–143.

Barrett, A. H., and A. E. Lilley. "A search for the 18-cm line of OH in the interstellar medium," *A. J.* **62**, 5–6.

Blythe, J. H. (*a*) "A new type of pencil-beam aerial for radio astronomy," *M. N.* **117**, 644–651.

————. (*b*) "Results of a survey of galactic radiation at 38 Mc/s," *M. N.* **117**, 652–662.

Bolton, J. G., and J. P. Wild. "On the possibility of measuring interstellar magnetic fields by 21-cm Zeeman splitting," *Ap. J.* **125**, 296–297.

Davies, R. D. "On the nature of the Cygnus X radio source . . . ," *M. N.* **117**, 663–679.

Dieter, N. H. "Observations of neutral hydrogen in M 33," *Pub. A. S. P.* **69**, 356–357.

Ellis, G. R. "Cosmic radio-noise intensities below 10 Mc/s," *J. Geophys. Res.* **62**, 229–234.

Heeschen, D. S. (*a*) "Neutral hydrogen in M 32, M 51, and M 81," *Ap. J.* **126**, 471–479.

————. (*b*) "Neutral hydrogen emission from the Hercules and Corona Borealis clusters of galaxies," *Pub. A. S. P.* **69**, 350–351.

Hulst, H. C. van de, E. Raimond, and H. van Woerden. "Rotation and density distribution of the Andromeda Nebula derived from observations of the 21-cm line," *B. A. N.* **14**, 1–16 (No. 480).

Krasovsky, V. I., and I. S. Shklovsky. "The possible influence of supernova explosions on the evolution of life on Earth" [in Russian], *Dokl. Akad. Nauk SSSR* **116**, 197–199.

Mayer, C. H., T. P. McCullough, and R. M. Sloanaker. "Evidence for polarized radio radiation from the Crab Nebula," *Ap. J.* **126**, 468–470.

Mills, B. Y., and O. B. Slee. "A preliminary survey of radio sources in a limited region of the sky at a wavelength of 3.5 m," *Austral. J. Phys.* **10**, 162–194.

Morgan, W. W., and N. U. Mayall. "A spectral classification of galaxies," *Pub. A. S. P.* **69**, 291–303.

Morris, D., H. P. Palmer, and A. R. Thompson. "Five radio sources of small angular diameter," *Observatory* **77**, 103–106.

Muller, C. A. "21-cm absorption effects in the spectra of two strong radio sources," *Ap. J.* **125**, 830–834.

Muller, C. A., and G. Westerhout. "A catalogue of 21-cm line profiles," *B. A. N.* **13**, 151–195 (No. 475).

Osterbrock, D. E. "Electron densities in the filaments of the Crab Nebula," *Pub. A. S. P.* **69**, 227–230.

Pikelner, S. B., and I. S. Shklovsky. "An investigation of the properties and energy dissipation of the galactic halo" [in Russian], *Astr. Zhur. SSSR* **34**, 145–158.

Razin, V. A. *Uch. Zap. Gork. Gos. Univ., Ser. Fiz.* **35**, 35.

Schmidt, M. (*a*) "Spiral structure in the inner parts of the galactic system derived from the hydrogen emission at 21-cm wavelength," *B. A. N.* **13**, 247–268 (No. 475).

————. (*b*) "The distribution of mass in M 31," *B. A. N.* **14**, 17–19 (No. 480).

Seeger, C. L., G. Westerhout, and R. G. Conway. "Observations of discrete sources, the Coma cluster, the moon, and the Andromeda Nebula at a wavelength of 75 cm," *Ap. J.* **126,** 585–587.

1957   Shain, C. A. "Galactic absorption of 19.7-Mc/s radiation," *Austral. J. Phys.* **10,** 195–203.

Shcheglov, P. V. "The spectrum of the Crab Nebula" [in Russian], *Astr. Zhur. SSSR* **34,** 675–677.

Shklovsky, I. S. "On the nature of the optical emission from the Crab Nebula" [in Russian], *Astr. Zhur. SSSR* **34,** 706–715.

Westerhout, G. (*a*) "The distribution of atomic hydrogen in the outer parts of the Galactic System," *B. A. N.* **13,** 201–246 (No. 475).

————. (*b*) "Intensités relatives des quatre principales radiosources observées sur la longeur d'onde 22 cm; note sur la radiosource Sagittarius A," *C. R.* **245,** 35–38.

Whitfield, G. R. "The spectra of radio stars," *M. N.* **117,** 680–691.

Woerden, H. van, W. Rougoor, and J. H. Oort. "Expansion d'une structure spirale dans le noyau du Système Galactique, et position de la radiosource Sagittarius A," *C. R.* **244,** 1691–1695.

Woltjer, L. "The polarization and intensity distribution in the Crab Nebula derived from plates taken with the 200-inch telescope by Dr. W. Baade," *B. A. N.* **13,** 301–311 (No. 478).

1958   Dewhirst, D. W. Symposium on radio astronomy, Paris.

Dieter, N. H. *Neutral hydrogen in M 33,* thesis (Radcliffe College).

Heeschen, D. S., and N. H. Dieter. "Extragalactic 21-cm line studies," *Proc. I. R. E.* **46,** 234–239.

Hill, E. R. "A search for radio emission at 3.5 m from the local Supergalaxy," *Austral. J. Phys.* **11,** 580–583.

Hsi Tsê-tsung. "A new catalog of ancient novae," *Smithson. Contr. Astrophys.* **2,** 109–130.

Mills, B. Y. Symposium on radio astronomy, Paris.

Mills, B. Y., A. G. Little, K. V. Sheridan, and O. B. Slee. "High-resolution radio telescope for use at 3.5 m," *Proc. I.R.E.* **46,** 67–84.

Oort, J. H., F. T. Kerr, and G. Westerhout. "The galactic system as a spiral nebula," *M. N.* **118,** 379–389.

Rishbeth, H. "Radio emission from the Vela–Puppis region," *Austral. J. Phys.* **11,** 550–563.

Shain, C. A. (*a*) Symposium on radio astronomy, Paris.

————. (*b*) "The radio emission from Centaurus A and Fornax A," *Austral. J. Phys.* **11,** 517–529.

————. (*c*) "The Sydney 19.7-Mc/s radio telescope," *Proc. I.R.E.* **46,** 85–88.

Sheridan, K. V. "An investigation of the strong radio sources in Centaurus, Fornax, and Puppis," *Austral. J. Phys.* **11,** 400–408.

Stankevich, K. S. "The investigation of radio emission from the galactic center on a wavelength of 91.6 cm" [in Russian], *Astr. Zhur. SSSR* **35,** 157–158.

Udaltsov, V. A., and V. V. Vitkevich. "On the intensity distribution of the discrete radio source Taurus A" [in Russian], *Astr. Zhur. SSSR* **35,** 713–721.

Westerhout, G. "A survey of the continuous radiation from the Galactic System at a frequency of 1390 Mc/s," *B. A. N.* **14,** 215–260 (No. 488).

1959   Drake, F. D. "A high-resolution radio study of the galactic center," *A. J.* **64,** 329.

1959   Hanbury Brown, R., and C. Hazard. "The radio emission from normal galaxies I. Observations of M 31 and M 33 at 158 Mc/s and 237 Mc/s," *M. N.* **119**, 297–308.

Hindman, J. V., and C. M. Wade. "The η Carinae nebula and Centaurus A near 1400 Mc/s I. Observations," *Austral. J. Phys.* **12**, 258–269.

Kerr, F. J., J. V. Hindman, and C. S. Gum. "A 21-cm survey of the southern Milky Way," *Austral. J. Phys.* **12**, 270–292.

Large, M. I., D. S. Mathewson, and C. G. T. Haslam. (*a*) "A high-resolution survey of the Andromeda Nebula at 408 Mc/s," *Nature* **183**, 1250–1251.

————. (*b*) "A high-resolution survey of the Coma cluster of galaxies at 408 Mc/s," *Nature* **183**, 1663–1664.

Muller, C. A. "21-cm observations on the Coma cluster," *B. A. N.* **14**, 339–344 (No. 493).

Shakeshaft, J. R., and J. E. Baldwin. "The radio emission from the direction of the 'Supergalaxy,' " *M. N.* **119**, 46–53.

1960   Rougoor, G. W., and J. H. Oort. "Distribution and motion of interstellar hydrogen in the Galactic System with particular reference to the region within 3 kpc of the center," *Proc. Nat. Acad. Sci.* **46**, 1–13.

GENERAL BIBLIOGRAPHY

A few suggestions for further reading are given here; additional references may be found in each of the works cited. Most of the important monographic material within the purview of this book has been included and, like the References, arranged in chronological order to emphasize the evolution of the field.   — R.B.R.

1950   International Scientific Radio Union. "Solar and galactic radio noise," *Spec. Rep.* **1** (U. R. S. I., Brussels).

M. Ryle. "Radio astronomy," in *Rep. Progr. Phys.* **13** (Physical Society, London), 184–246.

1951   H. C. van de Hulst. *A course in radio astronomy,* Harvard lectures (Leiden Observatory, Leiden), mimeographed. "Fundamentals of observational methods," 4–36; "Galactic and extragalactic radio waves," 141–175.

1952   A. C. B. Lovell and J. A. Clegg. *Radio astronomy* (Chapman and Hall, London; Wiley, New York), 177–206.

1954   International Scientific Radio Union. "Discrete sources of extraterrestrial radio noise," *Spec. Rep.* **3** (U. R. S. I., Brussels), 56 pp.

————. "Interstellar hydrogen," *Spec. Rep.* **5**, 47–72.

M. Waldmeier. "Radiowellen aus dem Weltraum," *Neujahrsbl. naturf. Ges. Zürich* **98**, Beiheft **4**, 80 pp.

1955   J. L. Pawsey and R. N. Bracewell. *Radio astronomy* (Clarendon Press, Oxford). "Technique for observations of extraterrestrial radio waves," 10–77; "Cosmic radio waves," 231–279.

I. S. Shklovsky. *Radioastronomiya; populyarny ocherk* [Radio astronomy; a popular sketch] (ed. 2; Gostekhizdat, Moscow), 1–54, 102–289.

A. Unsöld. *Physik der Sternatmosphären* (ed. 2; Springer, Berlin). "Radiofrequenzstrahlung und kosmische Ultrastrahlung," 714–781.

J. P. Wild. "Observational radio astronomy," in L. Marton, ed., *Advances in Electronics and Electron Physics* **7** (Academic Press, New York), 299–362.

1956　L. H. Aller. *Gaseous nebulae* (Chapman and Hall, London; Wiley, New York). "Gaseous nebulae as radio sources," 302–312.

R. Coutrez. *Radioastronomie,* Obs. Royal de Belgique Monographie **5** (R. Louis, Ixelles). "Radioémissions galactiques et extragalactiques," 278–320.

V. L. Ginzburg, ed. *Trudy pyatogo soveshchaniya po voprosam kosmogonii* [Conference on Problems of Cosmogony **5**], *Radioastronomiya* March 1955 (Academy of Sciences, U. S. S. R., Moscow), 395–561 [discrete sources; galactic and extragalactic radio emission].

G. A. Gurzadyan. *Radioastrofizika* (Academy of Sciences, Armenian S. S. R., Erevan). [Galactic radio emission, 54–68; discrete sources of cosmic radio waves, 69–102; origin of cosmic radio emission, 187–203; monochromatic radio emission, 204–239.]

1957　Commonwealth Scientific and Industrial Research Organization, Radiophysics Laboratory. *Symposium on radio astronomy,* September 1956 (C. S. I. R. O., Melbourne), 52–85.

J. Dufay. *Galactic nebulae and interstellar matter,* trans. A. J. Pomerans (Hutchinson's, London; Philosophical Library, New York). "Emission of radio waves," 103–112.

H. C. van de Hulst, ed. *Radio astronomy,* I. A. U. Symposium **4**, August 1955 (Cambridge University Press, Cambridge). "Spectral line investigations," 3–103; "Point sources," 107–207; "Galactic structure and statistical studies of point sources," 211–250.

P. Wellmann. *Radioastronomie,* Dalp-Taschenbücher **340** (L. Lehnen, Munich; A. Francke, Bern), 140 pp.

1958　R. Hanbury Brown and A. C. B. Lovell. *The exploration of space by radio* (Chapman and Hall, London; Wiley, New York). "Techniques of radio astronomy," 29–59; "Galactic and extragalactic radio emissions," 60–89; "The hydrogen line," 90–101.

Institute of Radio Engineers. *Proc. I. R. E.* **46** [No. 1, Radio astronomy issue], 1–373 *passim.*

1959　R. N. Bracewell, ed. *Paris symposium on radio astronomy,* I. A. U. Symposium **9**, August 1958 (Stanford University Press, Stanford). "Galactic and extragalactic radio sources," 295–401; "The large-scale structure of galaxies," 405–468; "Discrete sources and the universe," 471–538; "Mechanisms of solar and cosmic emission," 541–601.

R. D. Davies and H. P. Palmer. *Radio studies of the universe* (Routledge, London; Van Nostrand, Princeton), 46–111.

R. Hanbury Brown. "Discrete sources of cosmic radio waves," in *Handbuch der Physik* **53** (Springer, Berlin), 208–238.

B. Y. Mills. "Radio-frequency radiation from external galaxies," in *Handbuch der Physik* **53**, 239–274.

J. H. Oort. "Radio-frequency studies of galactic structure," in *Handbuch der Physik* **53**, 100–128.

1960　D. H. Menzel, ed. *The radio noise spectrum* (Harvard University Press, Cambridge). "Cosmic radio noise," 151–176.

Triennial, 1952–

International Astronomical Union. *Transactions* (Cambridge University Press, Cambridge). Draft reports and reports of the meetings of Commission 40 (Radio Astronomy).

Index

(Prepared by Richard B. Rodman)

429